Materials Science and Engineering

Materials Science and Engineering

Kaushik Kumar
B. Tech (Mech. Engg.), MBA, PhD
Birla Institute of Technology-Mesra, Ranchi

Apurba Kumar Roy
B.E. (Mech. Engg.), M.E. (Mech Engg.), PhD
Birla Institute of Technology-Mesra, Ranchi

Nehal Dash
B. Tech (Mech. Engg.), M.E. (D.M.E.)
Birla Institute of Technology-Mesra, Ranchi

I.K. International Publishing House Pvt. Ltd.
NEW DELHI

Published by
I.K. International Publishing House Pvt. Ltd.
4435-36/7, Ansari Road, Daryaganj
New Delhi–110 002 (India)
E-mail:info@ikinternational.com
Website: www.ikbooks.com

ISBN 978-93-85909-64-1

Published by Krishan Makhijani for I.K. International Publishing House Pvt. Ltd., 4435-36/7, Ansari Road, Daryaganj, New Delhi–110 002 and Printed by Rekha Printers Pvt. Ltd., Okhla Industrial Area, Phase II, New Delhi–110 020.

Dedicated to Our Parents

Preface

Engineering materials play a vital role in many applications in our day-to-day life. In order to understand the basic knowledge of the real concepts behind various theories, mechanisms, properties, etc., of the engineering materials this book will provide the readers a firm solution to gain the knowledge regarding the same.

The didactic approach of this book is perfectly suited to the engineering students as well as to physicists or chemists, who are not specialized in materials but who, nevertheless, wish to be aware of this discipline. The work should also be much appreciated by specialists in a particular aspect of materials science wishing to have a global view on the subject and to position their activity in a wider context. It will be equally useful to those who left university benches more than ten years ago and want to refresh themselves in the field of materials science.

The implication from the above is that it is intended to present a treatise on the basics of materials science that has a fundamental character. This may seem an impossible undertaking, as at the same time the book is meant to be used also in the beginning of a materials science and engineering study. Of course, then some important results of advanced theories have to be introduced and accepted without proof, but this is no serious obstacle in order to develop a sound basis of the basics of the field. If this book realizes these aspirations satisfactorily, then this book will be used by the reader also at later stages of his/her study, because a fundamental background may be quickly grasped on the basis of what this book offers. Also therefore the material contained in the book is much more comprehensive than what can normally be offered in an introductory course on materials science. But we do feel that there is a need for a book as the present one in the light of our experience with the existing introductory texts for the field of materials science which we consider as often to be too superficial and too phenomenological of nature.

This book contains of five sections with 13 chapters in total and in each of these chapters the primary aim is to deliver the basic knowledge of the engineering materials. Starting with Section I that's Kinetics, Thermodynamics and Crystallographic Aspects of Engineering Materials the history behind the engineering materials and also the applications of these materials. Chapter 1 provides a firm knowledge about an introduction to the engineering materials. In chapter 2, the atomic structures and bonding in solids is discussed. Chapter 3 elaborates on the structure of the crystalline solids. Chapter 4 discusses structure and composition of the phase diagrams of the engineering materials. Section II gives a firm idea on properties and strengthening of materials. Chapter 5 deals with the mechanical behaviour of the engineering materials and their properties such as electrical, thermal, magnetic, optical, etc. Chapter 6 explains the phenomenon of strengthening of engineering materials. Section III: Degradation and Failure of Engineering Materials deals with degradation of engineering materials, corrosion and the phenomenon of fatigue and creep respectively in Chapters 7, 8 and 9. Section IV covers the properties and processing of engineering materials. Chapter 10 provides a detailed discussion on nonmetallic engineering materials such as polymers, ceramics and glasses, etc. Chapters 11 and 12 explain processing, forming and an introduction to some of the engineering materials respectively. The degradation and the failure of engineering materials constitutes for the proper application and usage of these engineering materials, chapter 13 takes into account the economic, environmental and the social issues associated with the utilization of these engineering materials. Model Question Papers and Multiple-Choice Questions (sectionwise along with solutions) are also included in the book so that it provides the readers with an idea to emphasize on several topics from examination point of view. Glossary contains brief explanation/ definitions of important terms used in the book.

At the end, we would like to assure our readers that this book will provide an understanding of structure and mechanics of engineering materials like metals, ceramics, polymers, composites, etc., and speculating the exhibited physical and chemical properties.

Kaushik Kumar
Apurba Kumar Roy
Nehal Dash

Acknowledgements

First and foremost we thank the God. In the process of putting this book together we realized how true this gift of writing is for us. You have given us the power to believe in our passion and pursue our dreams. We could never have done this without the faith we have in you, the Almighty. We are thankful for this kindness.

We would like to thank our grandparents, parents and relatives for allowing us to follow our ambitions. Our families showed patience and tolerated us for taking yet another challenge which decreases the amount of time we could spend with them. They were our inspiration and motivation. We hope that one day they will read this book and understand why we spent more time in front of our laptops and computers than with them.

We would need to thank our colleagues, friends and students of different institutes and organizations and specially Birla Institute of Technology, Mesra. They always supported us throughout our career and also while writing this book. This book was not only inspired by them; but also improved by their active involvement. We look forward discussing this book with them at future gatherings as we are sure they will all need and read this soon.

Late Prof. Sanat Kumar Mukherjee of Birla Institute of Technology, Mesra, had shown us the importance of learning while working. We appreciate that he believed in us to provide leadership and knowledge to make this small effort a reality. Thanks for his wishes.

We owe a huge thanks to all of our technical reviewers. Their efforts helped to make this book complete and we couldn't have done it without them.

Throughout the process of writing this book, many individuals, from different walks of life, have taken time out to help us out. Last, but definitely not least, we would like to thank them all, our well wishers, for providing us encouragement.

Kaushik Kumar
Apurba Kumar Roy
Nehal Dash

Contents

Section III

Degradation and Faliure of Engineering Materials

Section IV

Properties and Processing of Engineering Materials

Section V

Economic, Environmental and Social Aspects of Engineering Materials

SECTION I

Kinetics, Thermodynamics and Crystallographic Aspects of Engineering Materials

1

Introduction

Learning Objectives

After going through this chapter the reader will be able to:

- List the four components involved in the design, production, and utilization of materials, and describe the relationships among them.
- Overview of history of engineering materials.
- Understand classification of materials and their applications.
- Understand criteria for material selection.

1.1 ENGINEERING MATERIALS: RESOURCES AND IMPLICATIONS

As engineering materials represent foundation of technology, it's not necessary to know how materials behave like they do and why they take issue in properties. This can be done with the atomistic understanding allowed by quantum physics that initially explained atoms and solids. The mixture of physics, chemistry, and the link between the properties of a fabric and its microstructure is the domain of Materials Science.

Why do we study materials? Many applied scientists and engineers, whether mechanical, civil, chemical, or electrical, will at one time or another be exposed to a design problem involving materials. Examples might include a transmission gear, the superstructure for a building, an oil refinery component, or an integrated circuit chip. Of course, materials scientists and engineers are specialists who are involved in the investigation and design of materials. Many times, a materials problem is one of selecting the right material from the many that are available. There are a few criteria on which an official choice is regularly based. As a matter of first importance, the in-administration conditions must be described, for these will direct the properties required of the material. On just uncommon events does a material have the most extreme or perfect mix of properties, therefore, it might be important to exchange off one trademark for another. The classic example involves strength and ductility; normally, a material having a high strength will have only a limited ductility. In such cases a reasonable compromise between two or more properties is necessary.

The second determination thought is any weakening of material properties that may happen amid administration operation. For instance, critical decreases in mechanical quality may come about because of presentation to hoisted temperatures or destructive situations. At long last, presumably the abrogating thought is that of financial matters: what will the completed item cost? A material might be found that has the perfect arrangement of properties yet is restrictively costly. Here, once more, some bargain is inescapable. The cost of a completed piece likewise incorporates any cost brought about amid creation to deliver the craved shape.

"Structure" is at this point a nebulous term that deserves some explanation. In brief, the structure of a material usually relates to the arrangement of its internal components. Subatomic structure involves electrons within the individual atoms and interactions with their nuclei. On an atomic level, structure encompasses the organization of atoms or molecules relative to one another. The next larger structural realm, which contains large groups of atoms that are normally agglomerated together, is termed "microscopic," meaning that which is subject to direct observation using some type of microscope. Finally, structural elements that may be viewed with the naked eye are termed "macroscopic.

In addition to structure and properties, two other important components are involved in the science and engineering of materials, namely, "processing" and "performance."With regard to the relationships of these four components, the structure of a material depends on how it is processed. Furthermore, a material's performance will be a function of its properties. Thus, the interrelationship between processing, structure, properties, and performance is as depicted in the schematic illustration shown. Throughout this text we draw attention to the relationships among these four components in terms of the design, production, and utilization of materials. A schematic interrelation between these four components is shown in Figure 1.1.

A primary application of materials science is matching the right material or combination of materials to the intended purpose and use of a specific product, such as a car. To do this, materials scientists must consider such things as the weight and strength of a certain material as well as its ability to conduct electricity or insulate the product from heat. They must also consider the material's chemical stability, corrosion resistance, and economy. This is the basic science part.

As students experiment with the different classes of materials, they will discover what terms like ductility mean and what makes these properties important in designing and producing stuff. Take the properties of metal, for example. The shared outer electrons of metal are wholly or partially responsible for high electrical conductivity, high thermal conductivity, and ductility. Ceramics exhibit the opposite properties as their localized, mostly ionic, bonding produces low electrical and low thermal conductivity and contributes to the extreme brittleness of ceramics. Students will also see as they experiment why one class of material is preferred over another for certain products and how they can change or "improve" certain materials.

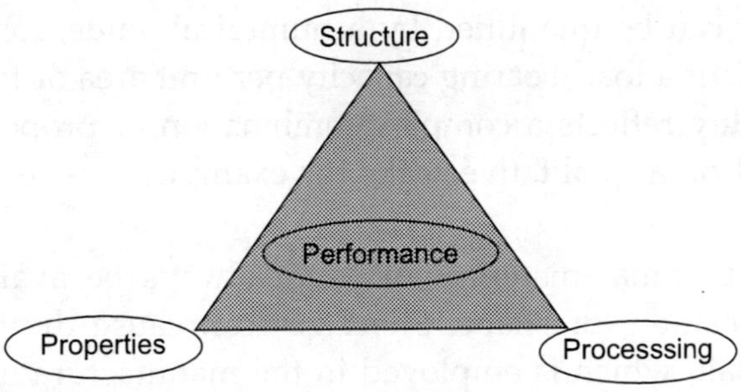

Figure 1.1 Interrelation between four components of materials science.

1.1.1 Selection of Materials

A key capacity of materials experts (e.g., researchers, metallurgists, ceramists, polymer specialists or engineers) is to incorporate helpful materials and produce them in mass amounts. Then again, civil, mechanical, chemical, and electrical engineers should habitually choose the materials to:

- implement designs for new products;
- improve existing products. In this case, new materials can cut production costs and rectify design shortcomings; and
- address emergency situations. These can include failure of the product with loss of consumer confidence, production problems, and shortages in the sources of supply requiring replacement materials.

A great many materials are conceivably accessible in grouped sizes and shapes, however, not rarely, the decision normally decreases to far less competitors. Among the issues that must be considered when settling on the choice to choose an engineering material are the following:

(i) **Essential property or properties of interest:** This is frequently the most important consideration. In rare cases there may only be one suitable material. For catalyst applications there is presently no good alternative to platinum. More commonly, however, there are materials choices. If, for example, mechanical property requirements are uppermost and a certain strength level is essential, many candidates will probably exist. Then other supplementary property requirements should be considered (e.g., density or weight per unit volume, toughness or resistance to fracture, environmental stability, resistance to elevated temperature softening, hardness and wear resistance, appearance) and rated according to importance.

(ii) **Fabricability:** If materials cannot be readily fabricated or assembled, they cease to be viable choices. Included under fabricability is the ability to be shaped by mechanical processes such as rolling, forging (hammering), and machining, and to be joined by welding, soldering, and other processes. Unlike some materials properties that can be quantified by a numerical value, for example, a melting point of 1083°C or a load-bearing capacity per unit area or tensile strength of 346 MPa, fabricability reflects a complex combination of properties. Therefore, it is usually ranked on a qualitative scale, for example, excellent, mediocre, or poor machinability.

(iii) **Availability:** Optimal materials may not always be available in the desired amount, composition, or shape. Shortages can cause dramatic price rises. An example is cobalt, which is employed in the manufacture of turbine blades and surgical prostheses. Numerous examples exist and include use of polyvinyl chloride (PVC) instead of brass and copper tubing, and polymer fibres instead of natural materials in clothing.

(iv) **Cost:** Materials selection is not possible without consideration of costs. Many basic engineering materials are sold as **commodities;** this means that their price is relatively fixed because there are numerous sources of supply. The increased value of a number of materials due to additional processing should be noted. Engineering materials such as mild (low-carbon) steel, wood, concrete, glass, and certain polymers have found such widespread use because they are relatively cheap. Initial purchase price is not the only cost consideration in materials selection, however. Subsequent labour, production, and assembly costs associated with the use of a particular material must also be taken into account.

In summary, a *concurrent* systems approach that synergistically integrates design, manufacturing, and materials selection must be adopted if optimum benefits are to be realized.

1.2 HISTORY OF MATERIALS AND ENGINEERING

Materials are thus imperative within the advancement of human development that the historians have recognized early times of progress by the name of most altogether utilised material, e.g., Stone Age, Bronze Age. This is often solely a perception created to showcase the importance of materials and their result on human development.

The assignment of progressive chronicled ages as the Stone, Copper, Bronze, and Iron ages mirrors the significance of materials to humanity. The relationship of a given material with the age or period that it characterizes is not just constrained to relics. The present atomic and data ages owe their presence to the misuse of two noteworthy components, uranium and silicon, individually. Despite the fact that current materials ages are to a great degree time packed with respect to the antiquated metal ages they share various normal qualities.

From the historic perspective, it can be said that human progress began with Stone Age where individuals utilized just characteristic materials, like stone, clay, skin, and wood for the purposes like to make weapons, instruments, cover, and so on. Nonetheless, the expanding requirement for better quality apparatuses delivered investigation that prompted Bronze Age, trailed by Iron Age. At the point when individuals discovered copper and how to make it harder by alloying, the Bronze Age began around 3000 BC. The utilization of iron and steel, a more grounded material that gave advantage in wars began at around 1200 BC. The following huge stride in human progress was the disclosure of a shoddy procedure to make steel around 1850 AD, which empowered the railways and the working of the present-day base of the mechanical world. A standout amongst the most noteworthy components of the *democratic material* is that number of clients simply exploded. In this way there has been a requirement for human and material assets for a considerable length of time, which as yet going solid. It's being said and concurred that we are in a matter of seconds in Space Age set apart by numerous mechanical improvements towards advancement of materials, bringing about more grounded and light materials like composites, electronic materials like semiconductors, materials for space voyage like high temperature ceramics, biomaterials, and so on.

It is instructive to appreciate some additional characteristics and implications of these materials. For example, imagine that time is frozen at 1500 BC and we focus on the Middle East, perhaps the world's most intensively excavated region with respect to archaeological remains. In Asia Minor (Turkey) the ancient Hittites were already experimenting with iron, while closeby to the east in Mesopotamia (Iraq), the Bronze Age was flowering. To the immediate north in Europe, the south in Palestine, and the west in Egypt, people were enjoying the benefits of the Copper and Early Bronze ages. Halfway around the world to the east, the Chinese had already melted iron and demonstrated a remarkable genius for bronze, a copper-tin alloy that is stronger and easier to cast than pure copper. Further to the west on the Iberian Peninsula (Spain and Portugal), the Chalcolithic period, an overlapping Stone and Copper Age held sway, and in North Africa survivals of the Late Stone Age were in evidence. Across the Atlantic Ocean the peoples of the Americas had not yet discovered bronze, but like others around the globe, they fashioned beautiful work in gold, silver, and copper, which were found in nature in the free state (i.e., not combined in oxide, sulphide, or other ores).

At the point when the 20th century unfolded the quantity of various materials controllably misused had, shockingly, not developed much past what was accessible 2000 years prior. A striking special case was steel, which introduced the Machine Age and reformed numerous aspects of life. Be that as it may, then a period followed in which there was an unstable increment in our comprehension of the basic nature of materials. The outcome was the development of polymeric (plastic), atomic, and electronic materials, new parts for metals and pottery, and the advancement of dependable approaches to process and make helpful items from them. On the whole, this advanced Age of Materials has penetrated the whole world and overshadowed the effect of past ages.

In the rundown, materials constitute establishment of innovation. The historical backdrop of human development advanced from the Stone Age to the Bronze Age, the

Iron Age, the Steel Age, and to the Space Age (contemporaneous with the Electronic Age). Every age is set apart by specific materials. The Iron Age brought instruments and utensils. The Steel Age brought railways, instruments, and the Industrial Revolution. The Space Age brought the materials far more grounded and light structures (e.g., composite materials). The Electronic Age brought semiconductors, and along these lines numerous assortments of electronic devices.

The advancement of numerous innovations that make our presence so agreeable has been personally connected with the openness of reasonable materials. A progression in the comprehension of a material sort is frequently the trailblazer to the stepwise movement of an innovation. For instance, cars would not have been conceivable without the accessibility of cheap steel or some other similar substitutes. In our contemporary time, refined electronic gadgets depend on parts that are produced using what are called semiconducting materials.

Figure 1.2 shows the advancement made in expanding the strength to density (or weight) proportion of materials. Two ramifications of these advances have been in enhanced aircraft design and energy savings in transportation frameworks.

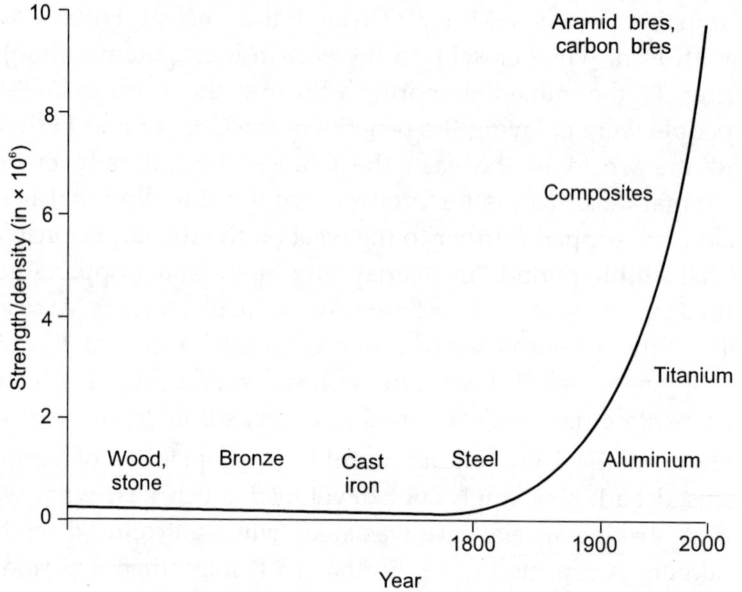

Figure 1.2 Graphical representation of strength to density ratio vs year of evolution.

By the time early polymers were introduced in the late 19th century, we had learnt enough about organic chemistry to manipulate materials at the molecular level. At this point, it became possible to design specific materials for specific needs. Designed materials constituted much of the accelerating pace of materials science. This launched us into the third and final era of materials history, which began in the 1950s. Designed materials are probably best illustrated by composites, which allow us to reinforce materials at the right places and in the right amounts to minimize weight and produce the desired mechanical properties. The graphite tennis racquet, golf club shaft, and fishing rod are all products of this designed materials revolution, as are the wings of new high-performance aircraft such as the Harrier.

Atomic structure and chemical composition were once major focus of materials science research. However, over the last few decades, this focus has changed dramatically as analytical chemistry, the electron microscope, X-ray diffraction, and a host of spectrometers have been developed that can analyze materials with accuracy. Because scientists can now understand chemical composition and the physical properties of materials, the major focus of materials science has shifted to understanding how materials can be improved and what new materials can be developed to meet society's needs. These scientific advances caused a revolution in knowledge in materials.

Examples of new materials abound. The space shuttle tile, which is used as a heat shield to protect the aluminium shell on the shuttle, is one example of this development of new and improved materials. When NASA (National Aeronautics and Space Administration) decided to build a space shuttle that would rocket into the orbit and eventually plunge through the atmosphere and land on the ground like an airplane, no known insulating material existed that would protect the flight crew from the fierce re-entry heat, be light enough to coat the entire craft, and be reused a number of times.

So, ceramists (materials scientists who work with ceramics) designed special tiles made from high-temperature glass fibres and sintered them to form a rigid, but almost unbelievably light structure. These tiles are glued to the shuttle with silicone rubber and now do an admirable job of keeping heat away from the crew. The ceramists designed the tiles from the "scratch" by adapting their knowledge of glass properties to meet the needs of the space shuttle. Further development continues as less bulky and more reliable materials are being developed to shield the Orient Express, a supersonic transport being developed for near-space travel over long distances around the earth.

Another example of the development of new materials is in biomedicine. The recent controversy over silicone breast implants shows how much care must be taken in choosing, testing, and using materials that are used inside human bodies. More successful examples of materials developed for human bodies are such things as hip, knee, and finger joint replacements made from composite materials.

A modern automobile is a good example of how materials have changed to keep pace with industry and culture. The American car of the 1950s was a durable machine and pretty well suited to its environment. Gas was cheap, metal was thick and lavishly used. The resulting car was heavy, but Americans demanded high performance for use on newly built freeways designed for speeds of at least 75 miles per hour. Engines were correspondingly large, with displacements approaching 500 cubic inches (that's about 8 litres). Americans tended to abuse these cars, and they were built to take abuse.

No one ever claimed these cars were fuel efficient. Then came the oil shock of 1973, and cars had to change radically, and still changing to meet the demand for more fuel-efficient transportation. In the quest for efficiency, car weight was reduced, and some changes were made in streamlining. Sheet metal used to build car bodies became much thinner and had to be much stronger. Unit body construction provided a way to produce stiffness without all the weight of a separate frame. Plastics and aluminium were used extensively in many parts of the car. The electronics revolution has provided onboard computers to manage the fuel and ignition systems of the engine, which now operate much closer to optimum parameters because of the need for both fuel efficiency and reduced emissions. Removing lead from gasoline stimulated use of better alloys for valves and valve seats. Spark plugs no longer exposed to lead deposits last 30,000 miles.

Materials engineers and scientists will be required in ever greater numbers as designs for everything from tennis racquets to space shuttles are driven harder by requirements for ever higher efficiency in manufacturing and by use and reuse of materials. And, increasingly, the disposal of used materials will require attention with emphasis on recycling these materials. "Environment-friendly" materials that can be produced, used, and disposed of without harmful effects to the biosphere is a field of study just beginning to emerge. This will require much research in our future. But, whether the need in 2010 is for 1000 materials scientists or 100,000, matters little. The greater need is for technologically literate citizens.

1.3 ENGINEERING MATERIALS AND SELECTED APPLICATIONS

All engineers need to think about materials. Indeed, even the most "insignificant", like programming or framework building or system engineering rely on the improvement of new materials. Expanding utilizations of framework building are in materials producing (mechanical designing) and complex ecological frameworks.

Advancement in designing implies utilization of another material for a particular application. For instance: plastic cans instead of metallic cans. So it is important that the expert engineer ought to know how to choose materials which best fit the demands of the configuration or design – monetary, stylish and aesthetic requests, and in addition requests of quality and toughness. Heretofore the planner must comprehend the properties of materials, and their restrictions. Accordingly, it is imperative that each architect must study and comprehend the ideas of Materials Science and Engineering. This empowers the engineer:

- to choose a material for a given use in view of contemplations of expense and execution;
- to comprehend the points of confinement of materials and the change of their properties with use;

- to have the capacity to make another material that will have some attractive and desirable properties; and
- to have the capacity to utilize the material for various applications.

In this segment a brief study of vital building materials and their properties is offered as a prologue to the rest of the book. By engineering materials we mean those that have survived the experience of testing what's more, determination to be consolidated into business items and structures. Like many other things, materials are classified in groups, so that we can handle their complexity. One can classify them based on many criteria, for example, crystal structures (arrangement of atoms and bonds between them), or properties, or use. Metals, ceramics, polymers, composites, semiconductors, and biomaterials constitute the main classes of present engineering materials. Different groupings of engineering solids exist yet our essential concern is with metals, ceramics, polymers and semiconductors.

1.3.1 Metals

High reflectivity, thermal conductivity (ability to conduct heat) and malleability (ability to be thinned by hammering) combined with low electrical resistance are some of the attributes we associate with metals. Materials in this group are composed of one or more metallic elements (such as iron, aluminium, copper, titanium, gold, and nickel), and often also nonmetallic elements (for example, carbon, nitrogen, and oxygen) in relatively small amounts. They also establish the electric fields that prevent the penetration of visible light, causing it to be reflected instead. In addition, metals are generally strong (high tensile strength), stiff (large modulus of elasticity), and tough (resistant to fracture). At the same time they are ductile and tend to deform in a plastic manner prior to breaking. Thus, metals are readily shaped and formed by mechanical forces. The reason for these desirable mechanical properties is traceable to an atomic bonding that allows atoms to slide readily past one another. These materials are portrayed by high temperature and electrical conductivity; solid yet deformable under connected mechanical burdens; misty to light (glossy if cleaned). These qualities are because of valence electrons that are withdrawn from molecules, and spread in an electron ocean that pastes the particles together, i.e., atoms are bound together by metallic bonds and weaker van der Waalls powers. Unadulterated metals are sufficiently bad for some applications, particularly basic applications. Along these lines metals are utilized as a part of compound shape, i.e., a metal blended with another metal to enhance the qualities, e.g., aluminium, steel, metal, gold. Figure 1.3 demonstrates the examination of the bar graph of room temperature thickness values for different metals, pottery, polymers, and composite materials.

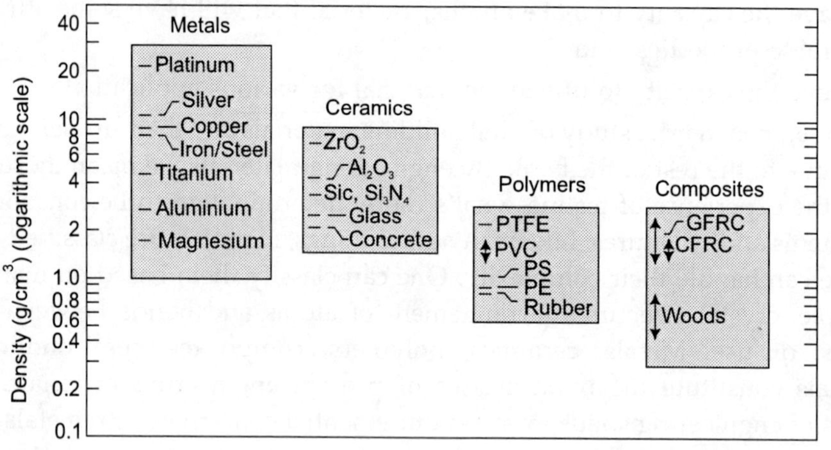

Figure 1.3 Bar chart of room temperature density values for various metals, ceramics, polymers, and composite materials.

1.3.2 Ceramics

These are inorganic compounds, and normally made both of oxides, carbides, nitrides, or silicates of metals. Ceramics production is normally somewhat crystalline and mostly nebulous. Particles in ceramics materials carry on generally like either positive or negative particles, and are bound by exceptionally solid Coulomb forces between them. These materials are portrayed by high quality under pressure, low pliability; typically protectors to warmth and power. Illustrations: glass, porcelain, numerous minerals. Ceramics are compounds between metallic and non-metallic elements; they are most frequently oxides, nitrides, and carbides. For example, some of the common ceramic materials include aluminium oxide (or alumina, Al_2O_3), silicon dioxide (or silica, SiO_2), silicon carbide (SiC), silicon nitride (Si_3N_4), and, in addition, what some refer to as the *traditional ceramics*—those composed of clay minerals (i.e., porcelain), as well as cement, and glass. With regard to mechanical behaviour, ceramic materials are relatively stiff and strong. Stiffnesses and strengths are comparable to those of the metals as shown in Figures 1.4 and 1.5.

Ceramic materials are found in crystalline frame, where atoms are geometrically masterminded in a dreary manner, and additionally in the formless, non-crystalline nature that we connect with glasses. Not at all like metals, for all intents and purposes no electrons are around to direct either charge or warmth, so earthenware production serves as great electrical and warm protectors. Their greatly high softening focuses means that their molecules are unequivocally clung to each other. Under connected stacking such atoms are not anticipated that would move past each other effortlessly. In this manner earthenware production and glasses are solid when packed, yet display next to no plastic extending when pulled by elastic powers. Or maybe, they have a

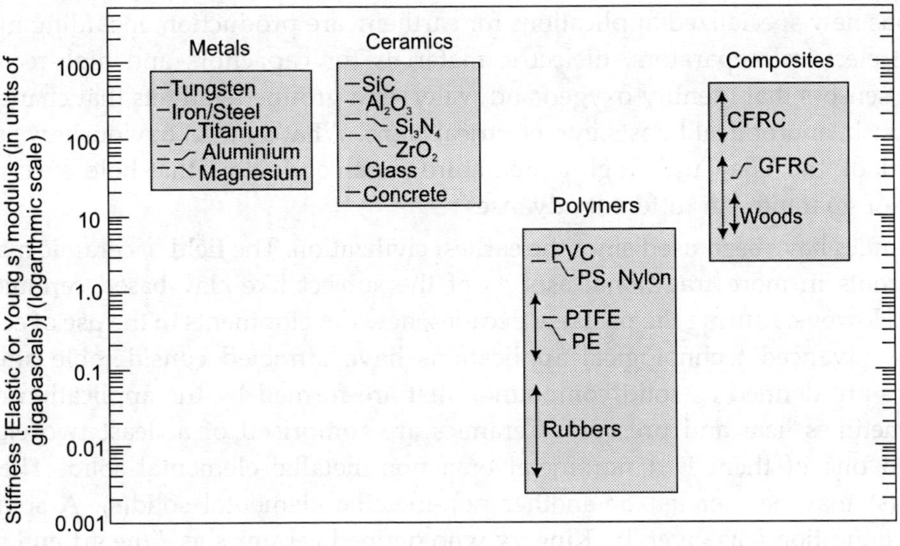

Figure 1.4 Bar-chart representation of room temperature stiffness (i.e., elastic modulus) values for various metals, ceramics, polymers, and composite materials.

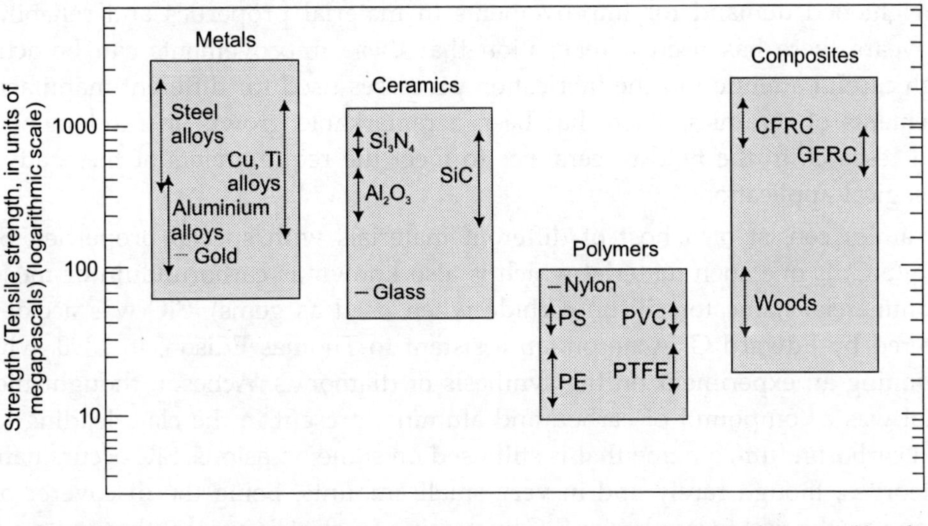

Figure 1.5 Bar-chart representation of room temperature strength (i.e., tensile strength) values for various metals, ceramics, polymers, and composite materials.

deadly inclination to crack in a weak way under pressure. Albeit strenuous endeavours have been made to toughen pottery, their utilization as auxiliary materials, besides cement, is constrained. Then again, applications have moved well past refractories, washroom installations, and white product. Last decades have seen the development of numerous new specialized applications for earthenware production, including magnets that are electrical separators, dielectric materials for capacitors and high-recurrence circuits, sensors that identify oxygen and water, and grouped gadgets that change over warm, push, and optical boosts into electrical signs. What's more, obviously we should not overlook the $YBa_2Cu_3O_7$ high-temperature superconductors that hold such a great amount of guarantee in different advances.

Ceramics have been used since the earliest civilization. The field of ceramic materials has its roots in more traditional aspects of the subject like clay based ceramics and glasses. However, during the past few decades, new developments in the use of ceramics in more advanced technological applications have attracted considerable attention. Ceramics are defined as solid compounds that are formed by the application of heat and sometimes heat and pressure. Ceramics are comprised of at least two elements provided one of them is a non-metal or a non-metallic elemental solid. The other element(s) may be metal(s) or another non-metallic elemental solid(s). A somewhat simpler definition was given by Kingery who defined ceramics as, "the art and science of making and using solid articles, which have as their essential component, and are composed in large part of, inorganic nonmetallic materials."

The increasing use of ceramics in advanced technological applications has resulted in a heightened demand for improvements in material properties and reliability. In recent years there has been a realization that these improvements can be achieved through careful attention to the fabrication processes used for different manufacturing requirements of ceramics. There has been a remarkable growth in fundamental and applied research in the field of ceramics to meet the requirements of these advanced technological applications.

Ceramics consist of a host of different materials with special properties. Silicon carbide (SiC) is one such material which is also known as carborundum or mossanite (the commercial name for silicon carbide when used as gems). SiC was accidentally discovered by Edward G. Acheson, an assistant to Thomas Edison, in 1890, when he was running an experiment on the synthesis of diamonds. Acheson thought the new material was a compound of carbon and alumina, present in the clay, leading him to name it carborundum, a name that is still used on some occasions. SiC occurs naturally in meteorites, though rarely and in very small amounts. Being the discoverer of SiC, Acheson was the first to synthesize SiC by passing an electric current through a mixture of clay and carbon. Nowadays, SiC is still produced via a solid state reaction between sand (silicon dioxide) and petroleum coke (carbon) at very high temperatures in an electric arc furnace. A purer product is produced with different advanced manufacturing processes. SiC, as a technical ceramic, is produced in two main ways. Reaction bonded SiC is made by infiltrating compacts made of mixtures of SiC and C with liquid Si (such as Poco Graphite samples). The silicon reacts with the carbon forming SiC. The reaction product bonds the SiC particles. Sintered SiC is produced from pure SiC powder with non-oxide sintering aids (such as Coors Tek sample).

1.3.3 Polymers

Man's quest for new and improved materials has been expanding with time and it can be said to be unending. The materials that have grown into familiarity and acceptance with the ages since the dawn of man's existence on the mother planet, the earth are: pieces of rocks and stones, sand, soil and various ceramic items; nails, horns, hydes, skins and bones of animals; wood, leaves and plant fibres, covering grass and straw, cotton, coir, jute, hemp, hair, wool and many other fibres of plant and animal origin; silk fibre of insect origin; natural adhesive / film - forming gums and resins (e.g. agar, algies, rosin, shellac, etc.); fossil fuels, viz., coal, lignite, natural gas and petroleum; glass and quartz, and metals and alloys extracted from minerals and ores. Polymers, as a class of materials with potential for use as rubbers, resins, plastics and composites, and as adhesives, laminates and coatings came nearly in the end of the chain of discoveries and developments of materials. Unfolding of the science of polymers and polymer - based materials really had its beginning and headway in the second and third decade of the 20th century.

Though introduced very late in the chain of materials, polymers occupy a major place and pivotal position in our materials map today. In application prospects and performance characteristics and in property range and diversity, they offer novelty and versatility that can hardly be matched by any other kind of materials. Polymers have gone deep, and far and wide in moulding the present-day human civilization and culture.

Even though scientists, particularly the chemists used to talk about polymers earlier to the early 20th century. There was a lot of confusion over the basic understanding of the structures of polymer molecules. It was a common experience for chemists working with polymers that most such materials were very viscous and tacky under melt or solution conditions. One could readily spread the melt or solution into thin films or draw them into fine filaments. In solution, they were recognized more as colloids or associated molecules. Attempts to find their molecular weights from dilute solutions in suitable solvents by cryoscopy often produced irreproducible, doubtful and uncertain and sometimes very high values. For natural rubber, rayon and cellulose derivatives, molecular weights ranging 45,000 - 50,000 or even higher were measured. Such high molecular weight values implied that the relevant polymer molecules were really very large; but this viewpoint was not favoured or accepted in view of a total lack of structural concept about such large or big molecules. The chemists continued to favour the concept of large associations of much smaller molecules of short-chain or cyclic structures. A state of growing dilemma and confusion imparted fresh impetus to the thinking about the size, shape, complexity, and behavioral patterns of the molecular systems called polymers.

Close your eyes, connect, and the odds are great that you have touched a polymer! More polymers are currently delivered on a volume premise than the majority of the metals consolidated, a consequence of their low normal thickness (≈1.5 Mg/m^ or 1.5

g/cm^ contrasted and ≈7.5 Mg/m^ for run of the mill metals). Their minimal effort and thickness, combined with straightforwardness in preparing and toughness, have been key purposes behind the substitution of polymers for metals, glasses, and common items. The outcome has been across the board utilization of polymers in apparel, fabrics, sustenance and drink compartments, bundling of different sorts, furniture, and sporting equipment.

Polymers as thermoplastics (nylon, polyethylene, polyvinyl chloride, elastic, and so forth.) comprise atoms that have covalent holding inside every particle and van der Waals constrains between them. Polymers as thermosets (e.g., epoxy, phenolics, and so on) comprise a system of covalent bonds. They depend on H, C and other non-metallic components. Polymers are formless, aside from a minority of thermoplastics. Because of the sort of holding, polymers are regularly electrical and warm protectors. In any case, directing polymers can be acquired by doping, and leading polymer-lattice composites can be obtained by the utilization of leading fillers. They break down at direct temperatures (100 – 400°C), and are lightweight.

1.3.4 Composite Materials

A composite material can be defined as a combination of two or more materials that results in better properties than those of the individual components used alone. In contrast to metallic alloys, each material retains its separate chemical, physical, and mechanical properties. The two constituents are a reinforcement and a matrix. The main advantages of composite materials are their high strength and stiffness, combined with low density, when compared with bulk materials, allowing for a weight reduction in the finished part. The reinforcing phase provides strength and stiffness. In most cases, the reinforcement is harder, stronger, and stiffer than the matrix. The reinforcement is usually a fibre or a particulate. Particulate composites have dimensions that are approximately equal in all directions. They may be spherical, platelets, or any other regular or irregular geometry. Particulate composites tend to be much weaker and less stiff than continuous fibre composites, but they are usually less expensive. Particulate reinforced composites usually contain less reinforcement (up to 40 to 50 volume per cent) due to processing difficulties and brittleness.

Continuous-fibre composites are often made into laminates by stacking single sheets of continuous fibres in different orientations to obtain the desired strength and stiffness properties with fibre volumes as high as 60 to 70 per cent. Fibres produce high-strength composites because of their small diameter; they contain far fewer defects (normally surface defects) compared to the material produced in bulk. As a general rule, the smaller the diameter of the fibre, the higher its strength, but often the cost increases as the diameter becomes smaller. In addition, smaller-diameter high-strength fibres have greater flexibility and are more amenable to fabrication processes such as weaving or forming over radii. Typical fibres include glass, aramid, and carbon, which may be continuous or discontinuous. The continuous phase is the matrix, which is a polymer, metal, or ceramic. Polymers have low strength and stiffness, metals have intermediate strength and stiffness but high ductility, and ceramics have high strength and stiffness but are brittle. The matrix (continuous phase) performs several critical functions, including maintaining the fibres in proper orientation and spacing and protecting them from abrasion and the environment. In polymer and metal matrix composites that form a strong bond between the fibre and the matrix, the matrix transmits loads from the matrix to the fibres through shear loading at the interface. In ceramic matrix composites, the objective is often to increase the toughness rather than the strength and stiffness; therefore, a low interfacial strength bond is desirable.

Composite materials are multiphase materials acquired by counterfeit blend of various materials to accomplish properties that the individual parts can't achieve. An illustration is a lightweight brake plate got by inserting SiC particles in Al-combination alloy matrix framework. Another example is reinforced cement concrete, a structural composite obtained by combining cement (the matrix, i.e., the binder, obtained by a reaction known as hydration, between cement and water), sand (fine aggregate), gravel (coarse aggregate), and, thick steel fibres. Also there are some natural composites which are available in nature, for example, wood. In general, composites are classified according to their matrix materials. The main classes of composites are metal-matrix, polymer-matrix, and ceramic-matrix. A composite is made out of two (or more) individual materials, which originate from the classification talked about above, viz., metals, ceramics, and polymers. The outline objective of a composite is to accomplish a blend of properties that is not shown by any single material, furthermore to fuse the best qualities of each of the segment materials. Countless sorts exist that are spoken to by various blends of metals, pottery, and polymers. Moreover, a few actually happening materials are likewise thought to be composites for instance, wood and bone. Nonetheless, a large portion of those we consider in our discussions are manufactured/synthetic (man-made) composites. One of the most common and familiar composites is fibre glass, in which small glass fibres are embedded within a polymeric material (normally epoxy or polyester). The glass fibres are relatively strong and stiff (but also brittle), whereas the polymer is ductile (but also weak and flexible). Thus, the resulting fibre glass is relatively stiff, strong, (Figures 1.4 and 1.5) flexible and ductile.

Materials can be classified as either isotropic or anisotropic. Isotropic materials have the same material properties in all directions, and normal loads create only normal strains. By comparison, anisotropic materials have different material properties in all directions at a point in the body. There are no material planes of symmetry, and normal loads create both normal strains and shear strains. A material is isotropic if the properties are independent of direction within the material. Bulk materials, such as metals and polymers, are normally treated as isotropic materials, while composites are treated as anisotropic. However, even bulk materials such as metals can become anisotropic, for example, if they are highly cold worked to produce grain alignment in a certain direction.

1.3.5 Semiconductors

Semiconductors are covalent in nature. Their nuclear structure is described by the most astounding possessed vitality/energy band (the valence band, where the valence electrons live vivaciously) full with the end goal that the vitality crevice between the highest point of the valence band and the base of the unfilled vitality band (the conduction band) is sufficiently little for some division of the valence electrons to be energized from the valence band to the conduction band by warm, optical, or different types of vitality. Their electrical properties depend to a great degree firmly on moment extents of contaminants. They are usually doped in order to enhance electrical conductivity. They are utilized as a part of the type of single precious stones without disengagements since grain limits and separations would debase electrical conduct. They are obscure to noticeable light however straightforward to the infrared. Examples: silicon (Si), germanium (Ge), and gallium arsenide (GaAs, a compound semiconductor). Dissimilar to alternative materials considered, which have changed our reality essentially in transformative ways, semiconductors have unleashed an innovative upheaval in numerous ranges. Semiconductors have electrical properties that are intermediate between the electrical conductors (viz., metals and metal alloys) and insulators (viz., ceramics and polymers) as shown in Figure 1.6. Furthermore, the electrical characteristics of these materials are extremely sensitive to the presence of minute concentrations of impurity atoms, for which the concentrations may be controlled over very small spatial regions. Semiconductors have made possible the advent of integrated circuitry that has totally revolutionized the electronics and computer industries (not to mention our lives) over the past three decades.

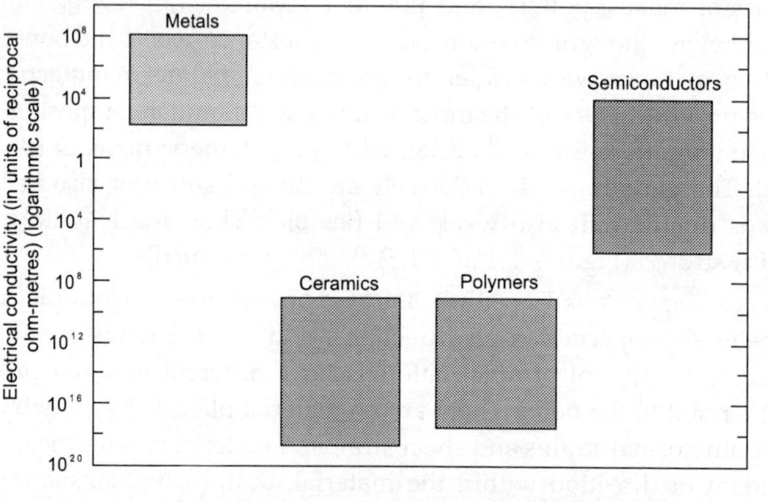

Figure 1.6 Bar chart representation of room temperature electrical conductivity ranges for semiconducting materials metals, ceramics, and polymers.

1.3.6 Biomaterials

These are any type of material that can be used for replacement of damaged or diseased human body parts. Primary requirement of these materials is that they must be biocompatible with body tissues, and must not produce toxic substances. Other important material factors are: ability to support forces; low friction, wear, density, and cost; reproducibility. Typical applications involve heart valves, hip joints, dental implants, intraocular lenses. Examples: Stainless steel, Co-28Cr-6Mo, Ti-6Al-4V, ultra high molecular weight polyethylene, high purity dense Al-oxide, etc. Biomaterials are employed in components implanted into the human body for replacement of diseased or damaged body parts. These materials must not produce toxic substances and must be compatible with body tissues (i.e., must not cause adverse biological reactions). All of the above materials metals, ceramics, polymers, composites, and semiconductors may be used as biomaterials.

1.3.7 Materials of Future: Smart Materials and Nanomaterials

Gathering of new and state-of-the-art materials now being created, and anticipated that would have huge impact on the present-day advances, particularly in the fields of drug, assembling and protection. Keen/intelligent material framework comprises some sort of sensor (recognizes an info) and an actuator (performs responsive and versatile capacity). Actuators might be called upon to change shape, position, normal recurrence, mechanical attributes because of changes in temperature, electric/attractive fields, dampness, pH, and so on. Four types of materials used as actuators are: shape memory alloys, piezoelectric ceramics, magnetostrictive materials, electro-/magnetorheological fluids. Materials/devices used as sensors in day-to-day applications: optical fibres, piezoelectric materials, micro-electromechanical systems (MEMS), nano-electro-mechanical systems (NEMSs), etc. Typical applications are by incorporating sensors, actuators and chip processors into system, researchers are able to stimulate biological human-like behaviour; fibres for bridges, buildings, and wood utility poles; (they also help in fast moving and accurate robot parts, high speed helicopter rotor blades); actuators that control chatter in precision machine tools; small microelectronic circuits in machines ranging from computers to photolithography prints; health monitoring detecting the success or failure of a product.

Smart (or intelligent) materials are a gathering of new and best in class materials presently being produced that will impact a hefty portion of our advances. The descriptive word "smart" suggests that these materials can sense changes in their surroundings and react to these progressions in foreordained conduct— qualities that are likewise found in living beings. Likewise, this "smart" idea is being reached out to rather modern frameworks that comprise both smart and customary materials.

Until very recent times the general procedure utilized by scientists to understand the chemistry and physics of materials has been to begin by studying large and complex structures, and then to investigate the fundamental building blocks of these structures that are smaller and simpler. This approach is sometimes termed "top-down" science. However, with the advent of scanning probe microscopes, which permit observation of individual atoms and molecules, it has become possible to manipulate and move atoms and molecules to form new structures and, thus, design new materials that are built from simple atomic-level constituents (i.e., "materials by design"). This ability to carefully arrange atoms provides opportunities to develop mechanical, electrical, magnetic, and other properties that are not otherwise possible. We call this the "bottom-up" approach, and the study of the properties of these materials is termed "nanotechnology"; the "nano" prefix denotes that the dimensions of these structural entities are on the order of a nanometre (10^{-9} m) – as a rule, less than 100 nanometres (equivalent to approximately 500 atom diameters).

1.3.8 Advanced Materials

These are materials utilized as a part of high-tech gadgets those work in view of generally unpredictable and modern standards (e.g., PCs, air/spacecrafts, electronic devices, and so on.). These materials are either conventional materials with upgraded properties or recently created materials with elite capacities. Consequently, these are generally costly. Ordinary applications: coordinated circuits, lasers, LCDs, fibre optics, warm security for space carry, and so on. Cases: metallic froths, between metallic mixes, multi-segment amalgams, attractive combinations, extraordinary pottery and high temperature materials, and so on.

A functionally graded material (FGM) is a two-component composite characterised by a compositional gradient from one component to the other. In contrast, traditional composites are homogeneous mixtures, and they therefore involve a compromise between the desirable properties of the component materials. Since significant proportions of an FGM contain the pure form of each component, the need for compromise is eliminated. The properties of both components can be fully utilised. For example, the toughness of a metal can be mated with the refractoriness of a ceramic, without any compromise in the toughness of the metal side.

Modern materials are developed through the invention of new or improved processes, for example, as a result of man-made materials/ingredients or human intervention, in other words not naturally occurring changes. They are altered to perform a particular function. Many smart and modern materials are developed for specialised applications but some eventually become available for general use.

Smart materials respond to differences in temperature or light and change in some way. They are called smart because they sense conditions in their environment and respond to those conditions. Smart materials appear to 'think' and some have a 'memory' as they revert to their original state. The term 'smart' can be ambiguous as in some cases it is difficult to distinguish between modern and smart.

Many naturally occurring food ingredients are smart in that they respond to heat and light and some changes are reversible. Such working characteristics are frequently exploited in food technology.

Modified starches respond to differences in temperatures, for example, they swell (thickening) in hot water or when heated, but return to a flow when cool. This working characteristic is used in pizza toppings. The topping thickens when heated in the oven and so does not run off the base, but on slight cooling the topping is runny again ready for eating. Other modified starches are used in instant desserts, which thicken without heating but do not return to their original state. Examples of modern food materials include genetically modified foods, antioxidants, modified enzymes, probiotic yoghurts/drinks, TVP, Quorn and Tofu.

Fibre and fabric technological developments have created a whole range of smart and modern textiles which can be used in many applications. These textiles have been used in functional sportswear, medical and safetywear and fashion clothing. Smart fabrics have been developed which can create a sense of well-being – they have anti-stress or calm-inducing properties. Aristoc, for example, has developed a range of well-being tights micro encapsulated with fragrant oils, moisturisers or vitamins. Smart textiles have a number of medical uses. Fabrics can be encapsulated with substances required by the body or antiseptics. Allergy control fabrics can be used in bedding for people with breathing problems caused by dust mites. Other smart textiles include sanitised fabrics for sportswear and socks which have antimicrobial protection. Antibacterial and antifungal fabrics have been used in clothing, linens, towels and carpets. Many synthetic fibres now have moisture management properties.

1.3.9 Need of Modern Materials

In spite of the fact that there have been colossal advancements throughout the decades in the field of materials science and designing, development of new innovations, and requirement for better exhibitions of existing advances requests substantially more from the materials field. Moreover, it is clear that new materials/advancements should have been ecologically amicable. Some commonplace needs, therefore, of present-day materials needs are recorded in the accompanying:

- At high temperatures, engine efficiency increases this requires high temperature structural materials.
- Nuclear energy applications that require solving problem with residues and advances in nuclear waste processing.
- Light materials for hypersonic flight applications.
- Construction related to civil that is metals with unbreakable windows.
- Structures like plastics which are corrosion resistant.
- Optical communications with optical fibres.

1.4 SUMMARY

This chapter has endeavoured to perform the part of materials inside more extensive authentic and societal settings, and also inside building what's more, assembling exercises. At the centre of the subject is the structure properties logical and scientific base. Concealing the centre are the designing worries of preparing, processing, performance and execution. For millennia, processing and performance of engineering solids were what mattered, and because the former was an art, the latter was variable in practice. Presently, structure-property connections are perceived as the way to comprehension the major way of materials, enhancing their properties, and making new materials. Hence, a hefty portion of the sections expand on double logical structure-designing property subjects. Further, in Section IV the book deals with the properties of engineering materials and their processing which constitutes of all the engineering materials like metals, polymers, ceramics, etc. In Section III (Chapter 8) the Corrosion of Metals are discussed in detail. Also later Chapter 12 (Section IV) explains more about the Engineering Materials.

EXERCISE

1. Why do we need to study Material Science? What is the importance of materials in our life?
2. Why is it more difficult to recycle polymers than metals?
3. How many products can you name that contain the following?
   (a) Metals, semiconductors, ceramics, and polymers.
   (b) Metals, ceramics, and polymers.
4. Materials-designated technologists (i.e., metallurgists, ceramists) have been called engineers' engineers. Why?
5. Sports equipment have undergone and continue to undergo great changes in the choice of materials used. Focus on the equipment used in your favourite sport and specify:
   (a) What materials change or substitution has occurred?
   (b) Reasons for the change.
6. It has been theorized that metals were first produced from oxide ores by ancient potters glazing and firing clayware. Suggest a possible scenario for such an occurrence.
7. Why does it take more energy to extract metals from oxide ores that are leaner in metal content?
8. Explain the importance of modern materials in day to life.
9. Is material selection an important concern for any production unit? Support your answer with relevant examples.
10. What are the several types of engineering materials? List some of their applications.

2

Bonding: Atomic Structure and Bonding in Solids

Learning Objectives

After going through this chapter the reader will be able to:

- Describe atomic electrons in single atom, defining atomic weight, isotopes and atomic mass units (amu).
- Describe the Pauli's principle and electronic configuration.
  (a) Explaining the periodic table and Bohr's atomic model.
  (b) Note on electrons in molecules and solids.
- (a) Briefly describe ionic, covalent, metallic, hydrogen, and van der Waals bonds.
  (b) Note which materials exhibit each of these bonding types.

2.1 INTRODUCTION

A portion of the essential properties of strong materials relies on geometrical nuclear gameplans, furthermore the associations that exist among constituent atoms. This part, by method for arrangement for resulting exchanges, considers a few basic and critical ideas—to be specific, nuclear atomic structure, electron designs and configurations in atoms and the periodic table, and the different sorts of essential and optional between nuclear bonds that hold together the particles involving a strong bond. These subjects are looked into quickly, under the supposition that a portion of the material is natural to the reader. It is all around acknowledged that atoms impact materials properties, yet which subatomic segments of atoms (e.g., electrons, cores comprising protons, neutrons) impact which properties is not all that self-evident. Before tending to this inquiry it is fundamental to survey a few basic ideas presented in essential chemistry courses. Elements are recognized by their atomic numbers and atomic weights. Inside every atom is a nucleus containing various positively charged protons that are equivalent to the atomic number (Z). Circulating about the nucleus are Z electrons that keep up electrical neutrality in the particle. The core additionally contains various neutrons; these are uncharged.

2.2 ATOMIC ELECTRONS IN SINGLE ATOMS

Every particle or atom comprises a little core of protons and neutrons, which is circled by moving electrons. Both electrons and protons are electrically charged, the charge size being which is negative in sign for electrons and positive for protons; neutrons are electrically unbiased. Masses for these subatomic particles are imperceptibly little; protons and neutrons have roughly same mass, which is essentially bigger than that of an electron.

2.2.1 Atomic Number (*Z*) and Atomic Weight

Every substance component is portrayed by the quantity of protons in the core, or the atomic number (Z). For an electrically neutral or complete atom, the nuclear number likewise breaks even with the quantity of electrons. This atomic number ranges in fundamental units from 1 for hydrogen to 92 for uranium. The atomic mass (*A*) of a particular particle might be communicated as the entirety of the masses of protons and neutrons inside the core. Despite the fact that the quantity of protons is the same for all molecules of a given component, the quantity of neutrons (*N*) might be variable. Therefore, particles of a few components have at least two distinctive nuclear masses, which are called isotopes. The atomic weight of a component compares to the weighted normal of the nuclear masses of the particle's actually happening isotopes. The atomic mass unit (amu) might be utilized for calculations of atomic weight.

A scale has been built up whereby 1 amu is characterized as of the nuclear mass of the most widely recognized isotope of carbon, Carbon 12. Inside this plan, the masses of protons and neutrons are marginally more prominent than solidarity, and

$$A \approx Z + N$$

The nuclear weight of a component or the sub-atomic weight of a compound might be indicated on the premise of amu per particle (atom) or mass per mole of material. In one mole of a substance there are 6.023 × 1023 (Avogadro's number) atoms or particles. These two nuclear weight plans are connected through the accompanying condition:

$$1 \text{ amu/atom (or molecule)} = 1 \text{ g/mol}$$

For instance, the nuclear weight of iron is 55.85 amu/iota, or 55.85 g/mol. Sometimes utilization of amu per particle or atom is advantageous; otherwise g (or kg)/mol is favoured.

Before appreciating the role of electrons in solids it is necessary to understand their behaviour in single isolated atoms. The simplest model of an atom assumes it to be a miniature solar system at whose centre is a positively charged nucleus (the sun) surrounded by a cloud of orbiting negatively charged atomic electrons (the planets). Charge is distributed so that the atom is electrically neutral. We now know these atomic electrons display a complex dynamical behaviour (the larger the Z, the more complex the behaviour) governed by the laws of quantum mechanics. The underlying philosophy and mathematical description of quantum theory are also involved. Nevertheless, the resulting concepts and laws that derive from them can be summarized in terms of a few relatively simple equations and rules that are discussed in turn.

2.2.2 Pauli's Principle

In all actuality, multi-electron atoms and hydrogen are more confusing than the straightforward Bohr model (Figure 2.1). Electrons circling nearer the core shields external electrons from drawing the atomic charge. This convolutes their movement adequately that diverse electrons are not at a similar vitality level; moreover, electron energies are not effectively figured. When all is said in done, more propelled speculations demonstrate that the electron flow inside all atoms is described by four quantum numbers n, I, m, and s. These emerge from answers for the observed Schrödinger condition, a foundation of the advanced quantum depiction of atoms. In particular, the three-dimensional movement of electrons is encapsulated in the quantum numbers n, I, and m.

The foremost quantum number is still n and it can expect just whole number qualities 1, 2, 3,.... Electrons are presently composed into shells. At the point when $w = 1$ we talk about the K electron shell, while for the L and M shells, $n = 2$ and $w = 3$, separately. As will be apparent after presentation of the other quantum numbers, there are 2 electrons in the K shell, 8 in the L shell, 18 in the M shell, etc. The precise energy emerging from the rotational movement of circling electrons is likewise quantized or compelled to expect particular qualities which are in the proportion of entire numbers. Acknowledgment of this reality is taken by allotting to the orbital quantum number estimations of 0, 1, 2, ... The state of electron orbitals is basically controlled by the quantum number. At the point when $l = 0$ we talk about s electron states. These electrons have no net precise energy, and as they move every which way with equivalent likelihood, the charge dispersion is circularly symmetrical about the core. For $n = 1, 2, 3, \ldots$, we have comparing p, d, f, ... states.

The third quantum number, m, indicates the introduction of the precise momenturn along a particular heading in space. Known as the magnetic quantum number, m goes up against the whole number values amongst $+l$ and $-l$, that is $-l, -l + 1, \ldots, +l - 1, +l$.

In conclusion there is the spin quantum number, s, in acknowledgement of the way that electrons turn as they all the while circle the core. Since there are just two introductions of turn precise energy, up or down, m accept $-1/2$ and $+1/2$ values.

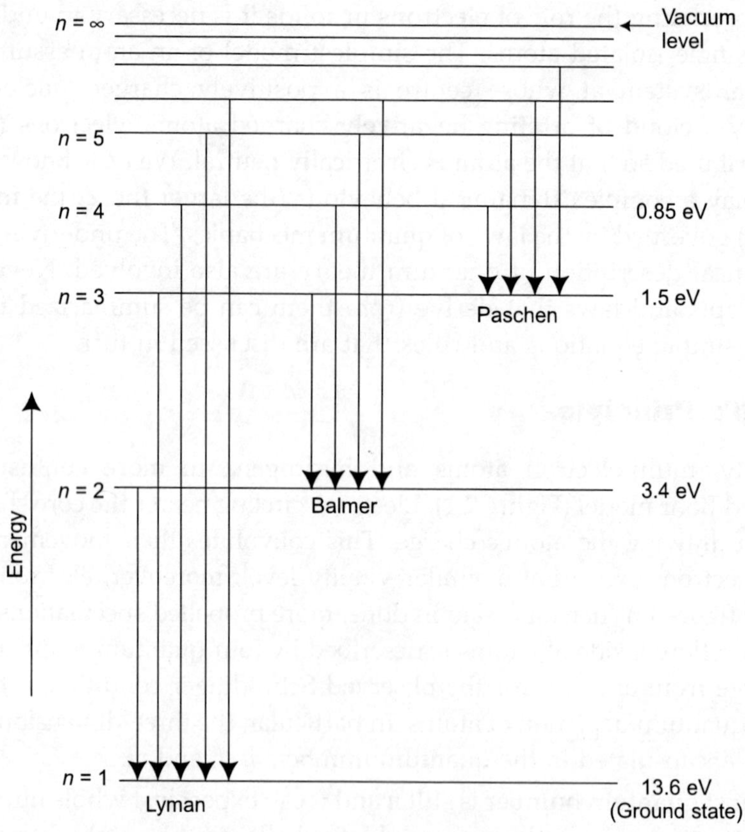

Figure 2.1 Electron energy levels in hydrogen atom.

Pauli's principle states that no two electrons in an atom can have the same four quantum numbers.

Let us apply Pauli's principle to an atom of sodium. As $Z = 11$ we have to specify a tetrad of quantum numbers (n, l, m, s) for each of the 11 electrons:

1*s* states [*K* shell) (1, 0, 0, 1/2) and (1, 0, 0, –1/2);

2*s* states (*L* shell) (2, 0, 0, 1/2) and (2, 0, 0, –1/2);

2*p* states (*L* shell) (2, 1, 0, +1/2), (2, 1, 0, –1/2), (2, 1, 1, +1/2), (2, 1, 1, –1/2), (2, 1, – 1, + 1/2), and (2, 1, –1, –1/2);

3*s* state (*M* shell) (3, 0, 0, +1/2).

Another way to identify the electron distribution in sodium is $1s^2\ 2s^2\ 2p^6\ 3s^1$ and similarly for other elements. In shorthand notation the integers and letters are the principal and orbital quantum numbers, respectively, and the superscript number tells how many electrons have the same *n* and *l* values.

Table 2.1 Electron configuration for some common elements

Element	Symbol	Atomic Number	Electron Configuration
Hydrogen	H	1	$1s^1$
Helium	He	2	$1s^2$
Lithium	Li	3	$1s^2 2s^1$
Beryllium	Be	4	$1s^2 2s^2$
Boron	B	5	$1s^2 2s^2 2p^1$
Carbon	C	6	$1s^2 2s^2 2p^2$
Nitrogen	N	7	$1s^2 2s^2 2p^3$
Oxygen	O	8	$1s^2 2s^2 2p^4$
Fluorine	F	9	$1s^2 2s^2 2p^5$
Neon	Ne	10	$1s^2 2s^2 2p^6$
Sodium	Na	11	$1s^2 2s^2 2p^6 3s^1$
Magnesium	Mg	12	$1s^2 2s^2 2p^6 3s^2$
Aluminium	Al	13	$1s^2 2s^2 2p^6 3s^2 3p^1$
Silicon	Si	14	$1s^2 2s^2 2p^6 3s^2 3p^2$
Phosphorus	P	15	$1s^2 2s^2 2p^6 3s^2 3p^3$
Sulphur	S	16	$1s^2 2s^2 2p^6 3s^2 3p^4$
Chlorine	Cl	17	$1s^2 2s^2 2p^6 3s^2 3p^5$
Argon	Ar	18	$1s^2 2s^2 2p^6 3s^2 3p^6$
Potassium	K	19	$1s^2 2s^2 2p^6 3s^2 3p^6 3s^1$
Calcium	Ca	20	$1s^2 2s^2 2p^6 3s^2 3p^6 3s^2$
Scandium	Sc	21	$1s^2 2s^2 2p^6 3s^2 3p^6 3d^1 4s^2$
Titanium	Ti	22	$1s^2 2s^2 2p^6 3s^2 3p^6 3d^2 4s^2$
Vanadium	V	23	$1s^2 2s^2 2p^6 3s^2 3p^6 3d^3 4s^2$
Chromium	Cr	24	$1s^2 2s^2 2p^6 3s^2 3p^6 3d^5 4s^1$
Manganese	Mn	25	$1s^2 2s^2 2p^6 3s^2 3p^6 3d^5 4s^2$
Iron	Fe	26	$1s^2 2s^2 2p^6 3s^2 3p^6 3d^6 4s^2$
Cobalt	Co	27	$1s^2 2s^2 2p^6 3s^2 3p^6 3d^7 4s^2$
Nickel	Ni	28	$1s^2 2s^2 2p^6 3s^2 3p^6 3d^8 4s^2$
Copper	Cu	29	$1s^2 2s^2 2p^6 3s^2 3p^6 3d^{10} 4s^1$
Zinc	Zn	30	$1s^2 2s^2 2p^6 3s^2 3p^6 3d^{10} 4s^2$
Gallium	Ga	31	$1s^2 2s^2 2p^6 3s^2 3p^6 3d^{10} 4s^2 4p^1$
Germanium	Ge	32	$1s^2 2s^2 2p^6 3s^2 3p^6 3d^{10} 4s^2 4p^2$
Arsenic	As	33	$1s^2 2s^2 2p^6 3s^2 3p^6 3d^{10} 4s^2 4p^3$
Selenium	Se	34	$1s^2 2s^2 2p^6 3s^2 3p^6 3d^{10} 4s^2 4p^4$
Bromine	Br	35	$1s^2 2s^2 2p^6 3s^2 3p^6 3d^{10} 4s^2 4p^5$
Kryton	Kr	36	$1s^2 2s^2 2p^6 3s^2 3p^6 3d^{10} 4s^2 4p^6$

2.2.3 The Periodic Table

Every component has been characterized by arrangement in the occasional table (Figure 2.2). Here, the components are arranged, with expanding nuclear number, in seven flat lines called periods. The gameplan is with the end goal that all components exhibited in a given section or gathering have comparative valence electron structures, and additionally concoction and physical properties. These properties change progressively, moving evenly over every period and vertically down every segment. The components situated in Group 0, the farthest right gathering, are the dormant gases, which have filled electron shells and stable electron setups. Group VIIA and VIA components are one and two electrons lacking, separately, from having stable structures. The Group VIIA components (F, Cl, Br, I, and At) are now and again named the incandescent light. The salt and the antacid earth metals (Li, Na, K, Be, Mg, Ca, and so on.) are named as Groups IA and IIA, having, separately, one and two electrons in abundance of stable structures. The components in the three long stretches, Groups IIIB through IIB, are named the move metals, which have somewhat filled *d* electron states and sometimes maybe a couple electrons in the following higher vitality shell. Bunches IIIA, IVA, and VA (B, Si, Ge, As, and so.) in plain view qualities that are middle between the metals and non-metals by temperance of their valence electron structures.

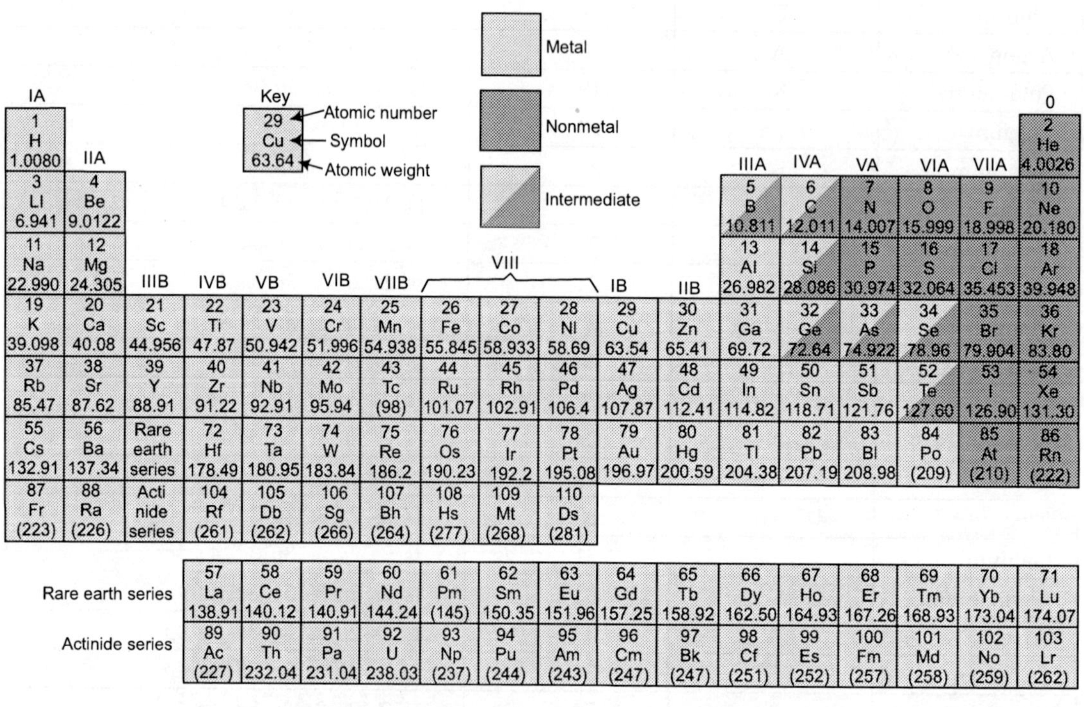

Figure 2.2 The periodic table of the elements.

As might be noted from the occasional table, a large portion of the components truly go under the metal order. These are some of the time named electropositive components, demonstrating that they are equipped for surrendering their few valence electrons to end up decidedly charged particles. Moreover, the components arranged on the right-hand side of the table are electronegative; that is, they promptly acknowledge electrons to shape contrarily charged particles, or now and then they impart electrons to different atoms. Figure 2.3 showcases electro-pessimism values that have been appointed to the different components orchestrated in the periodic table. When in doubt, electro-pessimism increments in moving from left to right and from base to beat. Particles will probably acknowledge electrons if their external shells are full, and in the event that they are less "protected" from (i.e., nearer to) the core.

IA	IIA	IIIB	IVB	VB	VIB	VIIB	VIII	VIII	VIII	IB	IIB	IIIA	IVA	VA	VIA	VIIA	0
1 H 2.1																	2 He -
3 Li 1.0	4 Be 1.5											5 B 2.0	6 C 2.5	7 N 3.0	8 O 3.5	9 F 4.0	10 Ne
11 Na 0.9	12 Mg 1.2											13 Al 1.5	14 Si 1.8	15 P 2.1	16 S 2.5	17 CI 3.0	18 Ar
19 K 0.8	20 Ca 1.0	21 Sc 1.3	22 TI 1.5	23 V 1.6	24 Cr 1.6	25 Mn 1.5	26 Fe 1.8	27 Co 1.8	28 Ni 1.8	29 Cu 1.9	30 Zn 1.6	31 Ga 1.6	32 Ge 1.8	33 As 2.0	34 Se 2.4	35 Br 2.8	36 Kr
37 Rb 0.8	38 Sr 1.0	39 Y 1.2	40 Zr 1.4	41 Nb 1.6	42 Mo 1.8	43 Tc 1.9	44 Ru 2.2	45 Rh 2.2	46 Pd 2.2	47 Ag 1.9	48 Cd 1.7	49 In 1.7	50 Sn 1.8	51 Sb 1.9	52 Te 2.1	53 I 2.5	54 Xe
55 Cs 0.7	56 Ba 0.9	57-71 La-Lu 1.2-1.2	72 Hf 1.3	73 Ta 1.5	74 W 1.7	75 Re 1.9	76 Os 2.2	77 Ir 2.2	78 Pt 2.2	79 Au 2.4	80 Hg 1.9	81 TI 1.8	82 Pb 1.8	83 BI 1.9	84 Po 2.0	85 At 2.2	86 Rn
87 Fr 0.7	88 Ra 0.9	89-102 Ac-No 1.1-1.7															

Figure 2.3 The electronegativity values for the elements.

Example 2.1 X, Y and Z are in the same period. Using the given information below, find relation between their metallic properties.

(i) Atomic number of Y is 12

(ii) Formula of compound produced by X and Y is YX_2

(iii) Z^{-2} and X^{-} have equal number of electrons

Solution:

(i) $_{12}Y$ has electron configuration: $1s^2 2s^2 2p^6 3s^2$

Y is in 3 period and (ii) A group. Thus, it has +2 value in compounds.

(ii) X in YX_2 compound is in (vii) A group and have −1 value in compound. Atomic number of X is 17.

(iii) Since their number of electrons are equal, Z is in 3 period and (vi) A group. Relation between metallic properties of elements;

$Y > Z > X$

Metallic property decreases when we go from left to right in the same period.

Example 2.2 Which ones of the following statements are true for element having atomic number 34?

(i) It is in *p* block.
(ii) It is a nonmetal.
(iii) Its valence electrons are all in *p* orbitals.

Solution: Electron configuration of element is: $1s^2 2s^2 2p^6 3s^2 3p^6 4s^2 3d^{10} 4p^4$

(i) Element is in *p* block, since the last electron is in *p* orbital. (i) is true.
(ii) In outer shell (4) it has 4 + 2 = 6 electrons. So it is a nonmetal. (ii) is true.
(iii) 4 valence electrons of element are in *p* orbital and 2 of them are in *s* orbital. So, (iii) is false.

Example 2.3 If an element has 15 filled and 1 half filled orbitals, which one of the following statements is false?

(i) Atomic number of element is 31
(ii) It is in *p* block of periodic table
(iii) It is transition metal
(iv) It is in 4 period
(v) It is in (iii) A group

Solution: We write orbitals on electron configuration; $1s^2 2s^2 2p^6 3s^2 3p^6 4s^2 3d^{10} 4p^1$ 1 + 1 + 3 + 1 + 3 + 1 + 5 + 1 = Filled orbitals and $4p^1$ is one half filled orbital.

(i) Number of electrons in neutral atom is equal to atomic number. Thus, if we sum the number of electrons in the given orbitals, we find atomic mass 31. (i) is true.
(ii) Since the last electron of the element is in *p* orbital, it is *p* block element. (ii) is true.
(iii) Since its last electrons are in *p* orbitals, it is not a transition metal. (If last electrons are in *d* orbitals, element becomes transition metal) (iii) is false.
(iv) Outer shell of element is 4, so it is i 4. period. (iv) is true.
(v) It has 3 valence electrons; 2 in 4*s* and 1 in 4*p*, so it is in (iii) A group. V is true.

Example 2.4 Ionization energies of *X* and *Y* are given in the table below.

	Ionization Energies (Kcal/mol)			
	$I.\ E_1$	$I.\ E_2$	$I.\ E_3$	$I.\ E_4$
X	118	1091	1652	2280
Y	215	420	3548	5020

Using the data given in the table, find which ones of the following statements are definitely true?

(i) Oxidation number of X is 1

(ii) Compound formed by Y and $_7N$ is; Y_3N_2

(iii) X^+ and Y^{+2} are isoelectronic with same noble gas

Solution:

(i) Since increase in the first IE_1 to the second IE_2 is higher than others, oxidation number of X is 1. (i) is true.

(ii) Since increase in the second IE_2 to the third IE_3 is higher than others, oxidation number of Y is 2. In compounds Y takes +2 value. N has electron configuration; N: $1s^2 2s^2 2p^3$ N accepts 3 electrons and has value −3 in compounds. So, N and Y form the following compound; Y_3N_2. (ii) is true.

(iii) X loses 1 electron and Y 2 loses 2 electrons to have a noble gas electron configuration. But, we cannot definitely say that they are isoelectronic.

Example 2.5 Which one of the following elements has the lowest first ionization energy?

(i) $1s^2 2s^2$

(ii) $1s^2 2s^2 2p^2$

(iii) $1s^2 2s^2 2p^4$

(iv) $1s^2 2s^2 2p^5$

(v) $1s^2 2s^2 2p^6 3s^2$

Solution:

(i) $1s^2 2s^2$ is in 2. period and (ii) A group

(ii) $1s^2 2s^2 2p^2$ is in 2. period and (iv) A group

(iii) $1s^2 2s^2 2p^4$ is in 2. period and (vi) A group

(iv) $1s^2 2s^2 2p^5$ is in 2. period and (vii) A group

(v) $1s^2 2s^2 2p^6 3s^2$ is in 3. period and (ii) A group

In the periodic table, ionization energy decreases from top to bottom and right to left. Thus, element given in (v) has the lowest first ionization energy.

Example 2.6 First ionization energies vs atomic mass graph of X, Y, Z, T and R is given below.

If Z is in 3rd period, which one of the following statements is false?

(i) Atomic number of X is 16

(ii) Y is in p block

(iii) Z and T are in same period

(iv) Electron configuration of R shows spherical symmetry property

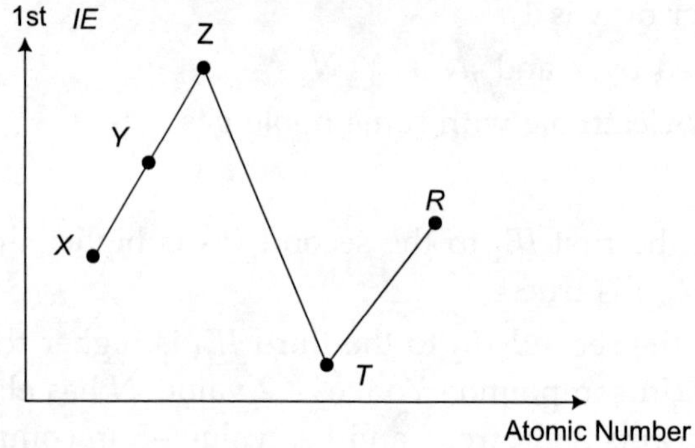

Solution: Since change in the first ionization energy of *Z* and *T* are too high, *Z* is a noble gas and *T* is in (i) A group. Thus, *Z* is a noble gas in the 3rd period and it has the following electron configuration:

Z: $1s^2 2s^2 2p^6 3s^2 3p^6$

Atomic number of Z is 18

(i) Atomic number of *X* = Atomic number of *Z* – 2

Atomic number of *X* is 16. (i) is true.

(ii) *Y* has atomic number 17 and has the following electron configuration;

Y: $1s^2 2s^2 2p^6 3s^2 3p^5$

As you can see *Y* is in *p* block. (ii) is true.

(iii) *Z* is in the third period but *T* is in the fourth period. (iii) is false.

(iv) Electron configuration of *R* is given below:

Z: $1s^2 2s^2 2p^6 3s^2 3p^6 4s^2$

Since all orbitals are filled, *R* shows spherical symmetry property. (iv) is true.

Example 2.7 Which one of the following statements is false for elements given in the periodic table?

(i) Electron affinity of *Y* is larger than others
(ii) *X* and *T* share electrons and form compound.
(iii) *Z* has the largest atomic radius
(iv) Ionic property of *ZY* compound is larger than others

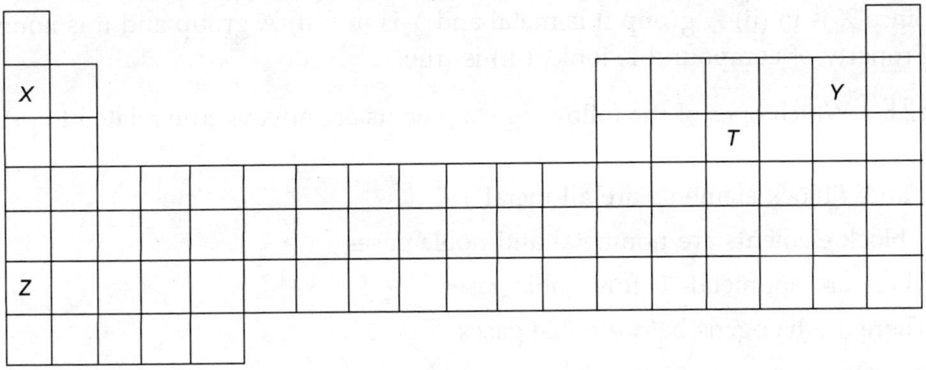

Solution:

(i) Electron affinity increases as we move from left to right and top to bottom in periodic table. Thus, Y has larger electron affinity. (i) is true.

(ii) Since X is in (i) A group, it is metal and T is in VA and nonmetal. So, X and T can form compound by electron transfer, not by electron sharing. (ii) is false.

(iii) Atomic radii increases as we move from right to left and top to bottom in periodic table. So, Z has largest atomic radii, (iii) is true.

(iv) Increasing in the electronegativity increases ionic property of compounds. Electronegativity increases from left to right and decreases from top to bottom in periodic table. Z is the one having smallest electronegativity and K is the one having highest electronegativity. Difference between electronegativities of these elements make compound have largest ionic property. (iv) is true.

Example 2.8 Which ones of the following statements are true for ${}_4X$, ${}_9Y$, ${}_{17}Z$?

(i) X and Z are in same group elements

(ii) X and Y are same period elements

(iii) X and Z form ionic compound

Solution: We first write electron configuration of elements to find group and period numbers.

${}_4X$: $1s^2 2s^2$

${}_9Y$: $1s^2 2s^2 2p^5$

${}_{17}Z$: $1s^2 2s^2 2p^6 3s^2 3p^5$

(i) Number of electrons in the last shell orbitals give group number:
Y and Z are in (vii) A group. (i) is true

(ii) Last shell number gives period number. So, X and Y are in second period. (ii) is true.

(iii) Since X is in (ii) A group it is metal and Y is in (vii) A group and it is nonmetal. Property of compound is ionic. (iii) is true.

Example 2.9 Which ones of the following statements are always true related to periodic table?

(i) s and d block elements are all metal

(ii) p block elements are nonmetal and noble gases

(iii) There are nonmetals before noble gases.

(iv) There are halogens before noble gases

Solution: H is in (i) A group but it is not metal. (i) is false.

p block elements are metals, nonmetals and noble gases. (ii) is false.

When we examine periodic table we see that there are always nonmetals before noble gases. (iii) is true.

H comes before He and it is not halogen. Thus, (iv) is false.

Example 2.10 Electron configuration of X^{+2} last with $2p^6$. Which one of the following elements have similar chemical properties with X element?

(i) $_8Y$

(ii) $_{20}Z$

(iii) $_4T$

Solution: Electron configuration of X^{+2} ion:

X^{+2}: $1s^2\,2s^2\,2p^6$ (X gives 2 electrons)

X: $1s^2\,2s^2\,2p^6\,3s^2$ (X is in (ii) A group)

Elements in same groups show similar chemical properties. Now we find group numbers of given elements:

$_8Y$: $1s^2\,2s^2\,2p^4$ (Y is in (vi) A group)

$_{20}Z$: $1s^2\,2s^2\,2p^6\,3s^2\,3p^6\,4s^2$ (Z is in (ii) A group)

$_4T$: $1s^2\,2s^2$ (T is in (ii) A group)

Z and T are in same group with X, thus they have similar chemical properties.

2.3 ATOMIC ELASTICITY (BOHR'S MODEL)

In the last part of the 19th century it was understood that numerous wonders including electrons in solids couldn't be clarified as far as traditional mechanics was concerned. What took after was the foundation of an arrangement of standards and laws that represent frameworks of nuclear and subatomic elements that came to be known as quantum mechanics. A comprehension of the conduct of electrons in molecules and crystalline solids fundamentally includes the exchange of quantum-mechanical ideas. Nonetheless, a definite investigation of these standards is past the extent of this book, hence, a basic and improved treatment is given. One early outgrowth of quantum mechanics was the disentangled Bohr atomic model, in which electrons are accepted to rotate around the nuclear core in discrete orbits, and the position of a specific electron is pretty much all around characterized as far as its orbital. This model of the particle is shown in Figure 2.4. Another essential quantum-mechanical guideline stipulates that the energies of electrons are quantized; that is, electrons are allowed to have just particular estimations of vitality. An electron may change vitality, however in doing so it must make a quantum hop either to a permitted higher vitality (with retention of vitality) or to a lower vitality (with outflow of vitality). Frequently, it is helpful to think about these permitted electron energies as being connected with vitality levels or states. These states don't shift constantly with vitality; that is, nearby states are isolated by limited energies. For instance, permitted states for the Bohr hydrogen particle are shown in Figure 2.5a. These energies are taken to be negative, though the zero reference is the unbound or free electron. Obviously, the single electron connected with the hydrogen iota will fill one and only of these states. Along these lines, the Bohr demonstrate speaks to an early endeavour to depict electrons in atoms, regarding both position (electron circles) and vitality (quantized vitality levels).

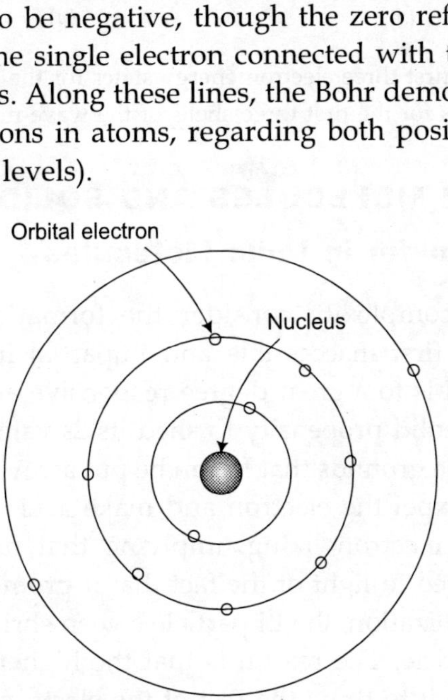

Figure 2.4 Schematic representation of the Bohr atom.

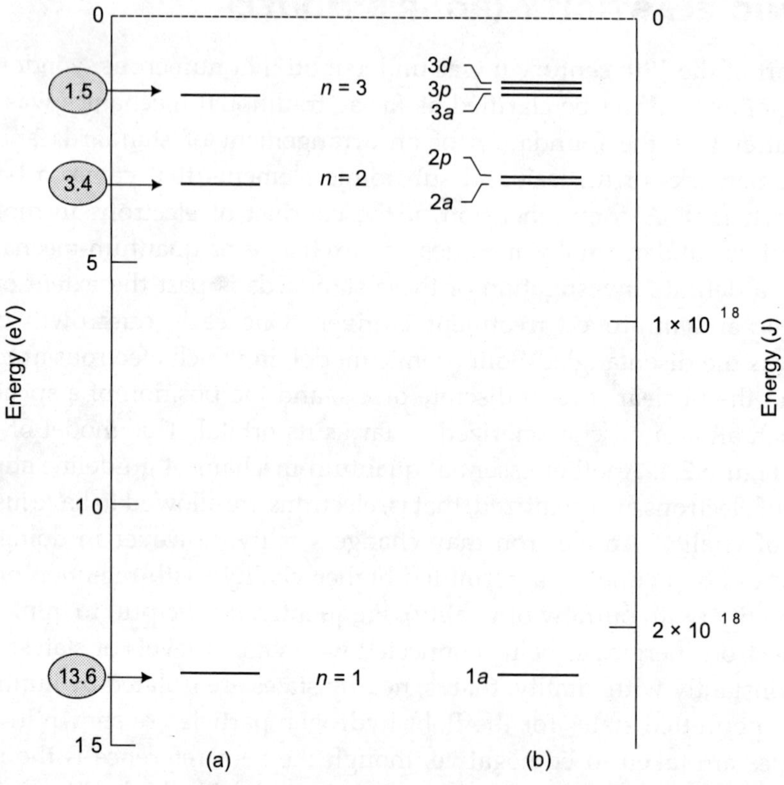

Figure 2.5 (a) The first three electron energy states for the Bohr hydrogen atom. (b) Electron energy states for the first three shells of the wave-mechanical hydrogen atom.

2.4 ELECTRONS IN MOLECULES AND SOLIDS

2.4.1 Electrons Transfer in Ionic Molecules

At a larger amount of complexity consider the formation a LiF particle through collaboration between at first inaccessible and impartial lithium and fluorine atoms (Fig. 2.6). As we know, Li is to a great degree responsive artificially (particularly with water) on account of the solid propensity to shed its 2*s* valence electron. Yet, this does not happen typically on the grounds that Li can be put away as metal. Ionization vitality of 5.4 eV is required to expel the electron and make a Li+ particle. Essentially, the F particle displays a solid electron liking, implying that fuse of an electron into its inadequate 2p shell is urged in light of the fact that it prompts to a vitality decrease of 3.6 eV. Strangely, amid ionization, the Li particle sweep shrivels from 0.157 to 0.078 nm as it turns into a Li+ particle. The reason is that the higher compelling charge on the core in the particle causes it to draw the rest of the electrons towards it. Then again, a lower powerful atomic charge in F causes the particle to extend to 0.133 nm from the first uncharged F nuclear sweep of 0.071 nm. Dissimilar to the symmetrical electronic charge circulation in the hydrogen atom, the electron thickness in the holding orbital is substantial near F and little close Li+. The turn around is valid for the counter holding orbital.

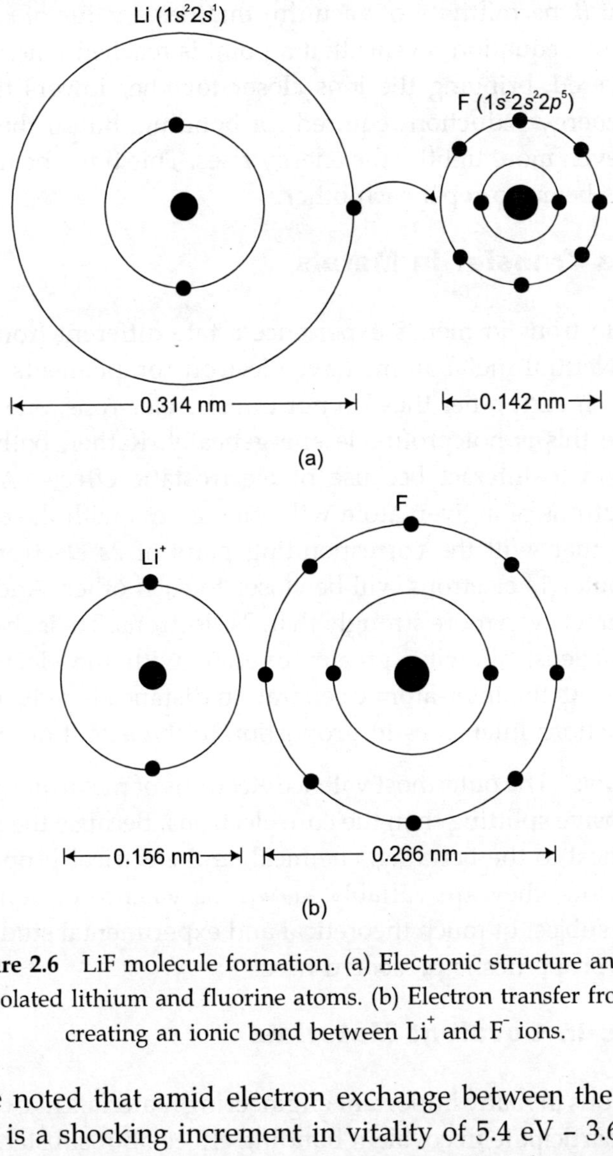

Figure 2.6 LiF molecule formation. (a) Electronic structure and size of the isolated lithium and fluorine atoms. (b) Electron transfer from Li to F creating an ionic bond between Li^+ and F^- ions.

It ought to be noted that amid electron exchange between these particles as they move close, there is a shocking increment in vitality of 5.4 eV – 3.6 eV = 1.8 eV! Why do they then respond to form molecules? The answer is that the particles are not as near to each other as they could be. Account must be taken of the electrostatic drive of fascination between the particles, which tumbles off with the separation (r) between them as r''. The potential vitality of communication (U) connected with the drive between these point charges is given by

$$U = \frac{-q^2}{4\pi\varepsilon_o r}$$

where, ε_o = constant (k-permittivity of vacuum) that has a value of 8.85×10^{-12} C/V-m. As r shrinks, the above equation reveals that a point is reached where $U = -1.8$ eV, and the energy is recovered; bringing the ions closer together lowers the energy further, leading to the net energy reduction required for bonding. But, if the Li and F ions are squeezed together even more tightly, the energy rises. This time the negatively charged, core electron clouds begin to repel each other.

2.4.2 Electrons Transfer in Metals

Core Electrons: Electrons in metals experience a fate different from that of electrons in ionic materials. Neutral metal atoms have electron complements. As the individual metal atoms approach each other they do not transfer electrons, one to the other alkali halide style because this is not profitable energetically. Rather, both the core and the outer electrons begin to interact because of electrostatic effects. At any distance of approach the 1*s* electrons of a given atom will interact less with the core 1*s* electrons of surrounding atoms than will the corresponding pairs of 2*s* electrons. The reason, of course, is that the outer 1*s* electrons will be closer to each other. And similarly, 2*p* and 3*s* electrons will interact even more strongly than 2*s* electrons. Their charge clouds extend further from the nucleus, allowing greater overlap with the electron clouds of the neighbouring atoms. As the inter-atomic separation distance is reduced, the strength of the electronic interactions intensifies in proportion to the extent of overlap.

Conducting Electrons: The outermost valence electrons of metal atoms undergo similar, but even more extensive splitting than the core electrons. Because the properties of these electrons are so critical to the bonding, chemical, and electrical properties of metals as well as semiconductors, they are variably known as valence or conduction electrons. They have been the subject of much theoretical and experimental study and a simplified description of their properties is presented here.

2.4.3 Electrons in Covalent Materials

Nonmetals: Electrons in many important engineering materials like diamond, Si, GaAs, polymers, and SiC participate in covalent bonding effects. But substances like H_2, which was treated above, F_2, and CH_4, together with hundreds of thousands of organic molecules, also exhibit covalent bonding. This type of bonding relies on neighbourly sharing of electrons. Bond energies and lengths have been determined for a greater number of covalently bonded atomic pairs and are listed in Table 2.2.

Table 2.2 Bonding energies (kcal/mol and kJ/mol) and bond lengths (nm)

Bond	Energy (kcal/mol)	Energy (kJ/mol)	Bond Length (nm)
C — C	81	340	0.154
C ═ C	148	620	0.134
C ≡ C	194	810	0.120
C — H	100	420	0.109
C — N	69	290	0.148
C — O	84	350	0.143
C ═ O	172	720	0.121
C — F	105	440	0.138
C — Cl	79	330	0.177
O — H	110	460	0.097
O — O	33	140	0.145
O ═ O	96	400	0.121
O — Si	103	430	0.16
N — O	43	180	0.12
N — H	103	430	0.101
H — H	105	440	0.074
Cl — Cl	57	240	0.199
H — Cl	103	430	0.127

Semiconductors: Elements in the fourth column of the periodic table are the most important examples of covalent solids from an engineering standpoint. Therefore, let us consider what happens to the electrons when silicon particles are united near the frame of the solid. Initially, electrons originating from a similar vitality level cover and split as depicted beforehand. Be that as it may, as particles move considerably the nearer electrons don't exclusively cover with electrons of a similar orbit. Or maybe, 3*s* and 3*p* electrons cover or blend fundamentally to yield the vitality band structure. This is known as hybridization and prompts to four hybridized electron circles that transmit ostensibly from every silicon molecule. Each of these electrons combines off with another like itself from neighbouring silicon atoms to concrete unequivocally coordinated covalent bonds. Finish electron sharing implies that eight electrons will now encompass every silicon atom and completely fill the greater part of its four valence band states. There are no more electrons to go around and, thus, the conduction band is vacant. Not at all like the circumstance for Na at the balance remove (TQ), there is currently a crevice in the electron vitality levels. There are no electron states and, hence, no electrons having energies inside the traverse characterized by the vitality crevice.

2.5 BONDING IN SOLIDS

A comprehension of a large number of the physical properties of materials is predicated on information of the nuclear powers that quandary the particles together. Maybe the standards of nuclear holding are best represented by considering the connection between two confined particles as they are brought into closeness from an interminable division. Everywhere the connections are insignificant, however as the molecules approach, each repels the other. These powers are of two sorts, alluring and appalling, and the greatness of each is an element of the detachment or interatomic separation.

The orbit of an alluring power F_A relies on the specific kind of holding that exists between the two molecules. The size of the alluring power shifts with the separation, as shown schematically in Figure 2.7a. At last, the external electron shells of the two atoms start to cover, and a solid shocking power F_R becomes possibly the most important factor. The net compel F_N between the two molecules is only the total of both alluring and terrible parts; that is,

$$F_N = F_R + F_A$$

Three unique sorts of primary or chemical bond are found in solids—ionic, covalent, and metallic. For every sort, the holding essentially includes the valence electrons; moreover, the nature of the bond relies on the electron structures of the constituent molecules. By and large, each of these three sorts of holding emerges from the inclination of the molecules to expect stable electron structures, similar to those of the inactive gasses, by totally filling the furthest electron shell. Auxiliary or physical strengths and energies are likewise found in numerous strong materials; they are weaker than the essential ones, yet regardless impact the physical properties of a few materials. The segments that take after clarify the few sorts of essential and auxiliary between nuclear or atomic bonds.

2.5.1 Primary Interatomic Bonds

Ionic Bonds: Ionic bonding may be the least demanding to depict and picture. It is constantly found in abundance that are made out of both metallic and non-metallic elements. Elements that are arranged at the flat limits of the occasional table. Molecules of a metallic element effortlessly surrender their valence electrons to the non-metallic atoms. In the process every one of the molecules procures steady or inactive gas setups and, furthermore, an electrical charge; that is, they get particles. Sodium chloride (NaCl) is an exemplary ionic material. A sodium molecule can accept the electron structure of neon (and a net single positive charge) by an exchange of its one valence 3*s* electron to a chlorine particle. After such an exchange, the chlorine particle has a net negative charge and an electron setup indistinguishable to that of argon. In sodium chloride, all the

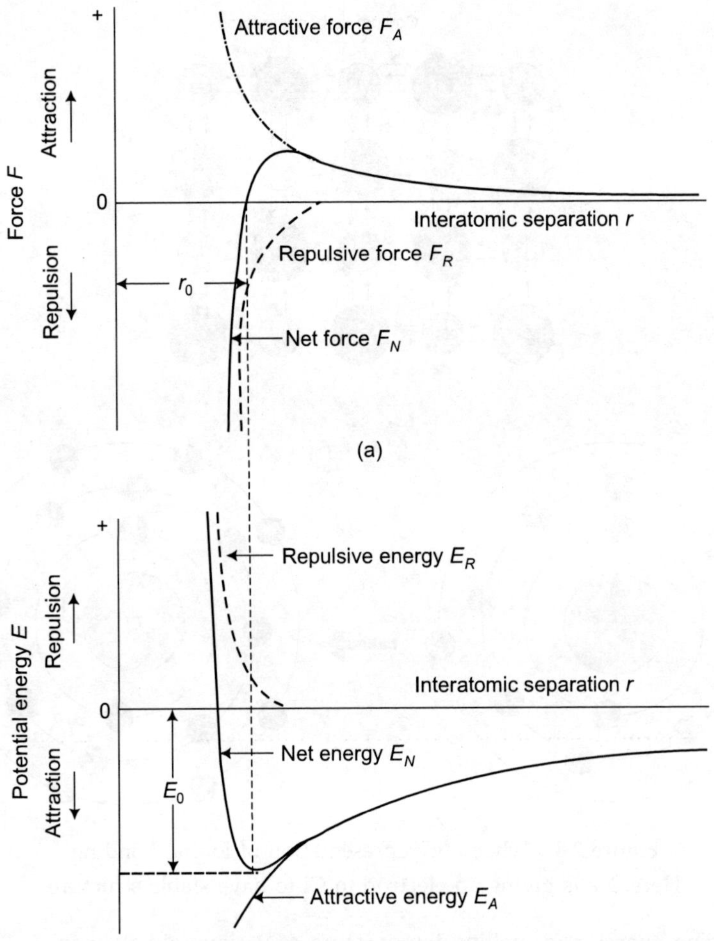

Figure 2.7 (a) The dependence of repulsive, attractive, and net forces on interatomic separation for two isolated atoms. (b) The dependence of repulsive, attractive, and net potential energies on inter-atomic separation for two isolated atoms.

sodium and chlorine exist as particles. This sort of holding is shown schematically in Figure 2.8.

Hence, we can summarize that this bond exists between two atoms when one of the atoms is negative (has an additional electron) and another is positive (has lost an electron). At that point there is a solid, coordinate Coulomb fascination. Essentially, ionic bonds are non-directional in nature. An illustration is NaCl. In the particle, there are more electrons around Cl, shaping Cl^- and less electrons around Na, framing Na^+. Ionic bonds are the most grounded bonds. In genuine solids, ionic holding typically exists alongside covalent holding.

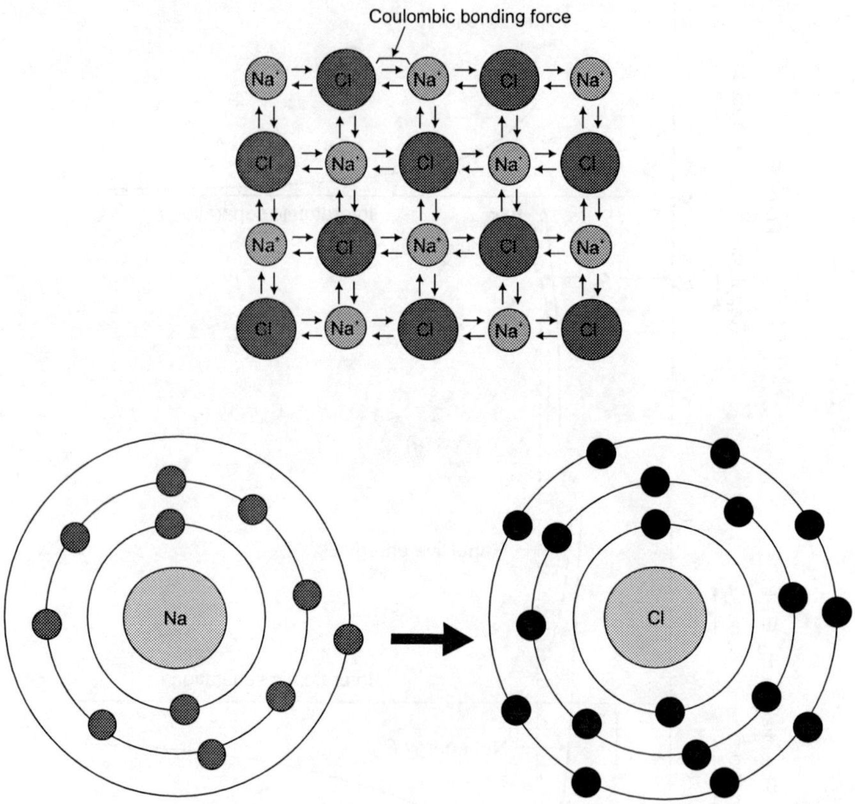

Figure 2.8 Schematic representation of ioning bonding. Here, Na is giving an electron to Cl to have stable structure.

Table 2.3 Bonding energy and melting temperature of various substances

Bonding Type	Substance	Bonding Energy		Melting Temperature (°C)
		kJ/mol	eV/Atom, Ion Molecule	
Ionic	NaCl	640	3.3	801
	MgO	1000	5.2	2800
Covalent	Si	450	4.7	1410
	C (diamond)	713	7.4	> 3550
	Hg	68	0.7	–39
Metallic	Al	324	3.4	660
	Fe	406	4.2	1538
	W	849	8.8	3410
van der Waals	Ar	7.7	0.08	–189
	Cl_2	31	0.32	–101
Hydrogen	NH_3	35	0.36	–78
	H_2O	51	0.52	0

Covalent Bonds: In covalent bonding, stable electron setups are expected by the sharing of electrons between neighbouring atoms. Two atoms that are covalently fortified will each contribute no less than one electron to the bond, and the mutual electrons might be considered to have a place with both particles. Covalent holding is schematically delineated in Figure 2.9 for an atom of methane (CH_4). The carbon particle has four valence electrons, though each of the four hydrogen molecules has a solitary valence electron. Every hydrogen iota can obtain a helium electron arrangement (two 1s valence electrons) when the carbon molecule offers with it one electron. Carbon now has four extra shared electrons, one from every hydrogen, for an aggregate of eight valence electrons, and the electron structure of neon. The covalent bond is directional; that is, it is between particular particles and may exist just in the course between one iota and another that takes part in the electron sharing.

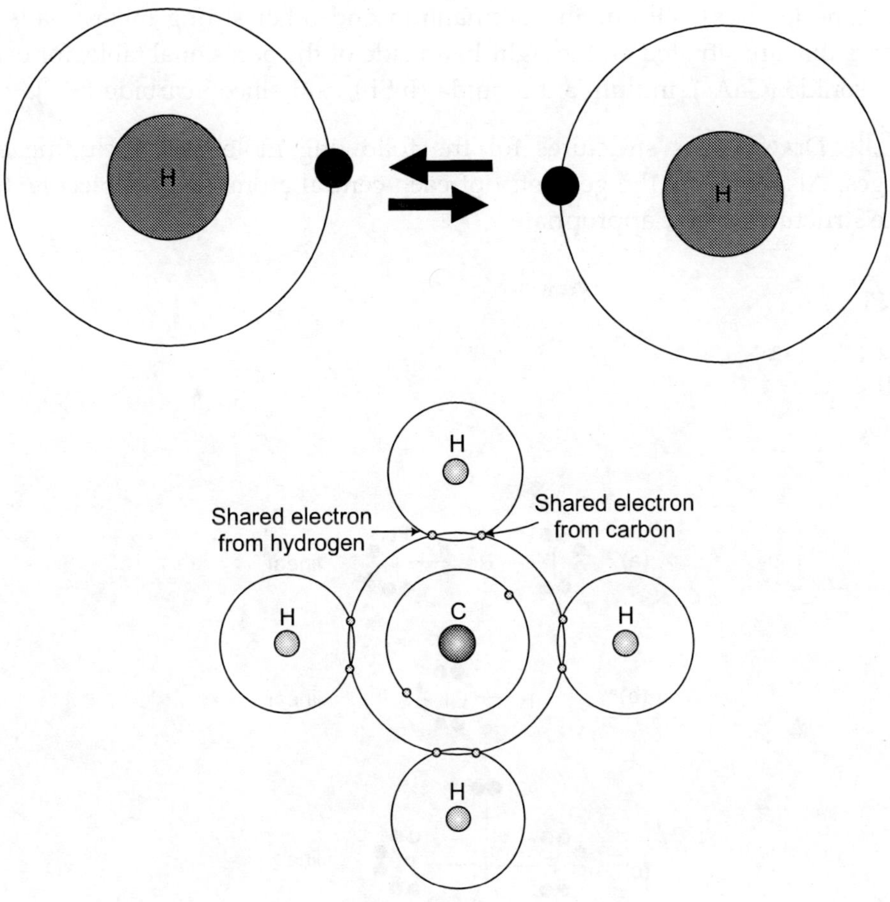

Figure 2.9 Schematic representation of a covalent bond in hydrogen molecule (sharing of electrons).

Hence, we can summarize covalent bonding that in this type of bonding electrons are shared between the atoms, to soak the valency. The most straightforward case is the H_2 particle, where the electrons invest more energy in the middle of the cores of two iotas than outside, along these lines delivering holding. Covalent bonds are stereo-particular, i.e., every bond is between a particular match of particles, which share a couple of electrons (of inverse attractive twists). Regularly, covalent bonds are exceptionally solid, and directional in nature. The hardness of precious stone is an after-effect of the way that every carbon molecule is covalently reinforced with four neighbouring particles, and every neighbour is fortified with an equivalent number of atoms to shape an inflexible three-dimensional structure.

Numerous non-metallic essential particles (H_2, Cl_2, F_2 and so forth.) and in addition atoms containing different molecules, for example, CH_4, HNO_3, H_2O and HF are covalently fortified. Moreover, this kind of holding is found in basic solids, for example, precious stone (carbon), silicon, and germanium and other strong mixes made out of components that are situated on the right-hand side of the occasional table, for example, gallium arsenide (GaAs), indium antimonide (InSb), and silicon carbide (SiC).

Example 2.11 Draw Lewis structures for the following molecules, including all formal charges. Also indicate the geometry of each central atom in the molecule. Include resonance structures where appropriate.

(a) BeI_2
(b) HCl
(c) IF_2
(d) ClF_3
(e) NO_3

Solution:

(a) 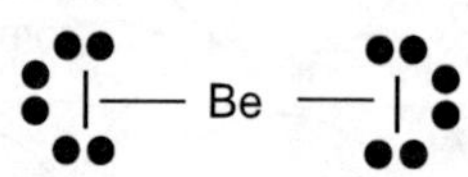

linear

(b) 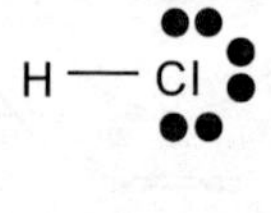

linear

(c) 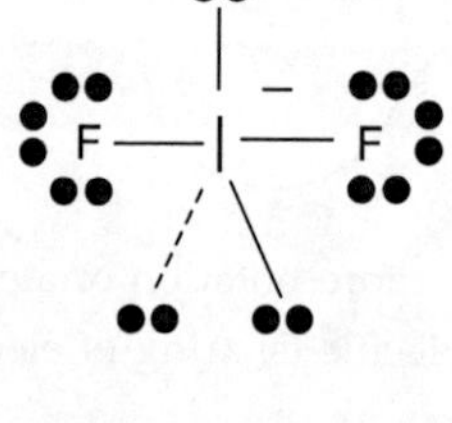

linear

(d) 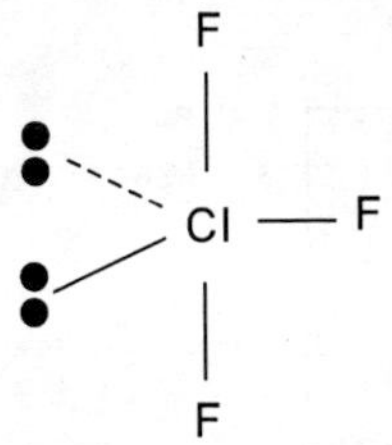 T-shaped

(e) 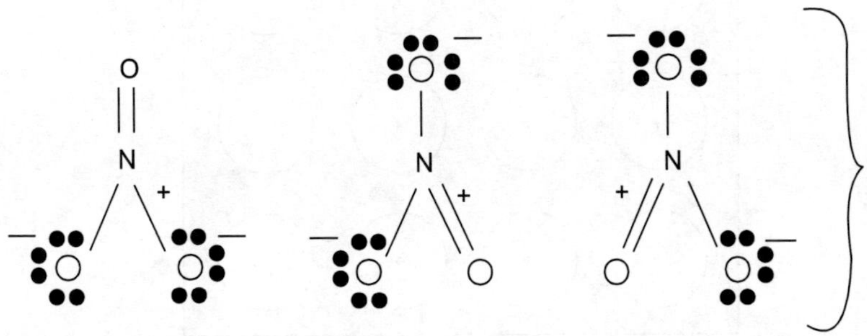 Trigonal planar

Metallic Bonds: Metallic bonding, the last essential bonding sort, is found in metals and their alloys. A moderately basic model has been recommended that practically approximates the bonding plan. Metallic materials have one, two, or at most, three valence electrons. With this model, these valence electrons are not bound to a specific molecule in particular and are pretty much allowed to float all through the whole metal. They might be considered having a place with the metal overall, or framing an "ocean of electrons" or an "electron cloud." The rest of the non-valence electrons and nuclear cores shape called ions core, have a net positive charge rise to in size to the aggregate valence electron charge per iota. Figure 2.10 is a schematic outline of metallic holding. The free electrons shield the decidedly charged particle centres from commonly horrible electrostatic strengths, which they would some way or another apply upon each other; consequently, the metallic bond is non-directional in character. Likewise, these free electrons go about as a "paste" to hold the particle cores together. Metallic bonding is found in the occasional table for Group IA and IIA elements and, truth be told, for every elemental metal. Some broad conduct of the different material sorts (i.e., metals, earthenware production, polymers) might be clarified by holding sort. For instance, metals are great channels of both power and warmth, as a result of their free electrons. By the method for difference, ionically and covalently reinforced materials are regularly electrical and warm separators, because of the nonappearance of vast quantities of free electrons.

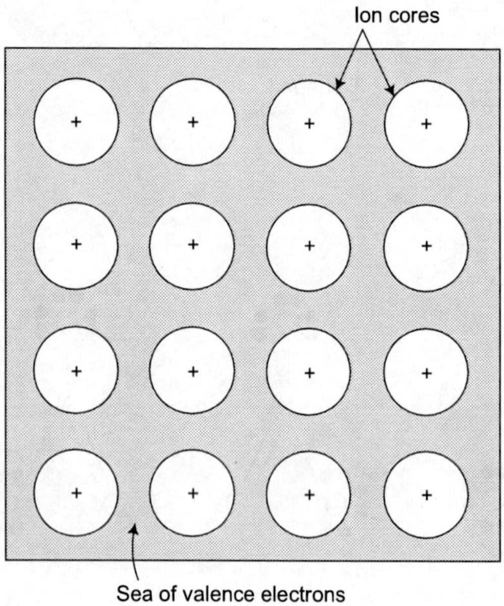

Figure 2.10 Schematic illustration of metallic bonding.

2.5.2 Secondary Inter-atomic Bonds (van der Waals Bonding)

Optional, van der Waals, or physical bonds are powerless in contrast with the essential or synthetic ones; holding energies are commonly on the request of just 10 kJ/mol (0.1 eV/molecule). Optional holding exists between for all intents and purposes all molecules or atoms, however, its nearness might be darkened if any of the three essential holding sorts is available. Auxiliary holding is proved for the latent gases, which have stable electron structures, and, what's more, between particles in subatomic structures that are covalently reinforced. Auxiliary holding powers emerge from nuclear or molecular dipoles. Generally, an electric dipole exists at whatever point there is some division of positive and negative segments of an iota or particle. The bonding comes about because of the coulombic fascination between the positive end of one dipole and the negative area of an adjoining one, as demonstrated in Figure 2.11. Dipole cooperations happen between actuated dipoles, between incited dipoles and polar particles (which have unchanging dipoles), and between polar atoms. Hydrogen holding, an extraordinary kind of helper holding, is found to exist between a couple of particles that have hydrogen as one of the constituents. These bonding frameworks are at present analyzed rapidly underneath.

Fluctuating Induced Dipole Bonds: A dipole might be made or prompted in a particle or atom that is regularly electrically symmetric; that is, the general spatial circulation of the electrons is symmetric as for the emphatically charged core, as appears in Figure 2.12a. All particles are encountering steady vibrational movement that can bring about

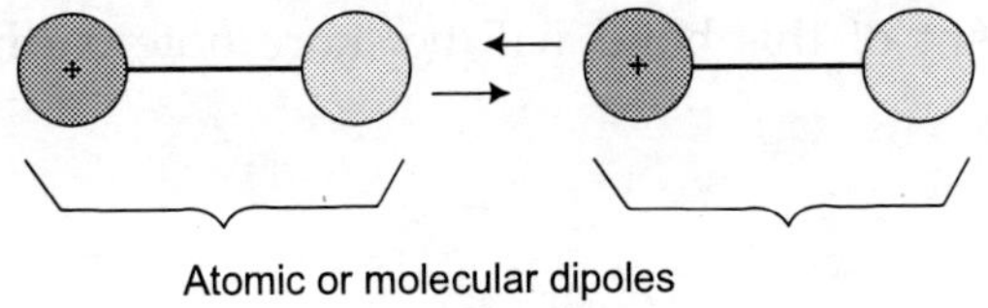

Figure 2.11 Schematic illustration of van der Waals bonding between two dipoles.

quick and brief mutilations of this electrical symmetry for a portion of the iotas or atoms, and the formation of little electric dipoles, as shown in Figure 2.12b. One of these dipoles can thus deliver a relocation of the electron conveyance of an adjoining particle or atom, which instigates the second one likewise to wind up a dipole that is then pitifully pulled in or attached to the main; this is one sort of van der Waals bonding. These alluring powers may exist between huge quantities of particles or atoms, which strengths are impermanent and vacillate with time.

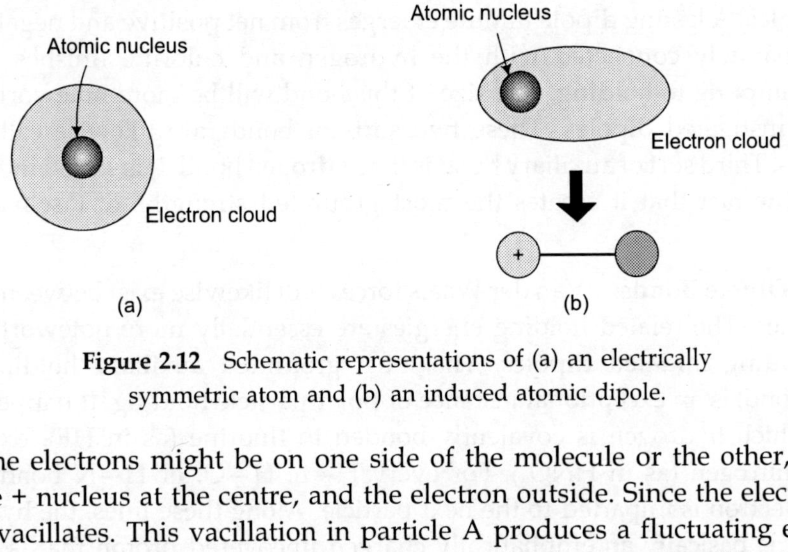

Figure 2.12 Schematic representations of (a) an electrically symmetric atom and (b) an induced atomic dipole.

Since the electrons might be on one side of the molecule or the other, a dipole is framed: the + nucleus at the centre, and the electron outside. Since the electron moves, the dipole vacillates. This vacillation in particle A produces a fluctuating electric field that is felt by the electrons of a nearby iota, B. Molecule B then spellbinds so that its external electrons are in favour of the atom nearest to the + side (or inverse to the – side) of the dipole in A.

Polar Molecule-induced Dipole Bonds: Perpetual dipole minutes exist in a few atoms by temperance of an unbalanced course of action of emphatically and contrarily charged districts; such particles are named polar particles. Figure 2.13 is a schematic representation of a hydrogen chloride atom; a changeless dipole minute emerges from net positive and negative charges that are individually connected with the hydrogen and chlorine finishes of the HCl molecule. Polar atoms can likewise instigate dipoles in contiguous nonpolar particles, and a bond will shape as a consequence of alluring powers between the two atoms. Besides, the extent of this bond will be more noteworthy than for fluctuating prompted dipoles.

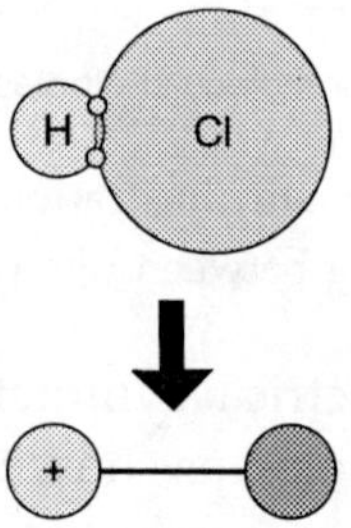

Figure 2.13 Schematic representation of a polar hydrogen chloride (HCl) molecule.

We can summarise this: It is another sort of optional bond which exists with asymmetric particles, additionally called polar atoms on account of positive and negative charged locales. A lasting dipole minute emerges from net positive and negative charges that are separately connected with the hydrogen and chlorine finishes of the HCl particle, prompting to holding. The size of this bond will be more noteworthy than for fluctuating instigated dipoles. These two sorts of bonds are likewise called van der Waals bonds. Third sort of auxiliary bond is the hydrogen bond. It is sorted independently in light of the fact that it creates the most grounded strengths of fascination in this class.

Permanent Dipole Bonds: Van der Waals forces will likewise exist between contiguous polar particles. The related holding energies are essentially more noteworthy than for bonds including initiated dipoles. The most grounded auxiliary holding sort, the hydrogen bond, is an exceptional instance of polar particle holding. It happens between atoms in which hydrogen is covalently bonded to fluorine (as in HF), oxygen (as in H_2O), and nitrogen (as in HNO_3). For every H–F, H–O, or H–N bond, the single hydrogen electron is imparted to the next particle. Along these lines, the hydrogen end of the bond is basically an emphatically charged uncovered proton that is unscreened by any electrons. This very decidedly charged end of the particle is fit for a solid appealing power with the negative end of a neighbouring atom, as exhibited in Figure 2.14 for HF. Generally, this single proton structures a scaffold between two contrarily charged iotas. The greatness of the hydrogen bond is by and large more noteworthy than that of alternate sorts of optional bonds and might be as high as 51 kJ/mol (0.52 eV/ particle), as appeared in melting and bubbling temperatures for hydrogen fluoride and water are strangely high in light of their low subatomic weights, as an outcome of hydrogen bonding.

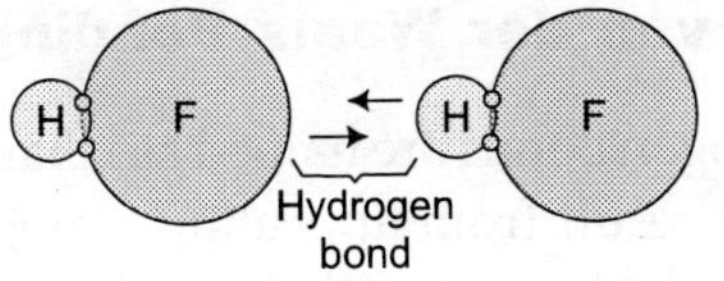

Figure 2.14 Schematic representation of hydrogen bonding in hydrogen fluoride (HF)

We can summarise this: It happens between particles as covalently fortified hydrogen iotas – for instance C-H, O-H, F-H – impart single electron to other molecule basically bringing about emphatically charged proton that is not protected by any electrons. This emphatically charged end of the atom is equipped for solid alluring power with the negative end of a nearby particle. The properties of water are impacted altogether by the hydrogen bonds/spans. The scaffolds are of adequate quality, and as a result water has the most astounding liquefying purpose of any atom of its size. Moreover, its warmth of vaporization is high.

2.6 SUMMARY

Electrons in Atoms and Periodic Table

This area began with an outline of the essentials of nuclear or atomic structure, displaying the Bohr. The Bohr shows electrons to be particles circling the core in discrete ways, in wave mechanics we consider them to be wave-like and treat electron position as far as a likelihood conveyance. Electron vitality states are determined as far as quantum numbers that offer ascent to electron shells and sub-shells. The electron arrangement of an iota compares to the way in which these shells and sub-shells are loaded with electrons in consistence with the Pauli guideline. The intermittent table of the components is created by course of action of the different components as indicated by valence electron setup.

Bonding Forces and Energies Primary Interatomic Bonds

Nuclear or atomic bonding in solids might be considered regarding alluring and horrible strengths and energies. The three sorts of essential bond in solids are ionic, covalent, and metallic. For ionic bonds, electrically charged particles are shaped by the transfer of valence electrons starting with one molecule sort then onto the next; powers are coulombic. There is a sharing of valence electrons between the neighbouring particles when holding is covalent. With metallic holding, the valence electrons shape an "ocean of electrons" that is consistently scattered around the metal particle centres and goes about as a type of paste for them.

Secondary Bonding or van der Waals Bonding

Both van der Waals and hydrogen bonds are named auxiliary, being powerless in contrast with the essential ones. They result from appealing strengths between electric dipoles, of which there are two sorts—instigated and changeless. For the hydrogen bond, very polar atoms frame when hydrogen covalently bonds to a non-metallic component, for example, fluorine.

EXERCISE

1. (a) Differentiate between atomic mass and atomic weight.
   (b) In one amu of a material contains how many grams?
2. Give the electron configurations for the following ions: P^{5+}, P^{3-}, Sn^{4+} and Se^{2-}.
3. To what group in the periodic table would an element with atomic number 112 belong?
4. (a) What electron sub-shell is being filled for the rare earth series of elements on the periodic table?
   (b) What electron sub-shell is being filled for the actinide series?
5. (a) Briefly cite the main differences between ionic, covalent, and metallic bonding.
   (b) State the Pauli exclusion principle.
6. What type(s) of bonding would be expected for each of the following materials: solid xenon, calcium fluoride, bronze, cadmium telluride, rubber, and tungsten?
7. Explain why hydrogen fluoride (HF) has a higher boiling temperature than hydrogen chloride (HCl) even though HF has a lower molecular weight?
8. Provide a reasonable physical argument for each of the following statements.
   (a) The higher the melting temperature of the solid, the greater the depth of the potential energy well.
   (b) Materials with deep energy wells are likely to have a more symmetrical potential energy curve.
   (c) Materials with high melting points tend to have low coefficients of thermal expansion.
   (d) Materials with high melting points tend to have large moduli of elasticity.
9. How many grams of Ni and Al are required to make 1 kg of the compound Ni_3Al?
10. What do all the elements in Group IIA of the periodic table have in common?
11. Which of the following elements is/are diatomic gases at ordinary temperature and pressure: neon, sodium, oxygen, bromine, carbon?
12. Which of the following element is easiest to ionize (has the lowest first ionization energy) sodium, potassium magnesium, calcium, fluorine.

3

Structure of Crystalline Solids

Learning Objectives

After going through this chapter the reader will be able to:

- Understand crystalline and noncrystalline structures.
- Have an overview of common crystal structures.
- Understand structures for FCC, BCC and HCP.
- List the concept of positions of atom, directions, and crystal structures planes.
- Gather idea on defects of crystalline solids.

3.1 INTRODUCTION

Why do we think about the structure of crystalline solids? The properties of a few materials are identified with their precious stone structures. For instance, unadulterated and undeformed magnesium and beryllium, having one precious crystal structure, are brittle (i.e., fracture at lower degrees of distortion) than are immaculate and undeformed metals, for example, gold and silver that have yet another gem structure. Moreover, noteworthy property contrasts exist amongst crystalline and noncrystalline materials having a similar structure. For instance, noncrystalline earthenware production and polymers ordinarily are optically straightforward; similar materials in crystalline (or semicrystalline) frame have a tendency to be obscure or, best case scenario, translucent.

In spite of the fact that the past part created models of solids in view of the electronic structure of atoms, it was to a great extent managed without suppositions in the matter of how particles were physically organized or situated in respect to each other. In this section a practically inverse perspective is received: the physical structure is our essential concern and the electronic structure is of optional intrigue. Obviously, at a more profound level the two sorts of structures are personally coupled. It is helpful to subdivide the structure of solids into two general classifications: crystalline and noncrystalline or amorphous.

Numerous solids have a crystalline structure that is portrayed by a requested geometric cluster of particles that extends interminably every which way in a dull manner. We don't consider materials like basic steel as being crystalline in the light of the fact that they are not clear, straightforward, faceted, shimmering, precise, etc. Regardless of outside appearances, materials as are metals, crystalline in light of the logical measure of an efficient inward plan of molecules. Then again, molecules are not situated in typically requested clusters in formless solids like silica glass and most hydrocarbon polymers.

Chapter 2 was concerned principally with the different sorts of nuclear holding, which are controlled by the electron structures of the individual molecules. The present one, i.e., chapter 3 is dedicated to the following level of the structure of materials, particularly, to a portion of the game plans that might be expected by molecules in the strong state. Inside this edge work, ideas of crystallinity and noncrystallinity are presented. For crystalline solids the thought of precious stone structure is introduced, determined as far as a unit cell. The three basic precious stone structures found in metals are then nitty gritty, alongside the plan by which crystallographic focuses, headings, and planes are communicated. Single crystals, polycrystalline, and noncrystalline materials are considered.

3.2 COMMON CRYSTAL STRUCTURES

3.2.1 Crystalline Materials

Strong materials might be ordered by normality with which particles or particles are organized as for each other. A crystalline material is one in which the particles are arranged in a rehashing or occasional exhibit over substantial nuclear separations; that is, long-run arrangement exists, with the end goal that upon hardening, the iotas will position themselves in a dull three-dimensional example, in which every molecule is clung to its closest neighbour particles. All metals, numerous artistic materials, and certain polymers frame crystalline structures under typical cementing conditions.

3.2.2 Crystalline Structures

Some of the properties of crystalline solids depend on the crystal structure of the material, the manner in which atoms, ions, or molecules are spatially arranged. There is an extremely large number of crystal structures having long range atomic order; these vary from relatively simple structures for metals to exceedingly complex ones, as displayed by some of the ceramic and polymeric materials. While portraying crystalline structures, particles (or particles) are considered strong circles having very much characterized breadths. This is *atomic hard sphere model* in which circles speaking to closest neighbour molecules touch each other. A case of the hard circle demonstrate for the nuclear game plan found in a portion of the basic metals is shown in Figure 3.1c. In this specific case every one of the molecules is indistinguishable. Infrequently the term grid is utilized as a part of the setting of gem structures. In this sense "lattice" implies a three-dimensional cluster of points agreeing with particle positions (or circle focuses).

3.2.3 Unit Cells

The atomic order in crystalline solids shows that little gatherings of particles frame a redundant example. Hence, in depicting precious stone structures, it is regularly helpful to subdivide the structure into little rehash elements called unit cells. Unit cells for most crystal structures are parallelepipeds or crystals having three arrangements of parallel confronts; one is drawn inside the total of circles (Figure 3.1c), which for this situation happens to be a shape. A unit cell is spoken to the symmetry of the precious stone structure, wherein all the particle positions in the crystal might be produced by interpretations of the unit cell basic separations along each of its edges.

In this manner, the unit cell is the fundamental auxiliary unit or building square of the gem structure and characterizes the gem structure by prudence of its geometry and the particle positions inside. Convenience generally manages that parallelepiped corners correspond with focuses of the hard sphere atoms. Moreover, more than a solitary unit cell might be decided for a specific gem structure; be that as it may, we for the most part utilize the unit cell having the largest amount of geometrical symmetry.

3.2.4 Metallic Crystal Structures

The nuclear holding in this gathering of materials is metallic and in this manner nondirectional in nature. Subsequently, there are negligible confinements with regards to the number and position of the closest neighbour molecules; this prompts to generally expansive quantities of closest neighbours and thick nuclear packings for most metallic crystal structures. Likewise, for metals, utilizing the hard circle show for the precious stone structure, every circle speaks to a particle centre. Table 3.1 presents the nuclear radii for various metals. Three moderately basic precious stone structures are found for a large portion of the normal metals: face-centered cubic, body-focused cubic, and hexagonal close-packed.

Table 3.1 Atomic radii and crystal structures for some metals

Metal	Crystal Structure[a]	Atomic Radius[b] (nm)	Metal	Crystal Structure	Atomic Radius (nm)
Aluminium	FCC	0.1431	Molybdenum	BCC	0.1363
Cadmium	HCP	0.1490	Nickel	FCC	0.1246
Chromium	BCC	0.1249	Platinum	FCC	0.1387
Cabalt	HCP	0.1253	Silver	FCC	0.1445
Copper	FCC	0.1278	Tantalum	BCC	0.1430
Gold	FCC	0.1442	Titanium (α)	HCP	0.1445
Iron (α)	BCC	0.1241	Tungsten	BCC	0.1371
Lead	FCC	0.1750	Zinc	HCP	0.1332

Face-centred Cubic Crystal Structure: The crystal structure found for some metals has a unit cell of cubic geometry, with atoms situated at each of the corners and the focuses of all the solid shape faces. It is appropriately called the face-centered cubic (FCC) precious stone structure. A portion of the recognizable metals having this gem structure are copper, aluminium, silver, and gold (see additionally Table 3.1). Figure 3.1a demonstrates a hard sphere model for the FCC unit cell, while in Figure 3.1b the atom centres are spoken to by little circles to give a superior point of view of molecule positions. The total of particles in Figure 3.1c speaks to a segment of precious stone comprising numerous FCC unit cells. These circles or particle centres touch each other over a face diagonal; the shape edge length and the atomic radius R are connected through

$$a = 2R\sqrt{2}$$

For the FCC crystal structure, each corner atom is shared among eight unit cells, whereas a face-centered atom belongs to only two. Therefore, one-eighth of each of the eight corner atoms and one-half of each of the six face atoms, or a total of four whole atoms, may be assigned to a given unit cell. This is depicted in Figure 3.1a, where only sphere portions are represented within the confines of the cube.

The cell comprises the volume of the cube, which is generated from the centres of the corner atoms as shown in the figure. Corner and face positions are really equivalent; that is, translation of the cube corner from an original corner atom to the centre of a face atom will not alter the cell structure. Two other important characteristics of a crystal structure are the **coordination number and the atomic packing factor (APF).** For metals, each atom has the same number of nearest neighbour or touching atoms, which is the coordination number. For face centered cubics, the coordination number is 12. This may be confirmed by examination of Figure 3.1(a); the front face atom has four corner nearest neighbour atoms surrounding it, four face atoms that are in contact from behind, and four other equivalent face atoms residing in the next unit cell to the front, which is not shown. The APF is the sum of the sphere volumes of all atoms within a unit cell (assuming the atomic hard sphere model) divided by the unit cell volume that is:

$$\text{APF} = \frac{\text{volume of atoms in a unit cell}}{\text{total unit}}$$

For the FCC structure, the atomic packing factor is 0.74, which is the most extreme pressing feasible for circles all having a similar measurement. Calculation of this APF is additionally included for instance issue. Metals ordinarily have generally huge nuclear pressing variables to expand the protecting given by the free electron cloud.

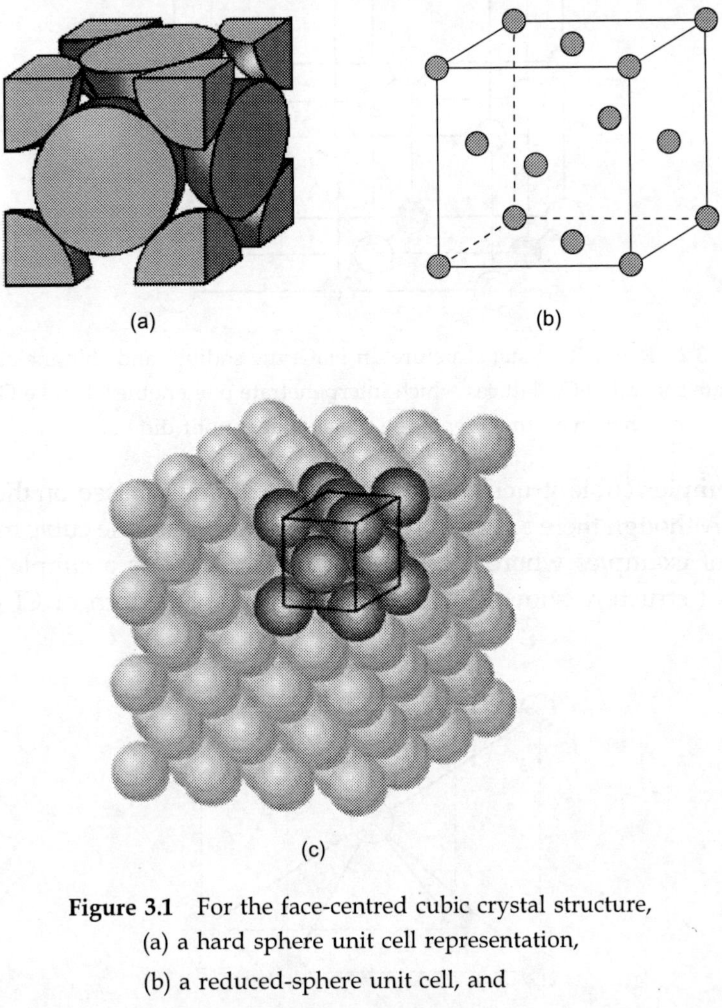

Figure 3.1 For the face-centred cubic crystal structure,
(a) a hard sphere unit cell representation,
(b) a reduced-sphere unit cell, and
(c) an aggregate of many atoms.

NaCI and CsCI: Let us again consider the FCC Bravais lattice. If we now faithfully replace each point by a NaCl molecule motif, we can generate the common rocksalt structure depicted in Figure 3.2. In a slight structural twist, the CI atom (ion) is positioned at the cell origin while the Na atom (ion) is located at the centre of an original FCC cube edge (or vice versa). Each atom has six nearest neighbours of the other kind. Many oxides (e.g., MgO, BeO), fluorides (e.g., LiF, NaF), and chlorides (e.g., KCl) crystallize with rocksalt structures.

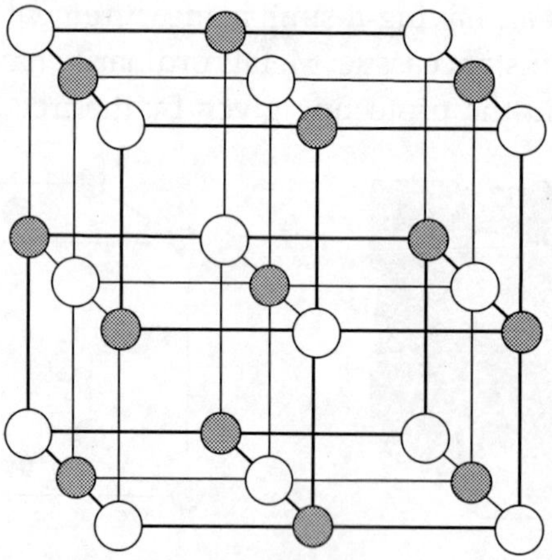

Figure 3.2 Rocksalt crystal structure. In NaCI the sodium and chlorine atoms populate separate FCC lattices which interpenetrate one another. Unlike GaAs, however, the orientation of the NaCI motif differs.

The least complex cubic structure containing two atoms is based on the simple cubic Bravais lattice. Although there are virtually no monoatomic simple cubic materials, there are a number of examples where a two-atom motif decorates a simple cubic Bravais lattice. The CsCl structure shown in Figure 3.3 repeats a pattern of CI (or Cs) at the

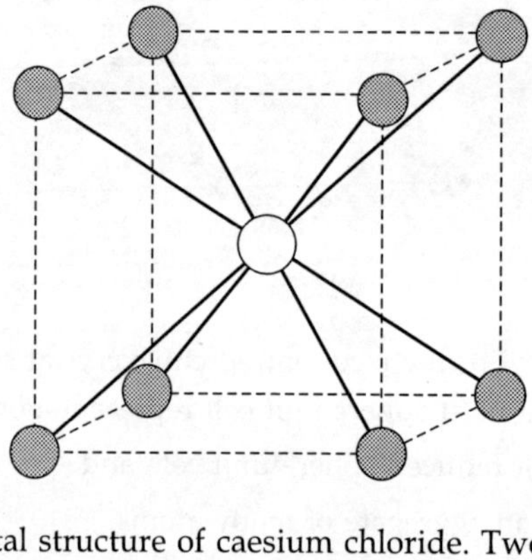

Figure 3.3 Crystal structure of caesium chloride. Two interpenetrating simple cubic lattices, one for Cs and the other for CI, generate this structure.

cube origin and Cs (or Cl) at the cube centre. Each atom is coordinated with eight atoms of the other kind in what is known as cubic coordination. Of course CsCl is an ionic soHd and if ions touch along the cube diagonal then $a\sqrt{3} = 2r_{cs} + 2r_{cl-}$. Other CsCl-type structures are exhibited by halide salts (e.g., Til, RbCl) and certain ordered binary metal alloys (e.g., CuZn or j8 brass, AlNi, BeCu).

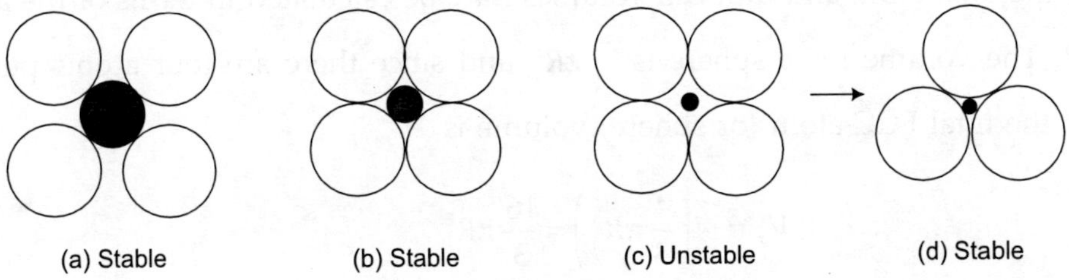

Figure 3.4 Stable and unstable coordinations between ions of different size. As the anion size remains constant, the cation successively shrinks in size (A)-(C), until an unstable configuration (C) is reached. This rearranges to a stable configuration (D).

Example 3.1 Calculate the volume of an FCC unit cell in terms of the atomic radius R.

Solution: In the FCC unit cell illustrated, the atoms touch one another across a face-diagonal the length of which is $4R$. Since the unit cell is a cube, its volume is a^3 where a is the cell edge length. From the right triangle on the face,

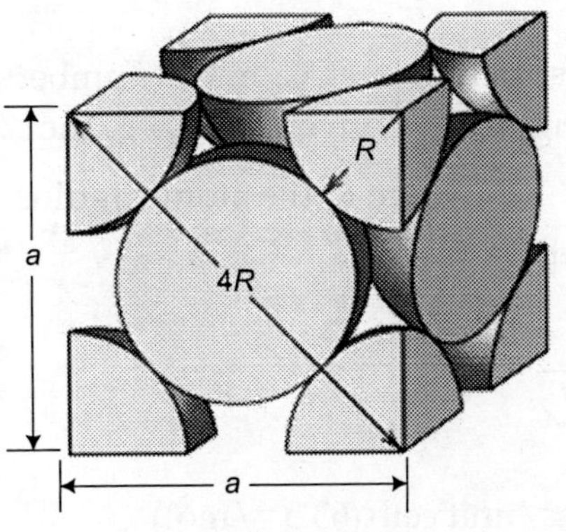

Figure 3.5

$$a^2 + a^2 = (4R)^2$$

We obtain $a = 2R\sqrt{2}$

The FCC unit cell volume V_c may be computed from

$$V_c = a^3 = (2R\sqrt{2})^3 = 16R^3\sqrt{2}$$

Example 3.2 Show that the atomic packing factor for the FCC crystal structure is 0.74.

Solution: The APF is defined as the fraction of solid sphere volume in a unit cell, or

$$\text{APF} = \frac{\text{Volume of atoms in a unit cell}}{\text{Total unit cell volume}} = \frac{V_s}{V_c}$$

Both the total atom and unit cell volumes may be calculated in terms of the atomic radius R. The volume for a sphere is $\frac{4}{3}\pi R^3$ and since there are four atoms per FCC unit cell, the total FCC atom (or sphere) volume is

$$V_s = 4\left(\frac{4}{3}\pi R^3\right) = \frac{16}{3}\pi R^3$$

From Example 3.1, the total unit cell volume is

$$V_c = a^3 = (2R\sqrt{2})^3 = 16R^3\sqrt{2}$$

Therefore, the atomic packing factor is,

$$\text{APF} = \frac{\text{Volume of atoms in a unit cell}}{\text{Total unit cell volume}} = \frac{V_s}{V_c} = \frac{\frac{16}{3}\pi R^3}{16R^3\sqrt{2}} = 0.74$$

Example 3.3 Copper has an atomic radius of 0.128 nm, an FCC crystal structure, and an atomic weight of 63.5 g/mol. Compute its theoretical density and compare the answer with its measured density.

Solution: Since the crystal structure is FCC, n, the number of atoms per unit cell, is 4. Furthermore, the atomic weight A_{Cu} is given as 63.5 g/mol. The unit cell volume V_c for FCC was determined as $16R^3\sqrt{2}$ where R, the atomic radius, is 0.128 nm.

Now substituting we get,

$$\rho = \frac{nA_{Cu}}{V_C N_A} = \frac{nA_{Cu}}{(16R^3\sqrt{2})\,N_A}$$

$$= \frac{(4\text{ atoms/unit cell})(63.5\text{ g/mol})}{[16\sqrt{2}(1.28\times10^{-8}\text{ cm})^3/\text{unit cell}](6.023\times10^{23}\text{ atoms/mol})}$$

$$= 8.89\text{ g/cm}^3$$

The literature value for the density of copper is 8.94 g/cm^3, which is in very close agreement with the foregoing result.

The Body-centred Cubic Crystal Structure: Another common metallic crystal structure also has a cubic unit cell with atoms located at all eight corners and a single atom at the cube centre. This is called a **body-centred cubic (BCC)** crystal structure. A collection of spheres depicting this crystal structure is shown in Figure 3.6c, whereas Figures 3.6a and 3.6b are diagrams of BCC unit cells with the atoms represented by hard sphere and reduced-sphere models, respectively.

Centre and corner atoms touch one another along cube diagonals, and unit cell length a and atomic radius R are related through

$$a = \frac{4R}{\sqrt{3}}$$

Chromium, iron, tungsten, and a few more metals recorded in Table 3.1 display a BCC structure. Two particles are connected with each BCC unit cell: one atom from the eight corners, each of which is shared among eight unit cells, and the single centre atom, which is completely contained inside its cell.

Likewise, corner and centre atoms positions are proportionate. The coordination number for the BCC precious stone structure is 8; every inside particle has as closest neighbours its eight corner atoms. Since the coordination number is less for BCC than FCC, so likewise is the atomic packing factor for BCC lower – 0.68 versus 0.74.

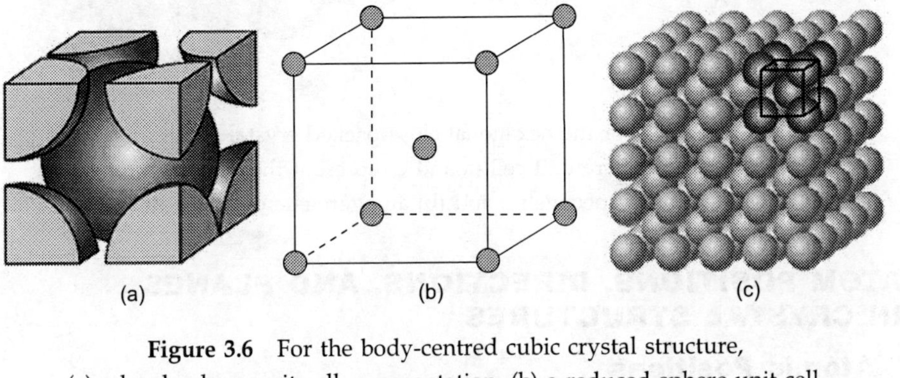

Figure 3.6 For the body-centred cubic crystal structure, (a) a hard sphere unit cell representation, (b) a reduced-sphere unit cell, and (c) an aggregate of many atoms.

The Hexagonal Close-packed Crystal Structure: Not all metals have unit cells with cubic symmetry; the final common metallic crystal structure to be discussed has a unit cell that is hexagonal. Figure 3.7a shows a reduced-sphere unit cell for this structure, which is termed **hexagonal close-packed (HCP)**; an assemblage of several HCP unit cells is presented in Figure 3.7b. The top and bottom faces of the unit cell consist of six atoms that form regular hexagons and surround a single atom in the centre. Another plane that provides three additional atoms to the unit cell is situated between the top and bottom planes. The atoms in this mid-plane have as nearest neighbours atoms in both of the adjacent two planes. The equivalent of six atoms is contained in each unit cell; one-sixth of each of the 12 top and bottom face corner atoms, one-half of each of the 2 centre face atoms, and all 3 mid-plane interior atoms. If a and c represent, respectively, the short and long unit cell dimensions of Figure 3.7a, the c/a ratio should be 1.633; however, for some HCP metals this ratio deviates from the ideal value. The coordination number and the atomic packing factor for the HCP crystal structure are the same as for FCC: 12 and 0.74, respectively. The HCP metals include cadmium, magnesium, titanium, and zinc; some of these are listed in Table 3.1.

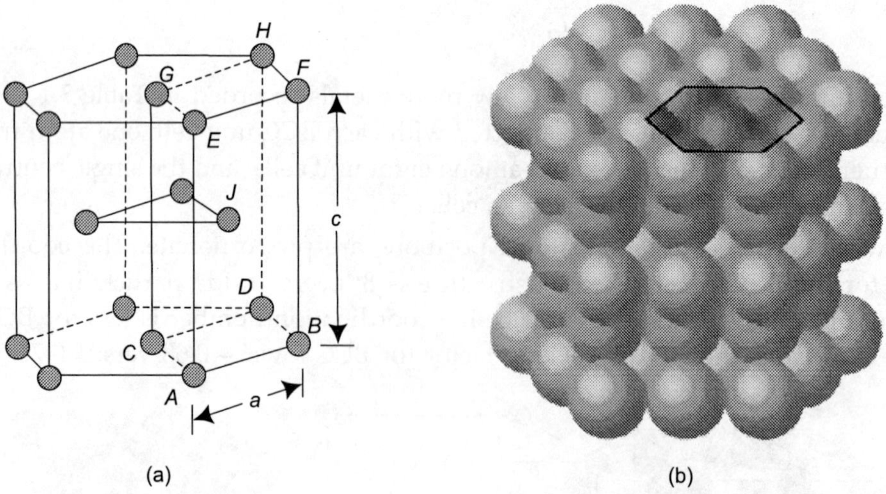

Figure 3.7 For the hexagonal close-packed crystal structure, (a) a reduced-sphere unit cell (a and c represent the short and long edge lengths, respectively), and (b) an aggregate of many atoms.

3.3 ATOM POSITIONS, DIRECTIONS, AND PLANES IN CRYSTAL STRUCTURES

3.3.1 Atomic Positions

Rather than continue to refer to "origin of unit cell," "cube diagonal," or "cube face" when speaking about atom positions, directions, and planes in a crystal, we have developed a shorthand notation to identify these geometric features. The notation is simply based on coordinate geometry and can be understood in most cases without resorting to analytic geometry or vector analysis. As the need arises, however, we will borrow some simple results from these latter mathematical methods. In what follows we strip away the atoms that were so carefully placed on Bravais lattice points in the previous section and just consider the Bravais points themselves. Let us consider the rectangular coordinate axes shown in Figure 3.8 and populate the space with SC, BCC, and FCC point lattices. Each point is given three indices to denote its x, y, and z coordinates in the space. Cube corners have a triad of integers, for example, (0, 0, 0), the origin, and (1, 0, 0), (0, 1, 0), (0, 0, 1), (1, 1,0), and so on. Cube centre indices are denoted by (½, ½, ½) whereas cube face-centred atoms are described by (0, ½ , ½), (½, 0, ½), (½,½, 0), (1, ½, ½), and so on. From the standpoint of crystallography, SC is characterized by (0, 0, 0), BCC by (0, 0, 0) and (½, ½, ½), and FCC by (0, 0, 0), (0, ½, ½), (½, 0, ½), and (½, ½, 0). These numbers are consistent with the respective one, two, and four atoms per unit cell noted earlier. Translation of either the single-, two-, or four-point pattern throughout the space will yield the indicated point lattices. In Figure 3.8 only the unit cube or cell is considered and all coordinates are positive. For neighbouring cells negative coordinates may arise. For example, BCC lattice cube centres in three adjacent cells might have coordinates of (½, – ½. ½), (½, ½, ½), and (½, 3/2, ½).

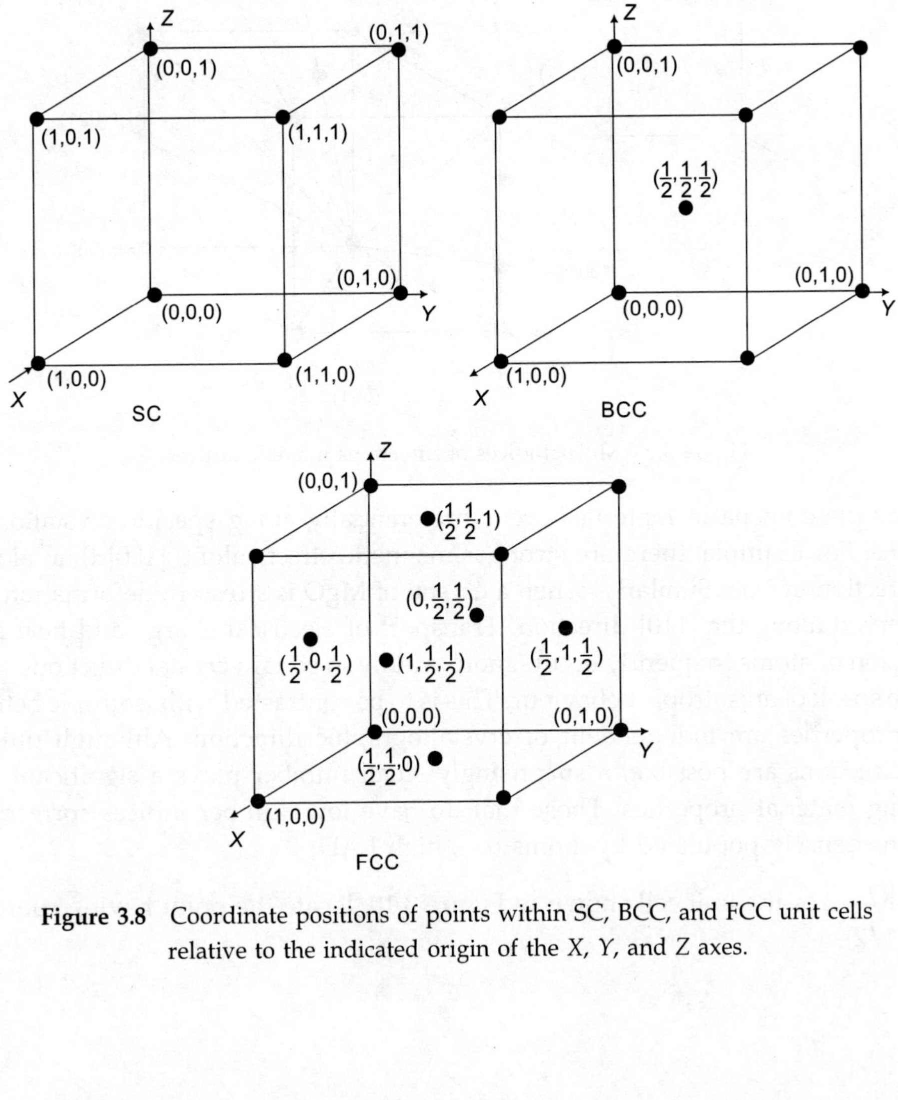

Figure 3.8 Coordinate positions of points within SC, BCC, and FCC unit cells relative to the indicated origin of the X, Y, and Z axes.

3.3.2 Directions

Directions are vectors that connect any two lattice points in a prescribed sense. The indices of a direction are given by the differences in x, y, z coordinates of the two points in question. For example, the direction connecting the outer cube centres noted at the end of the previous paragraph is (1/2, 3/2, 1/2) minus (1/2, –1/2, 1/2) or (0, 2, 0). By convention, the resultant direction components are reduced to smallest whole numbers and placed in brackets without commas; the so-called Miller indices for the direction are [010]. For the anti-parallel direction pointed towards the negative y axis the indices are [010]. The minus sign makes crystalline directions different mathematically, but not physically. A number of directions in cubic lattices are displayed in Figure 3.9.

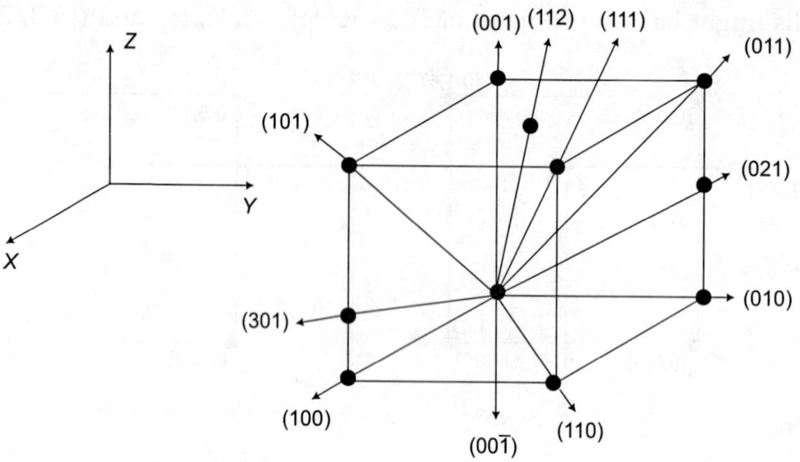

Figure 3.9 Miller indices of directions in cubic lattices.

Many phenomena in materials occur preferentially along specific crystallographic directions. For example, there are stronger magnetic effects along [100] than along the [111] direction in iron. Similarly, when a crystal of MgO is stressed, deformation effects are observed along the [110] direction. Transport of electrical charge and heat as well as diffusion of atoms frequently occurs more readily in certain crystal directions. In such cases we speak of anisotropic behaviour. This is to be contrasted with isotropic behaviour where properties are independent of crystallographic direction. Although numerous lattice directions are possible, a surprisingly small number plays a significant role in governing material properties. Those that do have low-number indices correspond to directions densely populated by atoms (i.e., high LAD).

Example 3.4 For the unit cell shown in Figure 3.10, locate the point having coordinates (1/4, 1, 1/2).

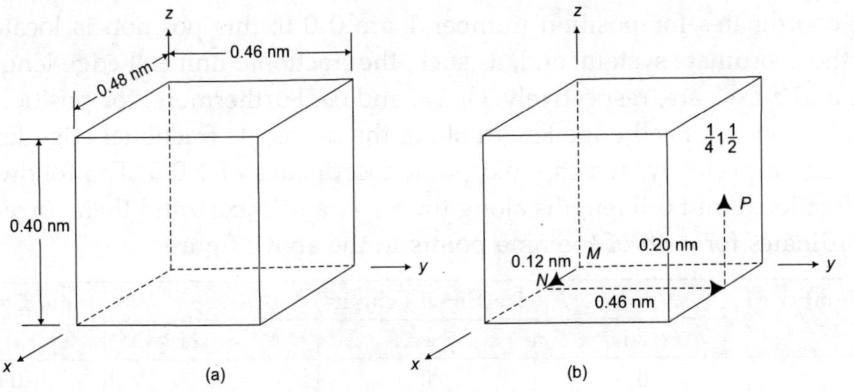

Figure 3.10

Solution: From sketch (a), edge lengths for this unit cell are as follows: a = 0.48 nm, b = 0.46 nm and c = 0.40 nm.

Furthermore, in light of the above discussion, fractional lengths are $q = 1/4$, $4 = 1$ and $s = 1/2$. Therefore, first we move from the origin of the unit cell (point M) qa = (1/4)(0.48) nm = 0.12 nm units along the x axis (to point N), as shown in the sketch (b). Similarly, we proceed rb = 1(0.46 nm) = 0.46 nm parallel to the y axis, from point N to point O. Finally, we move from this position, sc = (1/2)(0.40) nm = 0.20 nm units parallel to the z axis to point P as noted again in sketch (b). This point P then corresponds to the point (¼, 1, ½) coordinates.

Example 3.5 Specify point coordinates for all atom positions for a BCC unit cell.

Solution: Atom position coordinates correspond to the locations of the centres of all atoms in the unit cell, that is, the eight-corner atoms and single centre atom. These positions are noted (and also numbered) in the following figure.

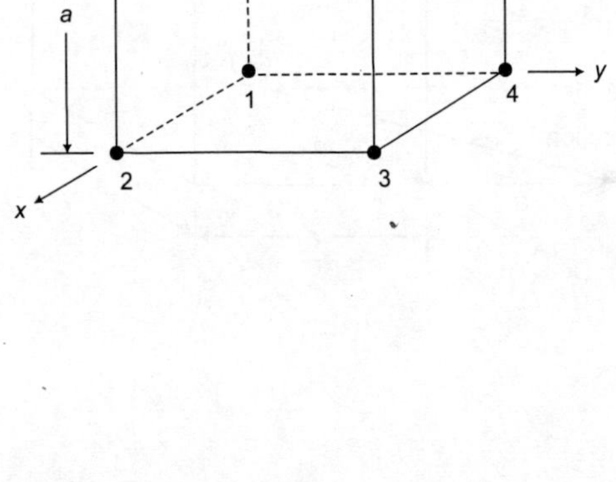

Point coordinates for position number 1 are 0 0 0; this position is located at the origin of the coordinate system, and, as such, the fractional unit cell edge lengths along the *x*-, *y*-, and *z*-axes are, respectively, 0*a*, 0*a*, and 0*a*. Furthermore, for position number 2, since it lies one unit cell edge length along the *x*-axis, its fractional edge lengths are *a*, 0*a*, and 0*a*, respectively, which yield point coordinates of 1 0 0. The following table presents fractional unit cell lengths along the *x*-, *y*-, and *z*-axes, and their corresponding point coordinates for each of the nine points in the above figure.

Point Number	Fractional Lengths			Point Coordinates
	x axis	y axis	z axis	
1	0	0	0	0 0 0
2	1	0	0	1 0 0
3	1	1	0	1 1 0
4	0	1	0	0 1 0
5	$\frac{1}{2}$	$\frac{1}{2}$	$\frac{1}{2}$	$\frac{1}{2}\ \frac{1}{2}\ \frac{1}{2}$
6	0	0	1	0 0 1
7	1	0	1	1 0 1
8	1	1	1	1 1 1
9	0	1	1	0 1 1

Example 3.6 Construction of Specified Crystallographic Direction

Draw a $[1\,\bar{1}\,0]$ direction within a cubic unit cell.

Solution: First construct an appropriate unit cell and coordinate axes system. In the accompanying figure the unit cell is cubic, and the origin of the coordinate system, point *O*, is located at one of the cube corners.

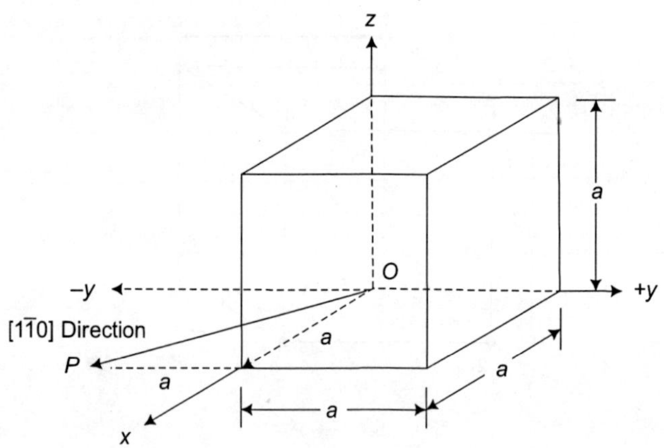

This problem is solved by reversing the procedure of the preceding example. For this $[1\bar{1}0]$ direction, the projections along the x-, y-, and z-axes are a, $-a$, and 0a, respectively. This direction is defined by a vector passing from the origin to point P, which is located by first moving along the x-axis a unit, and from this position, parallel to the y-axis -'a' units, as indicated in the figure. There is no z component to the vector, since the z projection is zero.

Example 3.7 Determination of Directional Indices for a Hexagonal Unit Cell

Determine the indices for the direction shown in the hexagonal unit cell of sketch (a) below.

Solution: In sketch (b), one of three parallelepipeds comprising the hexagonal cell is delineated—its corners are labelled with letters A through H, with the origin of the a_1–a_2–a_3–z, axes coordinate system located at the corner labelled C. We use this unit cell as a reference for specifying the directional indices. It now becomes necessary to determine projections of the direction vector on the a_1, a_2, and z axes. These respective projections are $a(a_1$ axis), $a(a_2$ axis) and $c(z$ axis), which become 1, 1, and 1 in terms of the unit cell parameters. Thus,

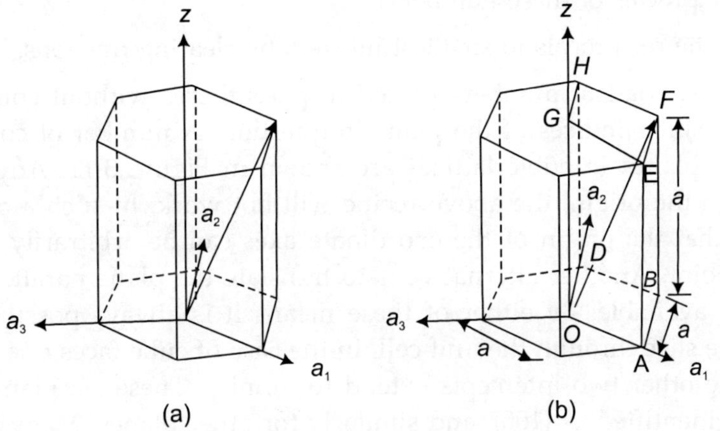

$$u' = 1 \quad v' = 1 \quad w' = 1$$

Also, we know that

$$u = \frac{1}{3}(2u' - v') = \frac{1}{3}[(2)(1) - 1] = \frac{1}{3}$$

$$v = \frac{1}{3}(2v' - u') = \frac{1}{3}[(2)(1) - 1] = \frac{1}{3}$$

$$t = -(u+v) = -\left(\frac{1}{3}+\frac{1}{3}\right) = -\frac{2}{3}$$

$$w = w' = 1$$

Multiplication of the above indices by 3 reduces them to the lowest set, which yields values for u, v, t, and w of 1, 1, – 2 and 3, respectively. Hence, the direction shown in the figure is $[11\bar{2}3]$.

3.3.3 Planes

Like points and directions, crystallographic planes are recognized by three Miller records or numbers. Any three focuses characterize a plane; any two directions will likewise characterize a plane in which both lie. Yet, a solitary course can lie in a wide range of planes. The accompanying basic formula can be utilized to recognize extraordinarily a given plane in cubic (and noncubic) crystals:

1. Express the intercepts of the plane on the three coordinate axes in the number of unit cell dimensions.
2. Take reciprocals of these numbers.
3. Reduce the reciprocals to smallest integers by clearing fractions.

The resulting triad of numbers placed in parentheses without commas, (hkl), is known as the Miller indices of the plane in question. A number of common as well as uncommon planes in cubic lattices are shown in Figure 3.11. Any time a plane passes through the origin, the above recipe will not work. In such a case it must be remembered that the origin of the coordinate axes can be arbitrarily shifted to any other lattice point. Another alternative is to translate the plane parallel to itself until intercepts are available. By either of these means it is always possible to have the involved plane slice through the unit cell. In the case of cube faces one intercept (e.g., x) is 1 and the other two intercepts extend to infinity. These six planes constitute a family that is identified as {100}, and similarly for other planes. Many phenomena in materials are associated with or occur preferentially on specific crystallographic planes. For example, brittle cleavage cracks readily nucleate and grow on (100) planes in MgO crystals. Silicon oxidizes more rapidly on the (111) plane than on the (100) plane. Crystal growth is frequently favoured on one plane relative to others. Microelectronic (e.g., transistor) and optoelectronic (e.g., laser) device fabrication can be reliably accomplished only on specific semiconductor planes. Regardless of application, the important planes are usually the most atomically dense ones (i.e., high APD). Such planes are also ones of low index.

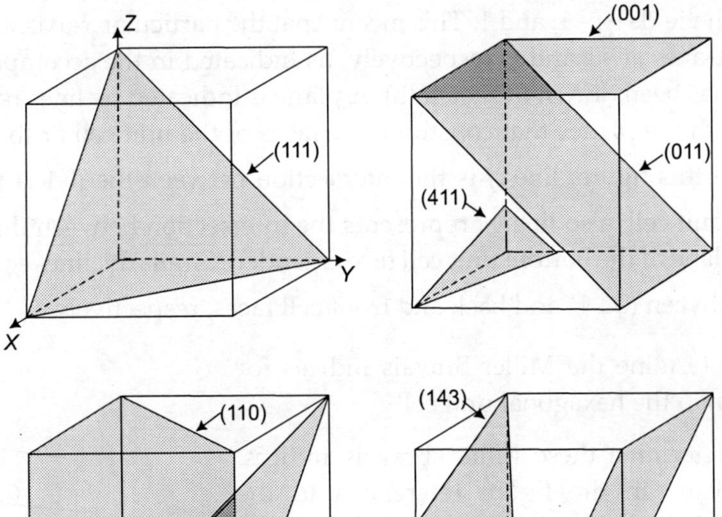

Figure 3.11 Miller indices of some planes in cubic lattices.
The suggested centre of the coordinate axes is shown in each case.

Example 3.8 Construction of Specified Crystallographic Plane

Construct a (011) plane within a cubic unit cell.

Solution:

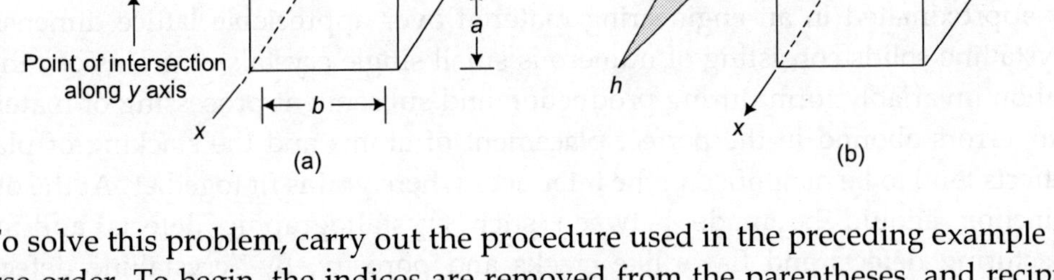

To solve this problem, carry out the procedure used in the preceding example in the reverse order. To begin, the indices are removed from the parentheses, and reciprocals

are taken, which yields ∞, – 1, and 1. This means that the particular x-axis while intersecting the y- and z-axes at – b and c, respectively, as indicated in the accompanying sketch (a). This plane has been drawn in sketch (b). A plane is indicated by lines representing its intersections with the planes that constitute the faces of the unit cell or their extensions. For example, in this figure, line ef is the intersection between the $(0\bar{1}1)$ plane and the top face of the unit cell; also line gh represents the intersection between this same $(0\bar{1}1)$ plane and the plane of the bottom unit cell face extended. Similarly, lines eg and fh are the intersections between $(0\bar{1}1)$ and back and front cell faces, respectively.

Example 3.9 Determine the Miller-Bravais indices for the plane shown in the hexagonal unit cell.

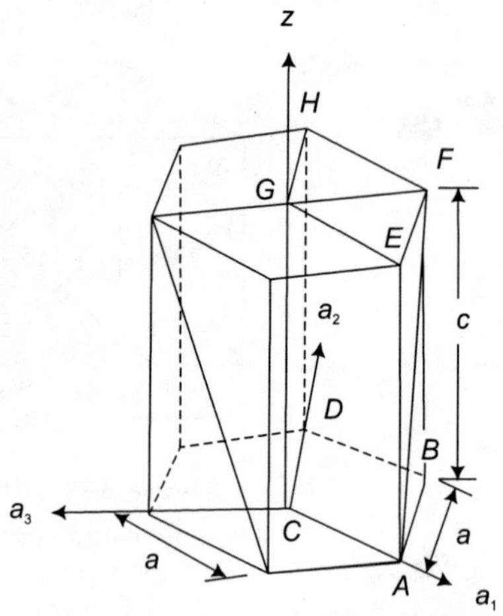

Solution: To determine these Miller–Bravais indices, consider the plane in the figure referenced to the parallelepiped labelled with the letters A through H at its corners. This plane intersects the a_1 axis at a distance a from the origin of the a_1–a_2–a_3–z coordinate axes system (point C). Furthermore, its intersections with the a_2 and z axes are – a and c, respectively. Therefore, in terms of the lattice parameters, these intersections are 1, – 1 and 1. Furthermore, the reciprocals of these numbers are also 1, – 1 and 1. Hence,

$$h = 1, k = -1 \quad \text{and} \quad l = 1$$

$$i = -(h + k) = -(1 - 1) = 0$$

Notice that the third index is zero which means that this plane parallels the a_3 axis. Upon inspection of the above figure, it may be noted that this is indeed the case.

3.4 DEFECTS IN CRYSTALLINE SOLIDS

Perfect crystal structures with all atoms situated in predictable lattice sites are rarely realized. Only in specially grown, oriented semiconductor single crystals is this ideal closely approximated in an engineering material over appreciable lattice dimensions. Polycrystalline solids consisting of numerous small single crystals or grains of random orientation invariably form during production and subsequent processing of materials. In them, errors abound in the perfect placement of atoms and the stacking of planes. Such effects tend to be magnified at the interfaces where grains fit together. At the outset a distinction should be made between such crystallographic defects and gross manufacturing defects and flaws like cracks and porosity. By "crystalline defect" is meant a lattice irregularity having one or more of its dimensions on the order of an atomic diameter. Classification of crystalline imperfections is frequently made according to geometry or dimensionality of the defect. Several imperfections are discussed in this chapter, including **point defects** (those associated with one or two atomic positions), linear (or one-dimensional) defects, as well as interfacial defects, or boundaries, which are two-dimensional. Impurities in solids are also discussed, since impurity atoms may exist as point defects. Finally, techniques for the microscopic examination of defects and the structure of materials are briefly described.

3.4.1 Point Defects

The simplest of the point defects is a vacancy, or vacant lattice site, one normally occupied from which an atom is missing (Figure 3.12). All crystalline solids contain vacancies and, in fact, it is not possible to create such a material that is free of these defects. The necessity of the existence of vacancies is explained using principles of thermodynamics; in essence, the presence of vacancies increases the entropy (i.e., the randomness) of the crystal.

The equilibrium number of vacancies N_v for a given quantity of material depends on and increases with temperature according to

$$N_v = N \exp\left(-\frac{Q_v}{kT}\right)$$

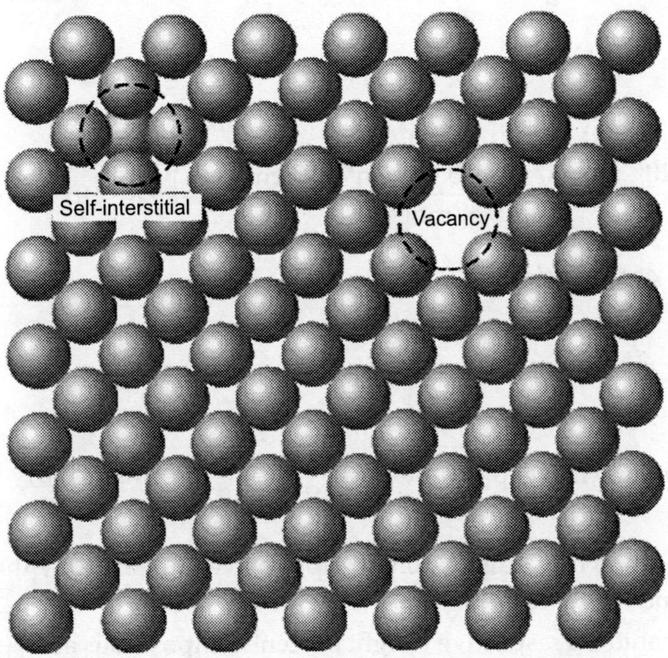

Figure 3.12 Two-dimensional representations of a vacancy and a self-interstitial.

where

N = Total number of atomic sites
Q_v = The energy required for the formation of a vacancy
T = Absolute temperature in kelvin
k = The gas or Boltzmann's constant = 1.38×10^{-23} J/atom-K
= 8.62×10^{-5} eV/atom-K

A self-interstitial is an atom from the precious stone that is swarmed into an interstitial site, a little void space that under standard conditions is not involved. This sort of deformity is likewise shown in Figure 3.12. In metals, a self-interstitial presents generally huge contortions in the encompassing cross section on the grounds that the iota is generously bigger than the interstitial position in which it is arranged. Thus, the arrangement of this deformity is not exceptionally likely, and it exists in little focuses, which are fundamentally lower than for opening.

Example 3.10 Calculate the equilibrium number of vacancies per cubic metre for copper at 1000°C. The energy for vacancy formation is 0.9 eV/atom; the atomic weight and density (at 1000°C) for copper are 63.5 g/mol and 8.4 g/cm^3, respectively.

Solution: The number of atomic sites per cubic metre for copper A_{Cu}, from its atomic weight its density and Avogadro's number N_A.

$$N = \frac{N_A \rho}{A_{Cu}}$$

$$= \frac{(6.023 \times 10^{23} \text{ atoms/mol})(8.4 \text{g/cm}^3)(10^6 \text{cm}^3/\text{m}^3)}{63.5 \text{ g/mol}} = 8 \times 10^{28} \text{ atoms/m}^3$$

Thus, number of vacancies at 1000°C is

$$N_v = N \exp\left(-\frac{Q_v}{kT}\right)$$

$$= (8 \times 10^{28} \text{atoms/m}^3) \exp\left[\frac{0.9 \text{ eV}}{(8.62 \times 10^{-5} \text{ eV/K})(1273\text{K})}\right] = 2.2 \times 10^{25} \text{ vacancies/m}^3$$

Example 3.11: Determine the composition, in atom percent, of an alloy that consists of 97 wt% aluminium and 3 wt% copper.

Solution: If we denote the respective weight percent compositions as $C_{Al} = 97$ and $C_{Cu} = 3$

$$C'_{Al} = \frac{C_{Al}\,A_{Cu}}{C_{Al}\,A_{Cu} + C_{Cu}\,A_{Al}} \times 100$$

$$= \frac{(97)\,(63.55\ g/mol)}{(97)\,(63.55\ g/mol) + (3)\,(26.98\ g/mol)} \times 100$$

$$= 98.7\ at\%$$

$$C'_{Cu} = \frac{C_{Cu}\,A_{Al}}{C_{Cu}\,A_{Al} + C_{Al}\,A_{Cu}} \times 100$$

$$= \frac{(3)\,(26.98\ g/mol)}{(3)\,(26.98\ g/mol) + (97)\,(63.55\ g/mol)} \times 100$$

$$= 1.30\ at\%$$

Point Defects in Ionic Compounds: In alkali halides and metal oxides where positive and negative particles populate cross section destinations, point imperfection structures are more unpredictable. Support of charge lack of bias is the reason. To envision the issues included let us consider Figure 3.13b which delineates an electrically unbiased cross section made out of monovalent (positive) cations and (negative) anions. Production of a cation opening means the nonappearance of a solitary positive charge or a successful negative lattice charge. This imperfection can't exist without anyone else's input in light of the fact that the cross section is no more extended electrically nonpartisan. In this way, a negative particle opening is made in the meantime and the related match (positive particle opportunity + negative particle opening) is known as a Schottky deformity. The part of Schottky deformities in respect to all atoms in the material is given by an equation like that for opportunities. In any case, the vitality is divided in the light of the fact that a couple of opportunities are included. Another defect, comprises a cation that bounces into an interstitial site. The subsequent cation opening interstitial particle is known as a Frenkel deformity. This deformity happens in AgCl. Alloying soluble base halides can convolute the deformity picture. For instance, if $CaCl_2$ is added to NaCl the two Cl-particles possess two typical anion locales, however the single Ca+ particle can involve one and only cation site. To make up for its additional charge an extra cation opportunity must be made. In such a case we should recognize two sorts of cation opening: those created inherently basically by warming and those presented extraneously by alloying. What's more, as we should later observe, semiconductor doping includes in the same manner as alloying ionic compounds.

3.4.2 Dislocations (Line Defects)

A *dislocation* is a linear or one-dimensional defect around which some of the atoms are misaligned. One type of dislocation is represented in Figure 3.13 an extra portion of a plane of atoms, or half-plane, the edge of which terminates within the crystal. This is named an edge dislocation. It is a direct imperfection that revolves around the line that is characterized along the end of the additional half-plane of molecules. This is once in a while named the separation line, which, for the edge disengagement in Figure 4.15, is opposite to the plane of the page. Inside the area around the disengagement line there is some localized lattice distortion.

The atoms over the separation line in Figure 3.13 are pressed together, and those beneath are pulled separated; this is reflected in the slight curvature and flow for the vertical planes of atoms as they curve around this additional half-plane. The magnitude of this distortion decreases with distance away from the dislocation line; at positions far evacuated, the crystal lattice is for all intents and purposes idealized. Now and then the edge separation in Figure 3.13 is spoken to by the image which likewise shows the position of the disengagement line. An edge separation may likewise be framed by an additional half-plane of atoms that is incorporated into the base part of the precious stone; its assignment is a *T*.

Another kind of separation, called a screw dislocation, might be considered as being framed by a shear push that is connected to create the mutilation as in Figure 3.14a: the upper front district of the gem is moved one nuclear separation to the right with respect to the base segment. The nuclear bending connected with a screw disengagement is

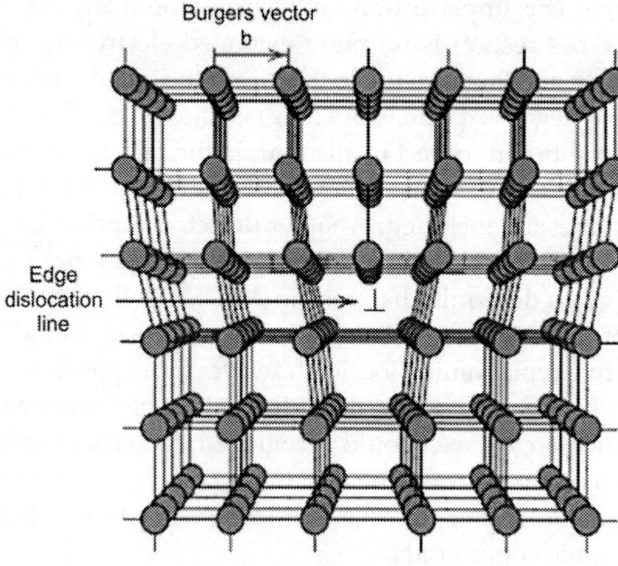

Figure 3.13 The atom positions around an edge dislocation; extra half-plane of atoms shown in perspective

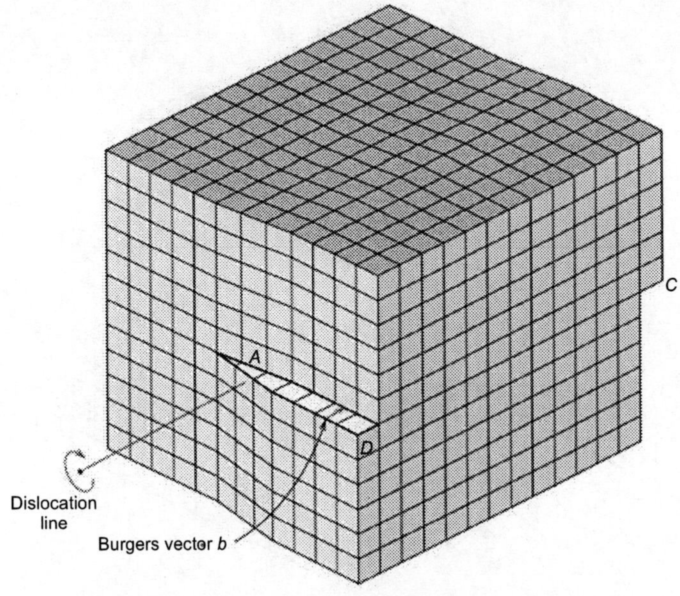

Figure 3.14(a) A screw dislocation within a crystal.

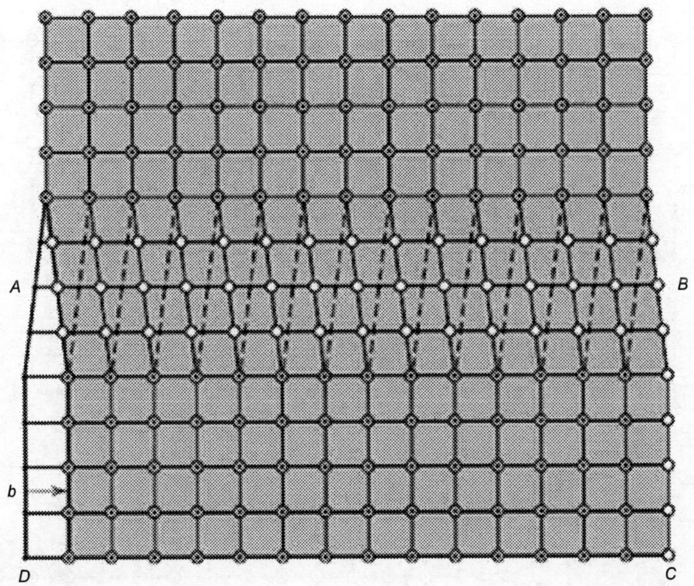

Figure 3.14(b) The screw dislocation in (a) as viewed from above. The dislocation line extends along line *AB*. Atom positions above the slip plane are designated by open circles, those below by solid circles.

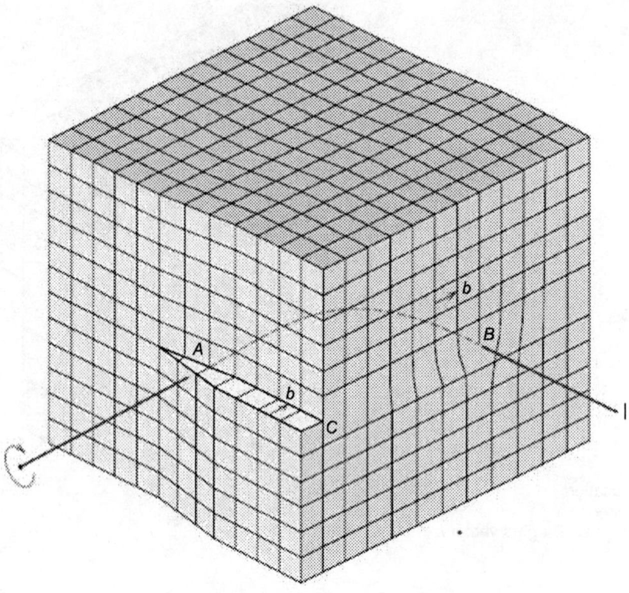

Figure 3.15(a) Schematic representation of a dislocation that has edge, screw and mixed character.

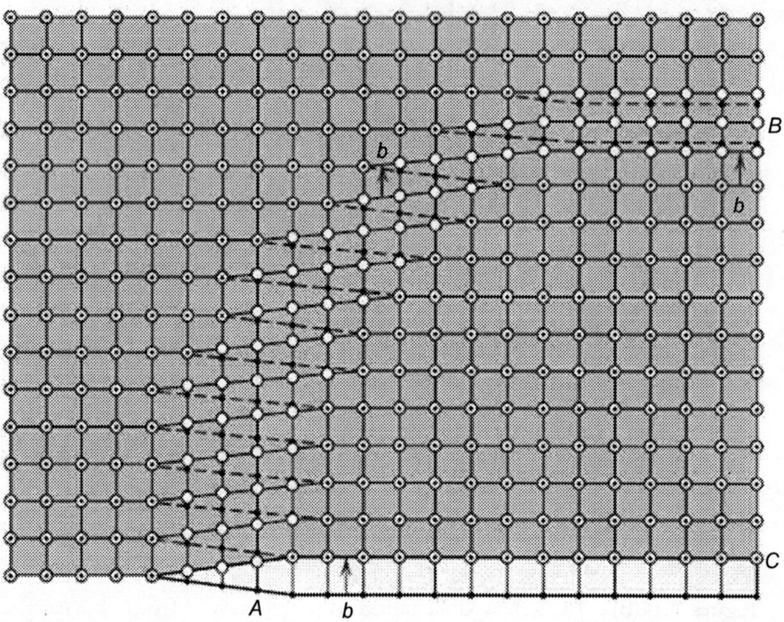

Figure 3.15(b) Top view, where open circles denote atom positions above the slip plane. At point *A*, the dislocation is pure screw, while at point *B*, it is pure edge. For regions in between where there is curvature in the dislocation line, the character is mixed edge and screw.

additionally direct and along a separation line, line AB in Figure 3.14b. The screw disengagement gets its name from the winding or helical way or slope that is followed around the separation line by the nuclear planes of atoms. Now and again the image of clockwise turn is utilized to assign a screw dislocation.

Most dislocations found in crystalline materials are presumably neither unadulterated edge nor immaculate screw, yet display segments of both sorts; these are named blended separations. Each of the three disengagement sorts are shown schematically in Figure 3.15; the lattice distortion that is created far from the two appearances is blended, having differing degrees of screw and edge character. The extent and heading of the cross section contortion connected with a separation is communicated as far as a Burgers vector, signified by a b. Burgers vectors are shown in Figures 3.13 and 3.14 for edge and screw separations, individually. Moreover, the nature of a disengagement (i.e., edge, screw, or mixed) is characterized by the relative introductions of separation line and Burgers vector. For an edge, they are perpendicular (Figure 3.13), while for a screw, they are parallel (Figure 3.14); they are neither opposite nor parallel for a mixed dislocation. Despite the fact that a dislocation alters course and nature inside a crystal (e.g., from edge to mixed to screw) Burgers vector will be the same at all focuses along its line. For instance, all places of the bended disengagement in Figure 3.15 will have the Burgers vector appeared. For metallic materials, the Burgers vector for a separation will point in a nearby stuffed crystallographic bearing and will be of extent equivalent to the between atomic spacing.

3.4.3 Interfacial Defects

Interfacial defects are limits that have two measurements and regularly isolate areas of the materials that have diverse crystal structures and additionally crystallographic introductions. These blemishes incorporate outside surfaces, grain limits, twin limits, stacking shortcomings, and phase boundaries.

External Surfaces: A standout amongst the most clear limits is the outer surface, along which the precious stone structure ends. Surface atoms are not attached to the most extreme number of closest neighbours, and are in this manner in a higher vitality state than the particles at inside positions. The obligations of these surface atoms that are not fulfilled offer ascent to a surface energy, communicated in units of vitality per unit area (J/m^2 or erg/cm^2). To decrease this energy, materials have a tendency to minimize, if at all conceivable, the aggregate surface territory. For instance, fluids expect a shape having a minimum range—the droplets get to be circular. Obviously, this is impractical with solids, which are mechanically inflexible and rigid.

Grain Boundaries: Another interfacial defect, the grain boundary is the boundary separating two small grains or crystals having different crystallographic orientations in polycrystalline materials. A grain boundary is represented schematically from an atomic perspective. Within the boundary region, which is probably just several atom distances wide, there is some atomic mismatch in a transition from the crystalline orientation of one grain to that of an adjacent one.

The atoms are bonded less consistently along a grain boundary (e.g., bond edges are longer), and thus, there is an interfacial or grain boundary energy like the surface energy portrayed previously. The greatness of this vitality is an element of the level of bewilderment, being bigger for high-edge limits. Grain boundaries are more synthetically responsive than the grains themselves as a result of this limit vitality. Moreover, contamination molecules regularly isolate along these limits as a result of their higher energy state. The aggregate interfacial energy is lower in huge or coarse-grained materials than in fine-grained ones, since there is less aggregate limit territory in the previous. Despite this scattered course of action of particles and absence of consistent holding along grain limits, a polycrystalline material is still exceptionally solid; durable powers inside and over the limit are available. Besides, the thickness of a polycrystalline example is for all intents and purposes indistinguishable to that of a solitary crystal of a similar material.

Twin Boundaries: A twin boundary is an extraordinary kind of grain limit crosswise over which there is a particular mirror lattice symmetry; that is, atoms on one side of the limit are situated in reflect picture places of the molecules on the opposite side (Figure 3.16). The locale of material between these limits is fittingly named a twin. Twins result from nuclear removals that are created from connected mechanical shear strengths (mechanical twins), furthermore amid strengthening thermal curing taking after twisting (toughening twins). Twinning happens on a positive crystallographic plane and in a particular bearing, both of which rely on the precious stone structure. Strengthening twins are normally found in metals that have the FCC crystal structure, while mechanical twins are seen in BCC and HCP metals.

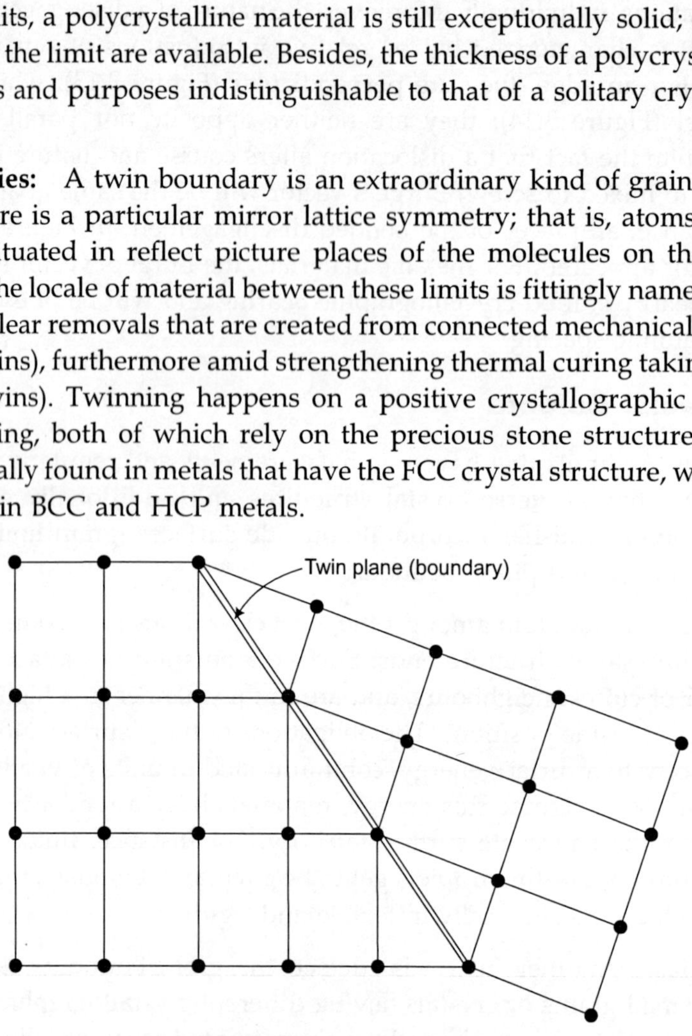

Figure 3.16 Schematic diagram showing a twin plane or boundary and the adjacent atom positions.

3.5 SUMMARY

Fundamental Concepts and Unit Cells

Atoms in crystalline solids are situated in deliberate and rehashed designs that are rather than the irregular and disarranged nuclear circulation found in non-crystalline or formless materials. Atoms might be spoken to as solid spheres, and, for crystalline solids, crystal structure is only the spatial course of action of these spheres. The different crystal structures are indicated regarding parallelepiped unit cells, which are portrayed by geometry and molecule positions inside.

Metallic Crystal Structures

Most normal metals exist in no less than one of the three moderately straightforward precious stone structures: face-centred cubic (FCC), body-centred cubic (BCC), and hexagonal close-packed (HCP). Two components of a crystal structure are coordination number (or number of closest neighbour particles) and atomic packing factor (the portion of strong circle volume in the unit cell). Coordination number and atomic packing factor are the same for both FCC and HCP crystal structures, each of which might be produced by the stacking of close-packed planes of atoms.

Point Coordinates, Crystallographic Directions and Crystallographic Planes

Crystallographic points, directions, and planes are determined as far as indexing plans. The reason for the assurance of every record is a facilitate coordinate axis system characterized by the unit cell for the specific crystal structure. The area of a point inside a unit cell is indicated utilizing arrangement that are fragmentary products of the cell edge lengths. Directional lists are processed as far as the vector projection on each of the organized tomahawks, while planar lists are resolved from the reciprocals of pivotal captures. For hexagonal unit cells, a four-record conspire for both bearings and planes is observed to be more helpful.

Vacancies and Self-Interstitials

Every strong material contains substantial quantities of defects or deviations from crystalline flawlessness. The few sorts of flaw are ordered on the premise of their geometry and size. Point deformities are those connected with may be a couple of nuclear positions, including opportunities (or empty lattice locales), self-interstitials (have atoms that involve interstitial destinations), and polluting influence atoms.

Impurities in Solids

A strong arrangement may shape when pollution atoms are added to a solid, in which case the first crystal structure is held and no new stages are framed. For substitutional solid arrangements, debasement atoms substitute for host molecules, and calculable solvency is conceivable just when nuclear distances across and electro-negativities for both particle sorts are comparative, when both components have a similar crystal structure, and when the polluting influence molecules have a valence that is the same as or not exactly the host material. Interstitial solid arrangements shape for generally little debasement particles that involve interstitial destinations among the host atoms.

Dislocations—Linear Defects

Separations are one-dimensional crystalline deformities of which there are two unadulterated sorts: edge and screw. An edge might be considered as far as the cross section mutilation along the end of an additional half-plane of molecules; a screw, as a helical planar slope. For blended separations, segments of both immaculate edge and screw are found. The size and bearing of cross section twisting connected with a separation are determined by its Burgers vector. The relative introductions of Burgers vector and disengagement line are opposite for edge, parallel for screw, and neither opposite nor parallel for mixing.

EXERCISE

1. What is the difference between atomic structure and crystal structure?
2. If the atomic radius of lead is 0.175 nm, calculate the volume of its unit cell in cubic metres.
3. For the HCP crystal structure, show that the ideal ratio c/a is 1.633.
4. Prove that the atomic packing factor for BCC is 0.68.
5. Show that the atomic packing factor for HCP is 0.74.
6. Show for the body-centred cubic crystal structure that the unit cell edge length a and the atomic radius R are related through $a = 4R\sqrt{3}$.
7. Indium has a tetragonal unit cell for which the a and c lattice parameters are 0.459 and 0.495 nm, respectively:
   (a) If the atomic packing factor and atomic radius are 0.693 and 0.1625 nm, respectively, determine the number of atoms in each unit cell.
   (b) The atomic weight of indium is 114.82 g/mol; compute its theoretical density
8. Beryllium has an HCP unit cell for which the ratio of the lattice parameters c/a is 1.568. If the radius of the Be atom is 0.1143 nm:
   (a) determine the unit cell volume, and
   (b) calculate the theoretical density of Be and compare it with the literature value.
9. Cobalt has an HCP crystal structure, an atomic radius of 0.1253 nm, and a c/a ratio of 1.623. Compute the volume of the unit cell for Co.

10. Caesium metal has a BCC structure with a lattice parameter of 0.6080 nm. What is the atomic radius?
11. Thorium metal has an FCC structure with a lattice parameter of 0.5085 nm. What is the atomic radius?
12. Rhodium has a lattice parameter of 0.3805 nm and the atomic radius is 0.134 nm. Does this metal have a BCC or FCC structure?
13. Niobium has a lattice parameter of 0.3307 nm and the atomic radius is 0.147 nm. Does this metal have a BCC or FCC structure?
14. Based on atomic weights and structural information show that gold and tungsten essentially have the same density. Calculate the density of each.
15. Calculate the density of GaAs if the lattice constant is 0.5654 nm.
16. Assuming energy for vacancy formation of 0.90 eV/atom, calculate the fraction of atom sites that are vacant for copper at its melting temperature of 1084°C (1357 K)
17. The density and atomic weight for Au are 18.63 g/cm^3 (at 900°C) and 196.9 g/mol respectively. Calculate the number of vacancies per cubic metre in gold at 900°C. The energy for vacancy formation is 0.98 eV/atom.
18. The atomic weight and density (at 800°C) for silver are, 107.9 g/mol and 9.5 g/cm^3 respectively. Calculate the energy for vacancy formation in silver, given that the equilibrium number of vacancies at 800°C (1073 K).
19. Cite the relative Burgers vector–dislocation line orientations for edge, screw, and mixed dislocations.
20. The grain boundary energy of a small angle grain boundary is less than for a high angle one. Why is this so?

4

Thermodynamics of Solids

Learning Objectives

After going through this chapter the reader will be able to:

- Understand the involvement of chemical reactions in thermodynamics of solids.
- Understand the concept of Gibbs phase rule and lever rule.
- Get an idea about binary phase diagram, the composition of an alloy.
- Understand the Fe-Fe_3C phase diagram in detail and note the significance of temperature variance.

4.1 INTRODUCTION

The concept of equilibrium is an important one in science. Statics, a typical course of study, manages the adjustment of powers that wins in mechanical structures and frameworks when they are in harmony. At the point when such state exists, it implies the structure or framework tends to move or change and, as a result, has minimized its mechanical or potential vitality. We may see the materials we are concerned with in this book as substance frameworks that are very often moderately immaculate, yet much of the time contain at least two distinct components. A few inquiries then emerge concerning synthetic, contrasted and mechanical frameworks: What do we mean by concoction balance? Are there equivalent strengths that equilibrate in substance balance? Is there a comparing vitality that is minimized at harmony? Completely created answers to these inquiries are given in science and thermodynamics courses, however here we can just stretch a few highlights subjectively.

Why do we read phase diagrams? One reason that a learning and comprehension of stage graphs is essential to identify the plan and control of heat treatment techniques; a few properties of materials are elements of their microstructures, and, thus, of their warm histories. Despite the fact that most stage charts speak to stable (or harmony) states and microstructures, they are helpful in comprehension the improvement and conservation of non-balance structures and their orderly properties. These properties are more alluring than those connected with the balance state. This is apropos outlined by the marvel of precipitation solidifying.

The comprehension of stage graphs for combination frameworks is critical in light of the fact that there is a solid relationship amongst the microstructure and mechanical properties, and the improvement of microstructure of an amalgam is identified with the attributes of its stage chart. Moreover, stage outlines give significant data about dissolving, throwing, crystallization, and other wonders. This section shows and talks about the accompanying themes: Chemical reactions connected with thermodynamics of solids, single segment frameworks, introduction to phase diagrams (phrasing connected with stage graphs and stage changes), structure and sythesis of stages.

4.2 CHEMICAL REACTIONS

4.2.1 Oxide Formation Reactions

There are many instances where chemical reactions play an important role in materials science and engineering. The oxidation of metals, high-temperature reactions in polymer mixtures, corrosion, and the deposition of hard ceramic coatings from gases are but a small sample of the examples that can be profitably described by the thermodynamics of chemical reactions. According to the rules of thermo-chemistry that apply to chemical reactions at equilibrium, we recall that the energies (enthalpies, free energies) of reactants are additive when account is taken of the number of moles involved, and similarly for the products. The net energy difference (ΔH, ΔG) is what remains when the energies of the products are subtracted from those of the reactants. To see how this works, consider the oxidation of metal M by the reaction:

$$M + O_2 = MO_2, \ \Delta G^0_{MO_2}$$

If MO_2 is produced under standard conditions (e.g., pure M, 1 atm O_2), then the change in free energy of oxide formation per mole, ΔG^0, is given by $\Delta G^0 = G_{MO_2} - G_M - G_{O_2}$.

As a generalization, the metal of an oxide that has a more negative ΔG^0 than a second oxide will reduce the latter, and be oxidized in the process.

4.2.2 Chemical Vapour Deposition Reactions

Chemical vapour deposition (CVD) is a chemical process used to produce high quality, high-performance, solid materials. The process is often used in the semiconductor industry to produce thin films. In typical CVD, the wafer (substrate) is exposed to one or more volatile precursors, which react and/or decompose on the substrate surface to produce the desired deposit. Frequently, volatile by-products are also produced, which are removed by gas flow through the reaction chamber. Micro-fabrication processes widely use CVD to deposit materials in various forms, including monocrystalline, polycrystalline, amorphous, and epitaxial.

There are various essential applications in which thin movies or coatings are incorporated from high-temperature gas blends. The target of these chemical vapour deposition (CVD) procedures is to coat chosen substrates with the required materials. For instance, consider the TiN, TiC, and Al_2O_3 compound coatings saved on tungsten carbide metal machining and cutting instruments. (The gold shading normally observed on cutting instruments and on tungsten carbide device supplements is because of a TiN covering.) Hundreds of a large number of hardware bits worldwide have been covered with either single or successively kept different layers of these materials.

Every covering has a hardness twofold that of solidified apparatus steel, a hard material in its own privilege. Just 5 to 10 μm (5000 to 10,000 nm) thick layers are saved. This covering thickness is sufficient to give extensive wear imperviousness to the hidden apparatus with the goal that it can last 5 to 10 times longer than uncoated instruments before requiring honing. A reactor used to store every one of the three coatings is indicated schematically in Figure 4.1. Input gases respond at the substrate surface shaping the compound store of intrigue while vaporous items are cleared out of the reactor. The substance responses depends on

$$2TiCl_4(g) + N_2(g) + 4H_2(g) = 2TiN(s) + 8HCl(g)$$

$$TCl_4(g) + CH_4(g) = TiC(s) + 4HCl(g)$$

$$2AlCl_3(g) + 3CO_2(g) + 3H_2(g) = Al_2O_3(s) + 3CO(g) + 6HCl(g)$$

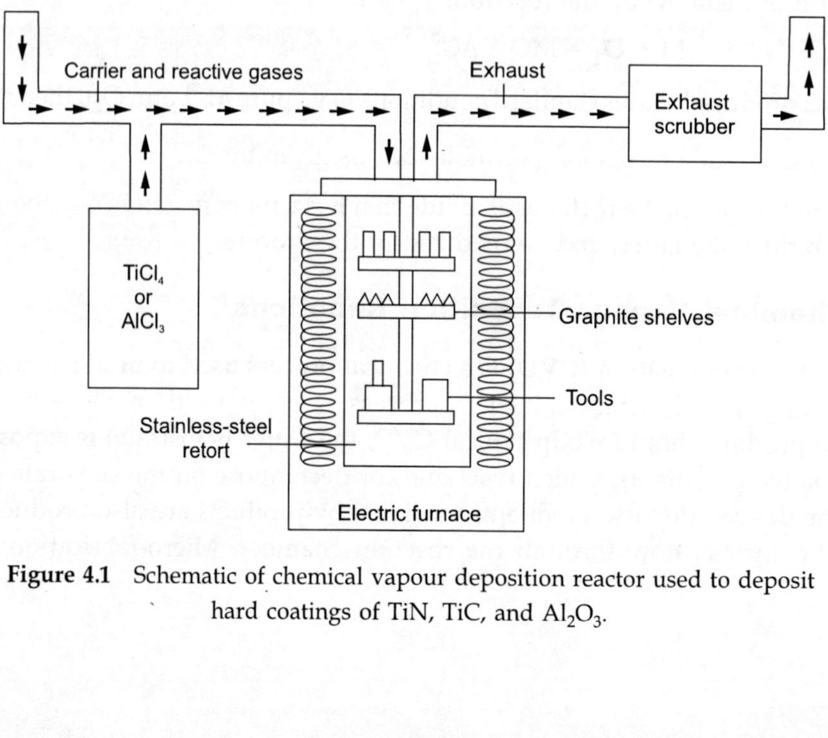

Figure 4.1 Schematic of chemical vapour deposition reactor used to deposit hard coatings of TiN, TiC, and Al_2O_3.

Every response happens at about 1000°C under surmised balance conditions. Additionally, the majority of the reactants and items, except for the covering meant by (s), are vaporous; this is a general element of CVD reactions.

4.3 COMPONENT AND SYSTEMS

4.3.1 Definitions of Component and System

It is important to set up an establishment of definitions and fundamental ideas identifying with amalgams, stages, and harmony before digging into the translation and use of stage outlines. The term part is every now and again utilized as a part of this discourse; segments are immaculate metals as well as mixes of which an amalgam is formed. For instance, in a copper-zinc metal, the parts are Cu and Zn. Solute and solvent, which are additionally normal terms, were characterized in Section 4.3. Another term utilized as a part of this setting is framework, which has two implications.

First, "system" may refer to a specific body of material under consideration (e.g., a ladle of molten steel). On the other hand, it might identify with the arrangement of conceivable amalgams comprising similar parts, yet without respect to combination structure (e.g., the iron-carbon framework). The idea of a strong arrangement was presented in Section 4.3. By method for survey, a strong arrangement comprises atoms of no less than two distinct sorts; the solute molecules possess either substitutional or interstitial positions in the solvent lattice, and the crystal structure of the dissolvable is kept up.

4.3.2 Solubility Limit

For some alloy systems and at some particular temperature, there is the most extreme grouping of solute particles that may break up in the solvent to form solid solution; this is known as a **solubility limit**. The expansion of solute in abundance of this dissolvability confine brings about the development of another strong arrangement or aggravate that has an unmistakably extraordinary sythesis. To outline this idea, consider the sugar-water ($C_{12}H_{22}O_{11}$-H_2O) system.

At first, as sugar is added to water, a sugar-water arrangement or syrup frames. As more sugar is presented, the arrangement turns out to be more focused, until as far as possible is come to, or the arrangement gets to be immersed with sugar. As of now the arrangement is not fit for dissolving any more sugar, and further increases just settle to the base of the compartment. In this manner, the framework now comprises two separate substances: a sugar-water syrup liquid solution and solid crystals of undissolved sugar.

This solubility limit of sugar in water relies on the temperature of the water and might be spoken to in graphical shape on a plot of temperature along the ordinate and creation (in weight percent sugar) along the abscissa, as shown in Figure 4.2. Along the composition axis, expanding sugar fixation is from left to right, and rate of water is perused from right to left. Since just two segments are included (sugar and water), the total of the focuses at any sythesis will meet 100 wt%.

As far as possible is spoken to as the almost vertical line in the figure. For organizations and temperatures to one side of the dissolvability line, just the syrup fluid arrangement exists; to one side of the line, syrup and strong sugar exist together. As far as possible at some temperature is the sythesis that compares to the crossing point of the given temperature arrange and as far as possible line. For instance, at 200°C the most extreme dissolvability of sugar in water is 65 wt%. As Figure 4.2 shows, as far as possible increments somewhat with rising temperature.

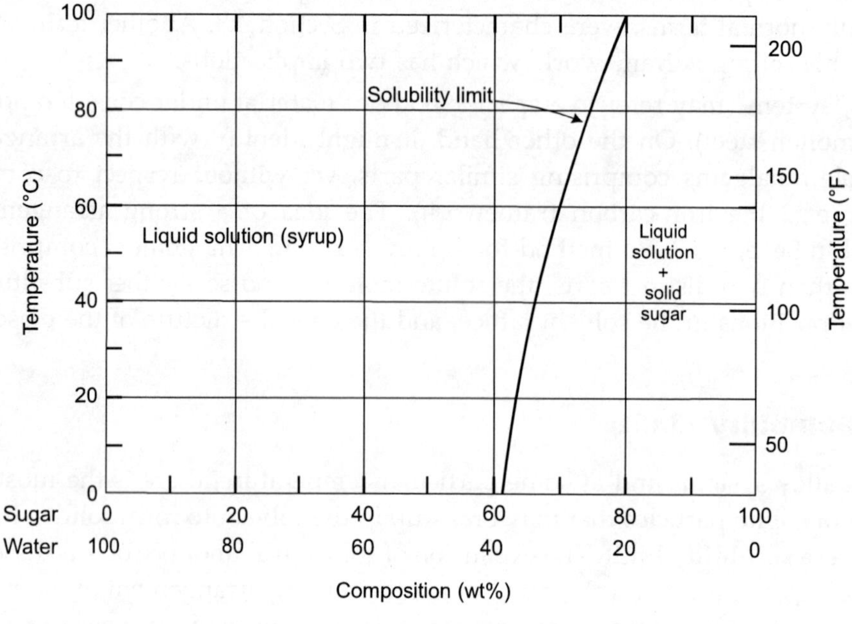

Figure 4.2 The solubility of sugar in a sugar-water syrup.

4.4 PHASE DIAGRAMS

4.4.1 Phases and Single-component Systems

By single-component systems we mean materials possessing one chemical species that is retained in all of its phases over the broad range of temperatures (T) and pressures (P) of interest. For example, a single type of either atom (carbon) or molecule (H_2O) can constitute a single component. Thus, carbon can exist as graphite, diamond, liquid carbon, or carbon vapour, and H_2O as water, ice, or steam. Each of these recognizable homogeneous forms (solid, liquid and gas) constitutes a **phase.** A convenient way of visualizing regions of phase stability in single-component systems as a function of T and P is through phase diagrams. The widely reproduced phase diagram for H_2O is shown in Figure 4.3. We know from experience that water, ice, and steam are individually stable over broad temperature and pressure ranges. The phase diagram, determined from careful experiment, broadly maps these limits of single-phase stability. In addition, the lines depict the special set of conditions under which two phases coexist in equilibrium. Thus, the boundary between liquid and vapour defines a set of pressure versus boiling point data, whereas that between the vapour and solid defines corresponding vapour pressure versus sublimation temperature data. Under a very restricted set of conditions, namely, 0.0075°C and 0.006 atm, all three phases of H_2O coexist in equilibrium at the triple point.

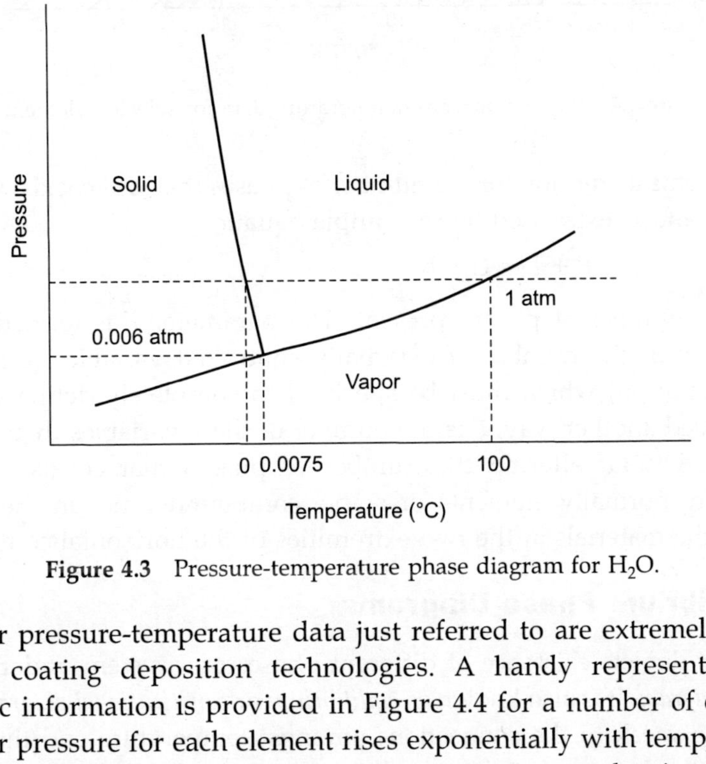

Figure 4.3 Pressure-temperature phase diagram for H_2O.

The vapour pressure-temperature data just referred to are extremely important in thin-film and coating deposition technologies. A handy representation of such thermodynamic information is provided in Figure 4.4 for a number of elements. Note that the vapour pressure for each element rises exponentially with temperature. This is not an entirely unexpected dependence because evaporation may be viewed as a reaction where the vapour (V) and condensed liquid (L) or solid (S) phases are in equilibrium.

4.4.2 Gibbs Phase Rule

The development of stage graphs and a portion of the standards representing the conditions for stage balances are administered by the laws of thermodynamics. One of these is the **Gibbs phase rule,** proposed by the 19th century physicist J. Willard Gibbs.

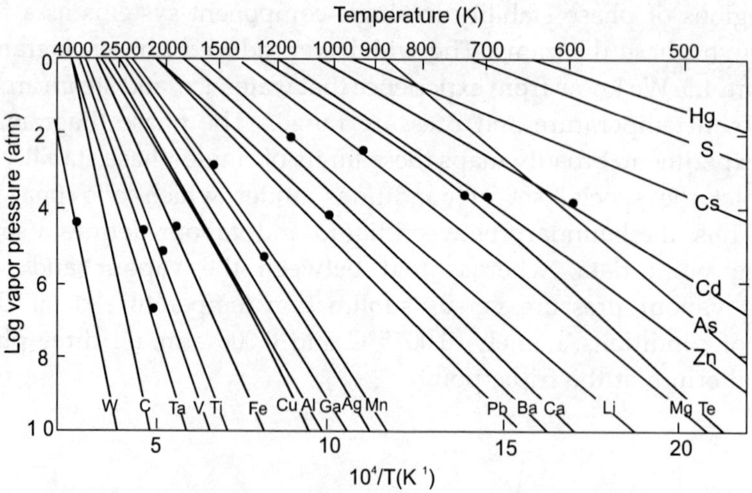

Figure 4.4 Vapour pressure-temperature data for selected elements.

This rule represents a criterion for the number of phases that will coexist within a system at equilibrium, and is expressed by the simple equation

$$P + F = C + N$$

where, P is the number of phases present. The parameter F is termed the *number of degrees of freedom* or the number of externally controlled variables (e.g., temperature, pressure, composition) which must be specified to completely define the state of the system. Expressed another way, F is the number of these variables that can be changed independently without altering the number of phases that coexist at equilibrium. Components are normally elements or stable compounds and, in the case of phase diagrams, are the materials at the two extremities of the horizontal compositional axis.

4.4.3 Equilibrium Phase Diagrams

A diagram that depicts existence of different phases of a system under equilibrium is termed *phase diagram*. It is also known as *equilibrium* or *constitutional diagram*. Equilibrium phase diagrams speak to the connections amongst temperature and the organizations and the amounts of stages at balance. All in all practice it is adequate to consider just strong and fluid.

These diagrams don't demonstrate the flow when one stage changes into another. Be that as it may, it delineates data identified with microstructure and phase structure of a specific framework in a helpful and compact way. Vital data, valuable for the researchers and specialists who are included with materials improvement, determination, and application in item outline, reachable from a stage graph can be compressed as takes after:

- To determine phases which are present at different compositions and temperatures under slow cooling (equilibrium) conditions.
- To show equilibrium solid solubility of one element/compound in another.
- To visualize temperature at which an alloy starts to solidify and the range of solidification.
- To depict the temperature at which different phases start to melt.
- Amount that can be obtained in each phase in a two-phase mixture.

A phase diagram is really a gathering of dissolvability farthest point curves. The phase fields in equilibrium diagrams rely on the specific system being delineated. Set of solubility curves that speaks to locus of temperatures above which all sytheses are liquid are called liquidus, while solidus speaks to set of solvency curves that signifies the locus of temperatures underneath which all compositions are solid.

Each phase diagram for at least two segments must demonstrate a liquidus and a solidus, and an interceding solidifying range, with the exception of immaculate framework, as softening of a stage happens over a scope of temperature. Whether the parts are metals or non-metals, there are sure areas on the stage outline where the liquidus and solidus meet. For an unadulterated segment, a contact point lies at the edge of the outline. The liquidus and solidus additionally meet at the other invariant positions on the graph. The liquidus and solidus additionally meet at the other invariant positions on the outline. Every invariant point speaks to an invariant response that can happen just under a specific arrangement of conditions between specific stages, so is the name for it!

Phase diagrams are classified based on the number of components in the system. Single component systems have unary diagrams, two-component systems have binary diagrams, three-component systems are represented by ternary diagrams, and so on. When more than two components are present, phase diagrams become extremely complicated and difficult to represent. This chapter deals mostly with binary phase diagrams.

4.4.4 Unary Diagrams

Much of the information about the control of the phase structure of a particular system is conveniently and concisely displayed in what is called a *phase diagram,* also often termed an *equilibrium diagram.* Now, there are three externally controllable parameters that will affect phase structure—viz., temperature, pressure, and composition— and phase diagrams are constructed when various combinations of these parameters are plotted against one another. Perhaps the simplest and easiest type of phase diagram to understand is that for a one-component system, in which composition is held constant (i.e., the phase diagram is for a pure substance); this means that pressure and temperature are the variables. This one-component phase diagram (or *unary phase diagram*) [sometimes also called a *pressure–temperature* (or *P–T*) *diagram*] is represented as a two-dimensional plot of pressure (ordinate, or vertical axis) versus temperature (abscissa, or horizontal axis). Most often, the pressure axis is scaled logarithmically.

We illustrate this type of phase diagram and demonstrate its interpretation using as an example the one for H_2O, which is shown in Figure 4.5. Here it may be noted that regions for three different phases—solid, liquid, and vapour—are delineated on the plot. Each phase will exist under equilibrium conditions over the temperature-pressure ranges of its corresponding area. Furthermore, the three curves shown on the plot (labelled *aO*, *bO*, and *cO*) are phase boundaries; at any point on one of these curves, the two phases on either side of the curve are in equilibrium (or coexist) with one another. That is, equilibrium between solid and vapour phases is along curve *aO*—likewise for the solid-liquid, curve *bO*, and the liquid vapour, curve *cO*. Also, upon crossing a boundary (as temperature and/or pressure is altered), one phase transforms to another. For example, at one atmosphere pressure, during heating the solid phase transforms to the liquid phase (i.e., melting occurs) at the point labelled 2 on Figure 4.5 (i.e., the intersection of the dashed horizontal line with the solid-liquid phase boundary); this point corresponds to a temperature of 0°C. Of course, the reverse transformation (liquid-to-solid, or solidification) takes place at the same point upon cooling. Similarly, at the intersection of the dashed line with the liquid-vapour phase boundary [point 3 (Figure 4.5), at 100°C] the liquid transforms to the vapour phase (or vaporizes) upon heating; condensation occurs for cooling. And, finally, solid ice sublimes or vaporizes upon crossing the curve labelled *aO*.

As may also be noted from Figure 4.5, all three of the phase boundary curves intersect at a common point, which is labelled *O* (and for this H_2O system, at a temperature of 273.16 K and a pressure of 6.04×10^{-3} atm). This means that at this point only, all of the solid, liquid, and vapour phases are simultaneously in equilibrium with one another. Appropriately, this, and any other point on a *P–T* phase diagram where three phases are in equilibrium, is called a *triple point;* sometimes it is also termed an *invariant point* inasmuch as its position is distinct, or fixed by definite values of pressure and temperature. Any deviation from this point by a change of temperature and/or pressure will cause at least one of the phases to disappear.

4.4.5 Binary Diagrams

Another type of extremely common phase diagram is one in which temperature and compositions are variable parameters, and pressure is held constant normally 1 atm. There are a few unique assortments; at present, we will discuss alloys—those that contain two components. In the event that more than two parts are available, stage outlines turn out to be amazingly convoluted and hard to speak to. A clarification of the standards representing and the understanding of stage outlines can be shown utilizing twofold compounds despite the fact that most amalgams contain more than two parts. Binary

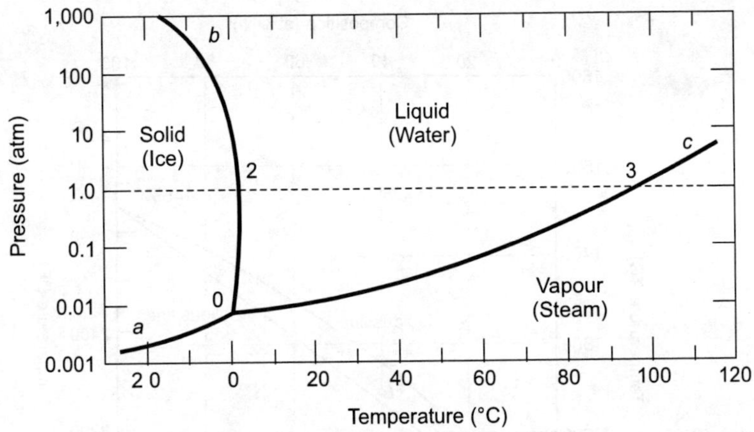

Figure 4.5 Pressure-temperature phase diagram for H_2O Intersection of the dashed horizontal line at 1 atm pressure with the solid-liquid phase boundary (point 2) corresponds to the melting point at this pressure (T = 0°C). Similarly, point 3, the intersection with the liquid-vapour boundary, represents the boiling point (T = 100°C).

phase diagrams are maps that speak to the connections amongst temperature and the creations and amounts of stages at balance, which impact the microstructure of an alloy. Numerous microstructures create from stage changes, the progressions that happen when the temperature is adjusted (normally upon cooling).This may include the move starting with one stage then onto the next, or the appearance or vanishing of a phase. Binary phase diagrams are useful in anticipating stage changes and the subsequent microstructures, which may have equilibrium or non-equilibrium character.

Binary Isomorphous Systems: Possibly the easiest type of binary phase diagram to understand and interpret is the type that is characterized by the copper-nickel system (Figure 4.6a). Temperature is plotted along the ordinate, and the abscissa represents the composition of the alloy, in weight per cent (bottom) and atom percent (top) of nickel. The composition ranges from 0 wt% Ni (100 wt% Cu) on the left horizontal extremity to 100 wt% Ni (0 wt% Cu) on the right. Three different phase regions, or fields, appear on the diagram, an alpha (α) field, a liquid (L) field, and a two-phase (α + L) field. Each region is defined by the phase or phases that exist over the range of temperatures and compositions delimited by the phase boundary lines.

The liquid L is a homogeneous liquid solution composed of both copper and nickel. The a phase (Fig. 4.6a) is a substitutional solid solution consisting of both Cu and Ni atoms, and having an FCC crystal structure. At temperatures below about 1080°C, copper and nickel are mutually soluble in each other in the solid state for all compositions. This complete solubility is explained by the fact that both Cu and Ni have the same crystal structure (FCC), nearly identical atomic radii and electronegativities, and similar valences.

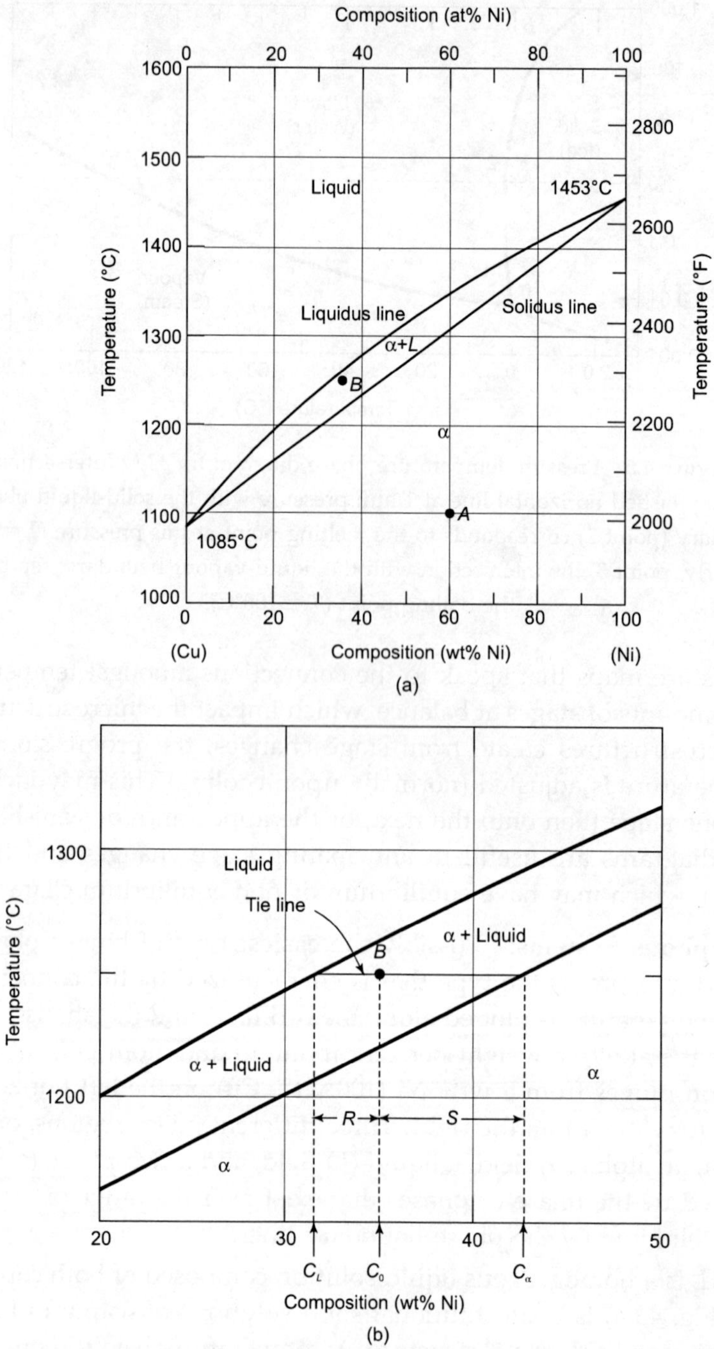

Figure 4.6 (a) The copper-nickel phase diagram. (b) A portion of the copper-nickel phase diagram for which compositions and phase amounts are determined at point *B*.

The copper-nickel system is termed **isomorphous** because of this complete liquid and solid solubility of the two components.

A few remarks are all together with respect to classification. In the first place, for metallic combinations, strong arrangements are usually assigned by lower case Greek letters (α, β and so forth.). Moreover, with respect to stage limits, the line isolating the L and α +L stage fields is named the liquidus line, as demonstrated in Figure 4.6a; the liquid phase is available at all temperatures and pieces over this line. The solidus line is situated between the α and $\alpha + L$ districts, beneath which just the solid phase exists.

For Figure 4.6a, the solidus and liquidus lines cross at the two furthest points; these relate to the softening temperatures of the immaculate segments. For instance, the dissolving temperatures of immaculate copper and nickel are 1085°C and 1453°C individually. Heating pure copper relates to moving vertically up the left-hand temperature hub. Copper stays strong until its softening temperature approaches. The strong to-fluid change happens at the softening temperature, and no further warming is conceivable until this change has been finished.

For any composition other than pure components, this melting phenomenon will occur over the range of temperatures between the solidus and liquidus lines; both solid and liquid phases will be in equilibrium within this temperature range. For example, upon heating an alloy of composition 50 wt% Ni-50 wt% Cu (Figure 4.6a), melting begins at approximately 1280°C; the amount of liquid phase continuously increases with temperature until about 1320°C, at which the alloy is completely liquid.

Example 4.1 Derive the lever rule.

Derivation

Consider the phase diagram for copper and nickel (Figure 4.6b) and alloy of composition C_0 at 1250°C and let C_α, C_L, W_α and W_L represent the same parameters as above. This derivation is accomplished through two conservations of mass expressions. With the first, since only two phases are present, the sum of their mass fractions must be equal to unity; that is,

$$W_\alpha + W_L = 1$$

For the second, the mass of one of the components (either Cu or Ni) that is present in both the phases must be equal to the mass of that component in the total alloy, or

$$W_\alpha C_\alpha + W_L C_L = C_0$$

Simultaneous solution of these two equations leads to the lever rule expressions for this particular situation,

$$W_L = \frac{C_\alpha - C_0}{C_\alpha - C_L}$$

$$W_\alpha = \frac{C_0 - C_L}{C_\alpha - C_L}$$

Example 4.2 For the alloys 5 wt% Sn-95 wt% Pb, 40 wt% Sn-60 wt% Pb, and 70 wt% Sn-30 wt% Pb, identify the phases present and determine their compositions and relative amounts at 330, 184, and 25°C. b. For a 1 kg specimen of the 70 wt% Sn-30 wt% Pb alloy, what weight of Pb is contained within the eutectic microstructure at 182°C?

Solution:

Temperature (°C)	Phases	Chemical composition (wt%Sn)	Phase amounts (weight fraction)	Comment
		5 Wt% Sn-95 wt% Pb		
330	L	5	1.0	All liquid
184	α	5	1.0	All solid
25	α	~ 2	(100 – 5)/(100 – 2) = 0.97	
	β	~ 100	(5 – 2)/(100 – 2) = 0.03	No eutectic microstructure
		40 wt% Sn-60 wt% Pb		
330	L	40	1.0	All liquid
184	α	19.2	(61.9 – 40)/(61.9 – 19.2) = 0.51	Proeutectic α
	L	61.9	(40 – 19.2)/(61.9 – 19.2) = 0.49	Eutectric liquid
25	α	~ 2	(100 – 40)/(100 – 2) = 0.61	
	β	~ 100	(40 – 2)/(100 – 2) = 0.39	
Or alternatively,				
182	α	19.2	(61.9 – 40)/(61.9–19.2) = 0.51	Proeutectic α
	α + β	61.9	(40–19.2)/(61.9-19.2) = 0.49	Eutectric liquid
		70 wt% Sn-30 wt% Pb		
330	L	70	1.0	All liquid
184	β	97.5	(70–61.9)/(97.5 – 61.9) = 0.23	Proeutectoid β
	L	61.9	(97.5–70)/(97.5 – 61.9) = 0.77	Eutectic liquid
25	α	~ 2	(100 – 70)/(100 – 2) = 0.31	
	β	~ 100	(70 – 2)/(100 – 2) = 0.69	
Or alternatively,				
182	β	~ 97.5	(70–61.9)/(97.5–61.9) = 0.23	Proeutectic β
	α+β	~ 61.9	(97.5–5.70)/(97.5–61.9) = 0.77	Eutectic mixture

In 1 kg of the 70 wt% Sn-30 wt% Pb alloy the eutectic mixture weighs 0.77 kg. Of this, 61.9% is Sn and 38.1% is Pb. Therefore, the amount of Pb = (0.77) (0.381) = 0.29 kg.

Example 4.3 The phase diagrams depicted are all thermodynamically impossible as drawn. Provide good physical arguments that specifically indicate what is wrong.

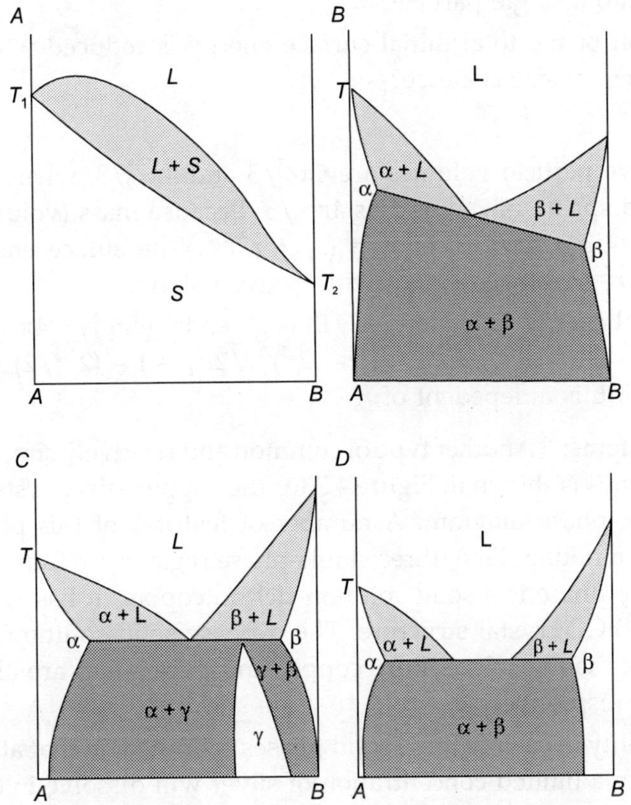

Solution:

(a) The hump in the lens-shaped $L + S$ region is incorrect. Below T_1 tie lines correctly intersect L and S fields. But above T_1 it is possible to draw tie lines that intersect only liquid fields on either side and not $L + S$ as labelled.

(b) A slanted eutectic isotherm is a contradiction in terms. It means the system is not in thermal equilibrium. Furthermore, horizontal tie lines to the slanted eutectic line connect phases that are inconsistent with the given labelling of the phase fields.

(c) On the eutectic isotherm four phases—*a*, *j*8, 7, and *L*—appear to coexist in equilibrium. But this is contrary to the phase rule (by phase rule equation $f = 2 + 1 - 4 = -1$) because the maximum number of phases in a two-component system is 3.

(d) The gap in composition along the eutectic isotherm is incorrect. It implies a variable composition of the eutectic liquid and negative degrees of freedom.

Example 4.4

(a) What is the energy reduction that occurs when spherical particles of radii $r_1$and r_2 coalesce into a single particle?

(b) What fraction of the total initial surface energy is reduced when two spherical powders of equal size coalesce?

Solution:

(a) The respective particle volumes are $4\pi r_1^3/3$ and $4\pi r_2^3/3$, whereas the volume of the coalesced sphere of radius r_c is $4\pi r_c^3/3$. Because mass (volume) is conserved, $4\pi r_c^3/3 = 4\pi r_1^3/3 + 4\pi r_2^3/3$, or $r_c = (r_1^3 + r_2^3)^{1/3}$. The suface energy reduction is therefore $(4\pi r_1^2\gamma + 4\pi r_2^2\gamma) - 4\pi r_c^2\gamma$, with r_c given above.

(b) The fractional energy reduction ($\Delta E/E$) is given by $(4\pi r_1^2\gamma + 4\pi r_2^2\gamma - 4\pi r_c^2\gamma)/(4\pi r_1^2\gamma + 4\pi r_2^2\gamma)$. As $r_1 = r_2 = r$, $\Delta E/E = \{1 - (2r^3)^{2/3}/2r^2\} = 1 - (2^{2/3}/2) = 0.206$ or 20.6%. Note that $\Delta E/E$ is indepedent of r.

Binary Eutectic Systems: Another type of common and relatively simple phase diagram found for binary alloys is shown in Figure 4.7 for the copper-silver system; this is known as a binary eutectic phase diagram. A number of features of this phase diagram are important and worth noting. First, three single-phase regions are found on the diagram: α, β and liquid. The phase is a solid solution rich in copper; it has silver as the solute component and an FCC crystal structure. The α-phase solid solution also has an FCC structure, but copper is the solute. Pure copper and pure silver are also considered to be α and β phases, respectively.

Thus, the solubility in each of these solid phases is limited, in that at any temperature below line *BEG* only a limited concentration of silver will dissolve in copper (for the α phase), and similarly for copper in silver (for the β phase). The solubility limit for the phase corresponds to the boundary line, labelled *CBA*, between the α/(α + β) and α(α + *L*) phase regions; it increases with temperature to a maximum [8.0 wt% Ag at 779°C] at point *B*, and decreases back to zero at the melting temperature of pure copper, point *A* [1085°C]. At temperatures below 779°C, the solid solubility limit line separating the α and α + β phase regions is termed **solvus line;** the boundary *AB* between the α and α + *L* fields is the **solidus line,** as indicated in Figure 4.7. For the β phase, both solvus and solidus lines also exist, *HG* and *GF*, respectively, as shown. The maximum solubility of copper in the β phase, point *G* (8.8 wt% Cu), also occurs at 779°C. This horizontal line *BEG*, which is parallel to the composition axis and extends between these maximum solubility positions, may also be considered a solidus line; it represents the lowest temperature at which a liquid phase may exist for any copper–silver alloy that is at equilibrium.

There are also three two-phase regions found for the copper–silver system (Figure 4.7): $\alpha + L$, $\beta + L$ and $\alpha + \beta$. The α- and β-phase solid solutions coexist for all compositions and temperatures within the $\alpha + \beta$ phase field; the α+ liquid and β+ liquid phases also coexist in their respective phase regions. Furthermore, compositions and relative amounts for the phases may be determined using tie lines and the lever rule as outlined previously.

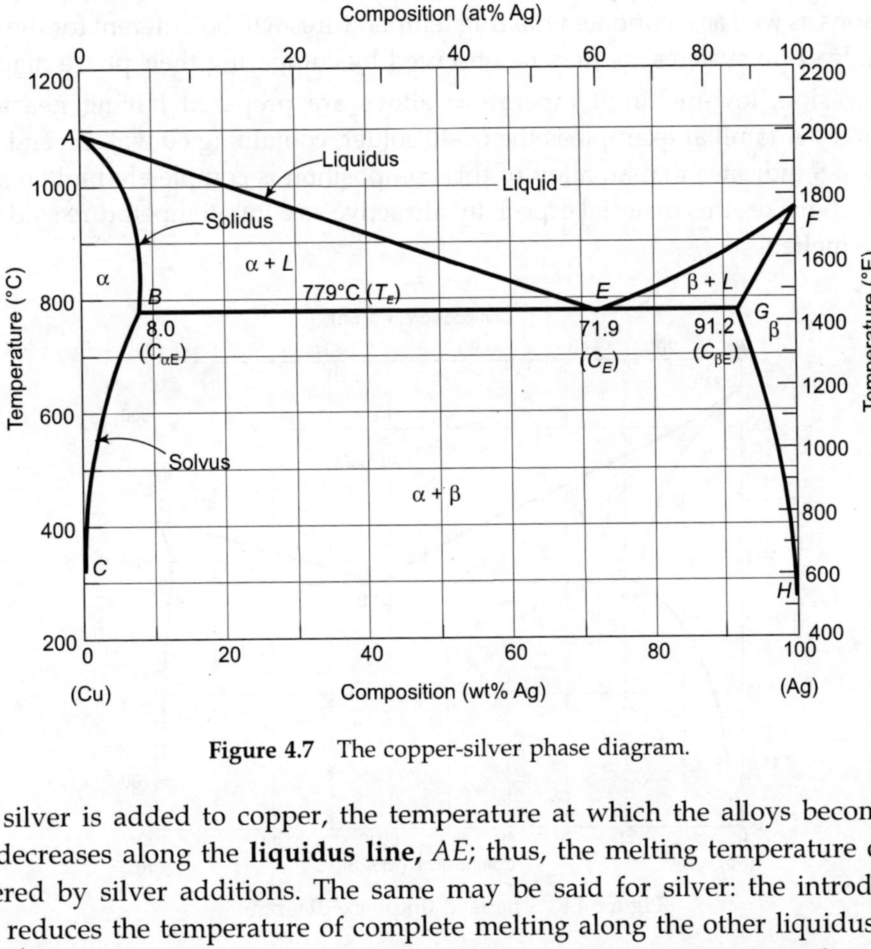

Figure 4.7 The copper-silver phase diagram.

As silver is added to copper, the temperature at which the alloys become totally liquid decreases along the **liquidus line,** *AE*; thus, the melting temperature of copper is lowered by silver additions. The same may be said for silver: the introduction of copper reduces the temperature of complete melting along the other liquidus line, *FE*. These liquidus lines meet at the point *E* on the phase diagram, through which also passes the horizontal isotherm line *BEG*. Point *E* is called an **invariant point,** which is designated by the composition C_E and temperature T_E for the copper-silver system, the values of C_E and T_E are 71.9 wt% Ag and 779°C respectively. An important reaction occurs for an alloy of composition as it changes temperature in passing through this reaction may be written as follows:

$$L(C_E) \underset{\text{heating}}{\overset{\text{cooling}}{\rightleftharpoons}} \alpha(C_{\alpha E}) + \beta(C_{\beta E})$$

Another common eutectic system is that for lead and tin; the phase diagram (Figure 4.8) has a general shape similar to that for copper-silver. For the lead–tin system the solid solution phases are also designated by α and β; in this case, represents a solid solution of tin in lead and, for β, tin is the solvent and lead is the solute. The eutectic invariant point is located at 61.9 wt% Sn and 183°C. Of course, maximum solid solubility compositions as well as component melting temperatures will be different for the copper-silver and lead-tin systems, as may be observed by comparing their phase diagrams.

On occasion, low-melting-temperature alloys are prepared having near-eutectic compositions. A familiar example is the 60-40 solder, containing 60 wt% Sn and 40 wt% Pb. Figure 4.8 indicates that an alloy of this composition is completely molten at about 185°C, which makes this material especially attractive as a low-temperature solder, since it is easily melted.

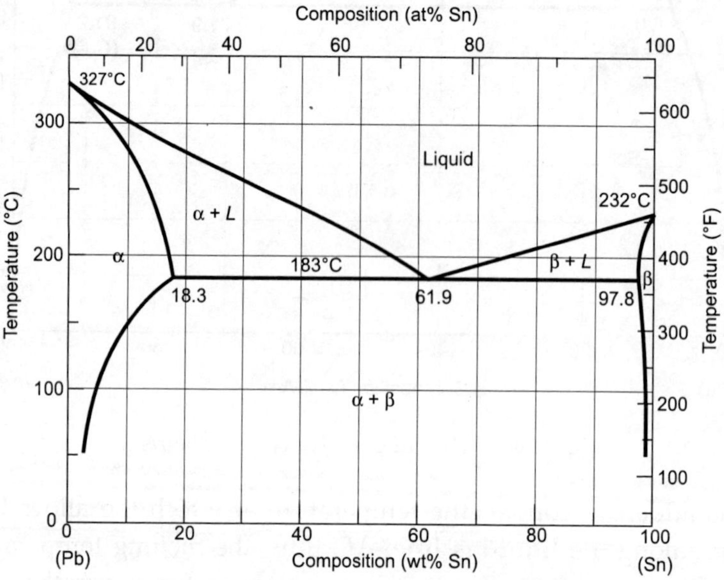

Figure 4.8 The lead-tin phase diagram.

Example 4.5 For a 40 wt% Sn-60 wt% Pb alloy at 150°C.

(a) What phase(s) is (are) present?

(b) What is (are) the composition(s) of the phase(s)?

Solution:

(a) Locate this temperature-composition point on the phase diagram (point *B* in Figure 4.9). Inasmuch as it is within α + β the region, both the phases will coexist.

(b) Since two phases are present, it becomes necessary to construct a tie line across the α + β phase field at 150°C as indicated in Figure 4.9. The composition of the phase corresponds to the tie line intersection with the α/α + β solvus phase boundary—about 10 wt% Sn-90 wt% Pb, denoted as C_α. Similarly, for the phase, which will have a composition of approximately 98 wt% Sn-2 wt% Pb (C_β).

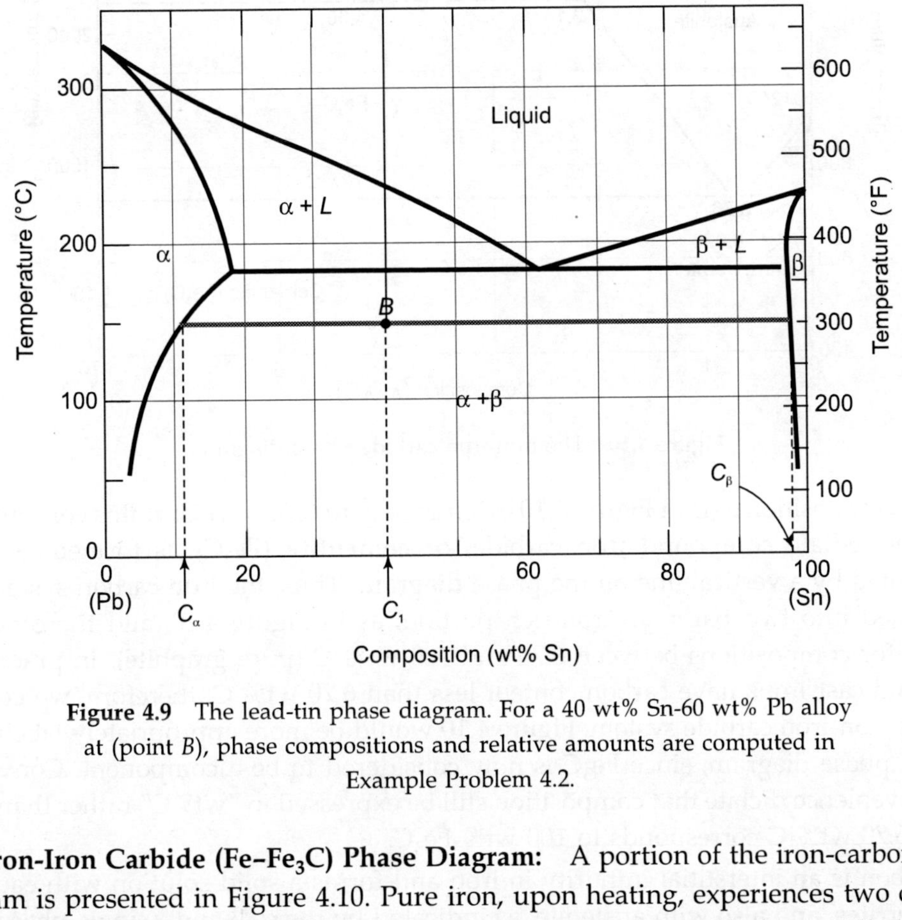

Figure 4.9 The lead-tin phase diagram. For a 40 wt% Sn-60 wt% Pb alloy at (point *B*), phase compositions and relative amounts are computed in Example Problem 4.2.

The Iron-Iron Carbide (Fe–Fe_3C) Phase Diagram: A portion of the iron-carbon phase diagram is presented in Figure 4.10. Pure iron, upon heating, experiences two changes in crystal structure before it melts. At room temperature the stable form, called **ferrite,** or α iron, has a BCC crystal structure. Ferrite experiences a polymorphic transformation to FCC **austenite,** or γ iron, at 912°C. This austenite persists to 1394°C, at which temperature the FCC austenite reverts to a BCC phase known as δ ferrite, which finally melts at 1538°C. All these changes are apparent along the left vertical axis of the phase diagram.

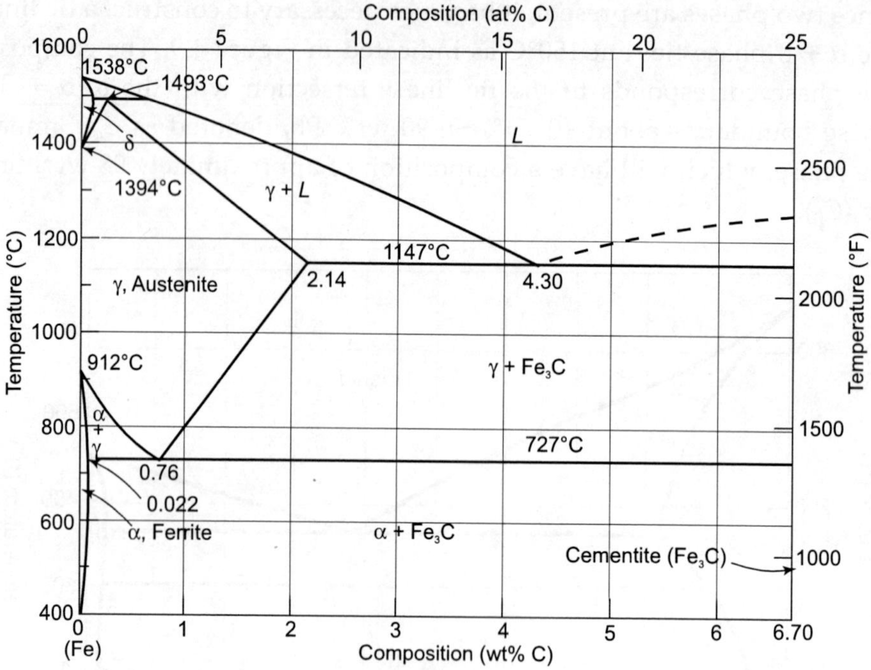

Figure 4.10 The iron-iron carbide phase diagram.

The composition axis in Figure 4.10 extends only to 6.70 wt% C; at this concentration the intermediate compound iron carbide, or **cementite** (Fe_3C), is formed, which is represented by a vertical line on the phase diagram. Thus, the iron-carbon system may be divided into two parts: an iron-rich portion, as in Figure 4.10, and the other (not shown) for compositions between 6.70 and 100 wt% C (pure graphite). In practice, all steels and cast irons have carbon content less than 6.70 wt% C; therefore, we consider only the iron-iron carbide system. Figure 4.10 would be more appropriately labelled the Fe-Fe_3C phase diagram, since Fe_3C is now considered to be a component. Convention and convenience dictate that composition still be expressed in "wt%C" rather than "wt% Fe_3C"; 6.70 wt% C corresponds to 100 wt% Fe_3C.

Carbon is an interstitial impurity in iron and forms a solid solution with each of α and δ ferrites, and also with austenite, as indicated by the α, β and γ single phase fields in Figure 4.10. In the BCC α ferrite, only small concentrations of carbon are soluble; the maximum solubility is 0.022 wt% at 727°C. The limited solubility is explained by the shape and size of the BCC interstitial positions, which make it difficult to accommodate the carbon atoms. Even though present in relatively low concentrations, carbon significantly influences the mechanical properties of ferrite. This particular iron-carbon phase is relatively soft, may be made magnetic at temperatures below 768°C, and has a density of 7.88 g/cm^3.

The austenite, or phase of iron, when alloyed with carbon alone, is not stable below 727°C), as indicated in Figure 4.10. The maximum solubility of carbon in austenite, 2.14 wt%, occurs at 1147°C). This solubility is approximately 100 times greater than the maximum for BCC ferrite, since the FCC interstitial positions are larger, and, therefore, the strains imposed on the surrounding iron atoms are much lower. As the discussions that follow demonstrate, phase transformations involving austenite are very important in the heat treating of steels. In passing, it should be mentioned that austenite is nonmagnetic.

The δ ferrite is virtually the same as α ferrite, except for the range of temperatures over which each exists. Since the δ ferrite is stable only at relatively high temperatures, it is of no technological importance and is not discussed further. Cementite (Fe_3C) forms when the solubility limit of carbon in ferrite is exceeded below 727°C) (for compositions within the α + Fe_3C phase region). As indicated in Figure 4.10, Fe_3C will also coexist with the γ phase between 727°C and 1147°C. Mechanically, cementite is very hard and brittle; the strength of some steels is greatly enhanced by its presence.

Cementite is only metastable; that is, it will remain a compound indefinitely at room temperature. However, if heated between 650°C and 700°C for several years, it will gradually change or transform into α iron and carbon, in the form of graphite, which will remain upon subsequent cooling to room temperature. Thus, the phase diagram in Figure 4.10 is not a true equilibrium one because cementite is not an equilibrium compound. However, inasmuch as the decomposition rate of cementite is extremely sluggish, virtually all the carbon in steel will be as Fe_3C instead of graphite, and the iron-iron carbide phase diagram is, for all practical purposes, valid.

The two-phase regions are labelled in Figure 4.10. It may be noted that one eutectic exists for the iron-iron carbide system, at 4.30 wt% C and 1147°C; for this eutectic reaction, the liquid solidifies to form austenite and cementite phases. Of course, subsequent cooling to room temperature will promote additional phase changes.

$$L = \underset{\text{heating}}{\overset{\text{cooling}}{\rightleftharpoons}} \text{ã} + Fe_3C$$ (Eutectic reaction for the iron-iron carbide system)

It may be noted that a eutectoid invariant point exists at a composition of 0.76 wt% C and a temperature of 727°C. This eutectoid reaction may be represented by the following equation. (Or, upon cooling, the solid phase is transformed into iron and cementite.)

$$\gamma(0.76\text{ wt\% C}) \underset{\text{heating}}{\overset{\text{cooling}}{\rightleftharpoons}} \alpha(0.022\text{ wt\% C}) + Fe_3C(6.7\text{ wt\% C})$$

Ferrous alloys are those in which iron is the prime component, but carbon as well as other alloying elements may be present. In the classification scheme of ferrous alloys based on carbon content, there are three types: iron, steel, and cast iron. Commercially pure iron contains less than 0.008 wt%C and, from the phase diagram, is composed almost exclusively of the ferrite phase at room temperature. The iron-carbon alloys that contain between 0.008 and 2.14 wt%C are classified as steels. In most steels the microstructure consists of both α and Fe_3C phases. Upon cooling to room temperature, an alloy within this composition range must pass through at least a portion of the phase field; distinctive microstructures are subsequently produced, as discussed below. Although a steel alloy may contain as much as 2.14 wt%C, in practice, carbon concentrations rarely exceed 1.0 wt%. Cast irons are classified as ferrous alloys that contain between 2.14 and 6.70 wt% C. However, commercial cast irons normally contain less than 4.5 wt% C.

Example 4.6 For a 99.65 wt% Fe-0.35 wt% C alloy at a temperature just below the eutectoid, determine the following:

(a) The fractions of total ferrite and cementite phases

(b) The fractions of the proeutectoid ferrite and pearlite

(c) The fraction of eutectoid ferrite

Solution:

(a) This part of the problem is solved by application of the lever rule expressions employing a tie line that extends all the way across the α + Fe_3C phase field. Thus, C_0' is 0.35 wt% C, and

$$W_{\alpha} = \frac{6.70 - 0.35}{6.70 - 0.022} = 0.95$$

$$W_{Fe_3C} = \frac{0.35 - 0.022}{6.70 - 0.022} = 0.05$$

(b) The fractions of proeutectoid ferrite and pearlite are determined by using the lever rule and a tie line that extends only to the eutectoid composition.

$$W_p = \frac{0.35 - 0.022}{0.76 - 0.022} = 0.44$$

$$W_{\alpha'} = \frac{0.76 - 0.35}{0.76 - 0.022} = 0.56$$

(c) All ferrite is either a proeutectoid or eutectoid (in the pearlite). Therefore, the sum of these two ferrite fractions will equal the fraction of total ferrite; that is,

$$W_{\alpha'} = W_{\alpha e} = W_{\alpha}$$

where, $W_{\alpha e}$ is the fraction of the total alloy that is eutectoid ferrite. The values $W_{\alpha'}$ and W_{α} are 0.56 and 0.95 respectively.

Hence, $W_{\alpha e} = W_{\alpha} = W_{\alpha'} = 0.95 - 0.56 = 0.39$

Example 4.7 For the lead-tin alloy calculate the relative amount of each phase present in terms of (a) mass fraction and (b) volume fraction. At 150°C take the densities of Pb and Sn to be 11.23 and 7.24 g/cm^3, respectively.

Solution:

(a) Since the alloy consists of two phases, it is necessary to employ the lever rule. If C_1 denotes the overall alloy composition, mass fractions may be computed by subtracting compositions, in terms of weight per cent tin, as follows:

$$W_\alpha = \frac{C_\beta - C_1}{C_\beta - C_\alpha} = \frac{98-40}{98-40} = 0.66$$

$$W_\beta = \frac{C_1 - C_\alpha}{C_\beta - C_\alpha} = \frac{40-10}{98-10} = 0.34$$

(b) To compute volume fractions it is first necessary to determine the density of each phase using by equation below. Thus,

$$\rho_a = \frac{100}{\dfrac{C_{Sn(\alpha)}}{\rho_{Sn}} + \dfrac{C_{Pb(\alpha)}}{\rho_{Pb}}}$$

where $C_{Sn(\alpha)}$ and $C_{Pb(\alpha)}$ denote the concentrations in weight per cent of tin and lead, respectively, in the α phase. From Example 4.5, these values are 10 wt% and 90 wt%. Incorporation of these values along with the densities of the two components lead to

$$\rho_\alpha = \frac{100}{\dfrac{10}{7.24\ \text{g/cm}^3} + \dfrac{90}{11.23\ \text{g/cm}^3}} = 10.64\ \text{g/cm}^3$$

Similar for the β phase:

$$\rho_\beta = \frac{100}{\dfrac{C_{Sn(\beta)}}{\rho_{Sn}} + \dfrac{C_{Pb(\beta)}}{\rho_{Pb}}}$$

$$= \frac{100}{\dfrac{98}{7.24\ \text{g/cm}^3} + \dfrac{2}{11.23\ \text{g/cm}^3}} = 7.29\ \text{g/cm}^3$$

Now it becomes necessary to employ the below given equations to determine V_α and V_β as

$$V_\alpha = \frac{\dfrac{W_\alpha}{\rho_\alpha}}{\dfrac{W_\alpha}{\rho_\alpha} + \dfrac{W_\beta}{\rho_\beta}}$$

$$= \frac{\dfrac{0.66}{10.64\,\text{g/cm}^3}}{\dfrac{0.66}{10.64\,\text{g/cm}^3} + \dfrac{0.34}{7.29\,\text{g/cm}^3}} = 0.57$$

$$V_\beta = \frac{\dfrac{W_\beta}{\rho_\beta}}{\dfrac{W_\alpha}{\rho_\alpha} + \dfrac{W_\beta}{\rho_\beta}}$$

$$= \frac{\dfrac{0.34}{7.29\,\text{g/cm}^3}}{\dfrac{0.66}{10.64\,\text{g/cm}^3} + \dfrac{0.34}{7.29\,\text{g/cm}^3}} = 0.43$$

4.5 PHASES: STRUCTURES AND COMPOSITION

4.5.1 Equilibrium Phases: Solid Solutions

Substitutional Solid Solution: Binary phase diagrams usually contain an assortment of single-phase materials known as solid solutions. In the substitutional solid solution alloy the involved solute and solvent atoms are randomly mixed on lattice sites. The single-phase solid alloys that extend across the entire phase diagram in the Cu-Ni and Ge-Si systems are good examples of random substitutional solid solutions. Both the atoms are randomly distributed on FCC lattice sites in Cu-Ni and on diamond cubic sites in Ge-Si. In either case it makes no difference which component one starts with; the atoms mix in all proportions. A *necessary* condition for single-phase solid solution formation across the entire phase diagram is that both components have the same crystal structure. It is not a *sufficient* condition, however, because there are combinations like Ag-Cu (both FCC) and Fe-Mo (both BCC) that do not form an extensive range of solid solutions. Instead **terminal solid solutions,** so named because they appear at the ends of the phase diagram, form. The terminal α and β phases in the Pb-Sn diagram are examples. They are both substitutional solid solutions and display the limited solubility often exhibited by such phases. Introduction of foreign atoms into the lattice, whether by design (as dopants or solutes) or accident (as impurities), will always create dilute solid solutions which are often substitutional in nature. In fact, all "pure" materials are in effect dilute terminal solid solutions because it is thermodynamically impossible to remove all impurities.

Interstitial Solid Solution: When undersized alloying elements dissolve in the lattice they sometimes form interstitial solid solutions. Important examples include carbon and nitrogen in BCC iron, the former being the α phase on the Fe-Fe_3C phase diagram. Rather than sit on substitutional sites, these atoms occupy the interstice in the centre of the α-Fe cube faces, e.g., (½, 0, ½). In FCC γ-Fe, carbon atoms assume the large interstitial site in the centre of the unit cell. Based on a hard sphere model it is not too difficult to show that considerably less room is available for carbon to squeeze into in the α-Fe interstitial site relative to the γ-Fe site. More severe lattice bond straining in the former means that less carbon will dissolve interstitially in α-Fe compared with γ-Fe, in accord with solubility limits defined by the Fe-Fe_3C phase diagram.

Intermediate Solid Solution: Unlike terminal solid solutions that extend inwards from the outer pure components, intermediate phase fields are found within the phase diagram. Two-phase regions border either side of an intermediate phase. In the case of Cu-Zn, α and η are terminal phases, β, γ, δ and ε are intermediate phases. These single phases are generally stable over a relatively wide composition range.

4.5.2 Compounds

Another very common type of solid phase that appears in phase diagrams is the compound. Elementary inorganic chemistry provides a basis for understanding compounds in many binary systems. For example, we know that salt compounds form when elements, drawn from opposite ends of the Periodic Table, react. Further, such compounds possess well-defined stoichiometries and melting points. Cementite (Fe_3C) is such a compound. The oxides of iron, magnetite (Fe_3O_4), and Wüstite (FeO) are also compounds, but they exist over a range of oxygen content and correspondingly have varying melting points. Among the most technically important compounds are those formed between column 3*A* and 5*A* elements, for example, GaAs. These compound semiconductors play an important role in optoelectronics. A closer look at GaAs reveals that it can be either slightly Ga-rich or slightly As-rich, although this is not visible on the scale of the diagram. In general, most of the compounds we are concerned with have a slightly variable stoichiometry.

4.6 SUMMARY

Thermodynamics is definite about denying the impossible and defining limits for behaviour that is possible. But, for a possible event, thermodynamics is noncommittal as to the extent to which it will occur, or even whether it will occur at all. For example, amorphous SiO_2 should crystallize and Fe_3C should thermodynamically decompose into iron and graphite at 300 K, but both can be retained indefinitely in metastable equilibrium. This is true of many other solid-state materials and reactions whether they are chemical in nature or involve physical transformations of state. Only reactions that lower the free energy of the system can proceed; and if the free energy of the system reaches a minimum and cannot be lowered further, true thermodynamic equilibrium is attained.

Phase Equilibria One-Component (or Unary) Phase Diagrams Binary Phase Diagrams Interpretation of Phase Diagrams

Equilibrium phase diagrams are an advantageous and succinct method for speaking to the most stable connections between stages in alloy systems. This talk started by considering the unary (or pressure-temperature) stage graph for a one phase diagram. Solid-, liquid-, and vapour-phase regions are found on this type of phase diagram. For binary systems temperature and composition are factors, though outer weight is held steady. Regions, or phase areas, are characterized on these temperature versus creation plots inside which possibly maybe a couple stages exist. For a combination of indicated organization and at a known temperature, the stages exhibit, their creations, and relative sums under balance conditions might be resolved. Inside two-stage areas, tie lines and the lever run must be utilized for phase composition and mass fractions calculations, individually.

Gibbs Phase Rule

The Gibbs phase rule was introduced; it is a simple equation that relates the number of phases present in a system at equilibrium with the number of degrees of freedom, the number of components, and the number of non-compositional variables.

Binary Isomorphous Systems

A few types of stage outlines were talked about for metallic frameworks. Isomorphous graphs are those for which there is finished dissolvability in the strong stage; the copper-nickel framework shows this conduct.

Binary Eutectic Systems

In an eutectic response, as found in some compound frameworks, a fluid stage changes isothermally to two diverse strong stages after cooling. Such a response is noted on the copper-silver and lead-tin stage outlines. Finish strong dissolvability for all pieces does not exist; rather, strong arrangements are terminal—there is just a constrained solvency of every component in the other.

Iron-Iron Carbide (Fe-Fe_3C) Phase Diagram

Significant consideration was given to the iron-carbon framework, and particularly, the iron-iron carbide stage graph, which mechanically is a standout amongst the most imperative. The improvement of microstructure in numerous iron-carbon combinations and steels relies on the eutectoid response in which the FCC austenite period of arrangement 0.76 wt% C changes isothermally to the BCC ferrite stage (0.022 wt% C) and the intermetallic compound, cementite (Fe_3C).

EXERCISE

1. An initially solid material becomes a liquid as the temperature is raised above the melting point. Explain the probable signs of the enthalpy, entropy, and free energy changes. What are the signs of these quantities as the liquid transforms back to the solid?
2. What is the minimum temperature at which it would be thermodynamically possible for carbon to reduce silica according to the reaction

   $SiO_2 + 2C = Si + 2CO$?

   Assume the effective concentrations are unity and the pressure is 1 atm.
3. Consider the sugar-water phase diagram.
   (a) in 1000 g of water at 80°C how much sugar will dissolve?
   (b) What will be the composition of the saturated liquid solution (in wt% sugar) at 20°C? If the saturated liquid solution in part (a) is cooled to 20°C, some of the sugar will precipitate out as a solid.
   (c) How much of the solid sugar will come out of solution upon cooling to 20°C?

   (For the HCP crystal structure, show that the ideal ratio c/a is 1.633)
4. At 100°C what is the maximum solubility
   (a) of Pb in Sn?
   (b) of Sn in Pb?
5. What thermodynamic condition must be met for a state of equilibrium to exist?

6. At a pressure of 0.1 atm, find
   (a) the melting temperature for ice, and
   (b) the boiling temperature for water.
7. Consider a specimen of ice that is at and 10 atm pressure. Considering unary system, phase diagram the pressure-temperature phase diagram for determining the pressure to which the specimen must be raised or lowered to cause it
   (a) to melt, and
   (b) to sublime.
8. Is it possible to have a copper-silver alloy that, at equilibrium, consists of a phase of composition 92 wt% Ag-8 wt% Cu, and also a liquid phase of composition 76 wt% Ag-24 wt% Cu? If so, what will be the approximate temperature of the alloy? If this is not possible, explain why.
9. A lead-tin alloy of composition 30 wt% Sn-70 wt% Pb is slowly heated from a temperature of 150°C.
   (a) At what temperature does the first liquid phase form?
   (b) What is the composition of this liquid phase?
   (c) At what temperature does complete melting of the alloy occur?
   (d) What is the composition of the last solid remaining prior to complete melting?
10. A 40 wt% Pb-60 wt% Mg alloy is heated to a temperature within the α+ liquid-phase region. If the mass fraction of each phase is 0.5, then estimate:
    (a) The temperature of the alloy
    (b) The compositions of the two phases
11. A 60 wt% Pb-40 wt% Mg alloy is rapidly quenched to room temperature from an elevated temperature in such a way that the high temperature microstructure is preserved. This microstructure is found to consist of the phase and Mg_2Pb, having respective mass fractions of 0.42 and 0.58. Determine the approximate temperature from which the alloy was quenched.
12. What is the principal difference between congruent and incongruent phase transformations?
13. Compute the mass fractions of ferrite and cementite in pearlite.
14. (a) What is the distinction between hypoeutectoid and hypereutectoid steels?
    (b) In a hypoeutectoid steel, both eutectoid and proeutectoid ferrite exist. Explain the difference between them. What will be the carbon concentration in each?
15. What is the carbon concentration of an iron-carbon alloy for which the fraction of total cementite is 0.10?
16. What is the proeutectoid phase for an iron-carbon alloy in which the mass fractions of total ferrite and total cementite are 0.86 and 0.14, respectively? Why?

17. Consider 6.0 kg of austenite containing 0.45 wt% C, cooled to below 727°C.

(a) What is the proeutectoid phase?

(b) How many kilograms each of total ferrite and cementite form?

(c) How many kilograms each of pearlite and the proeutectoid phase form?

(d) Schematically sketch and label the resulting microstructure.

18. Compute the mass fractions of proeutectoid ferrite and pearlite that form in an iron-carbon alloy containing 0.35 wt% C.

19. The mass fractions of total ferrite and total cementite in an iron-carbon alloy are 0.91 and 0.09, respectively. Is this a hypoeutectoid or hypereutectoid alloy? Why?

20. Is it possible to have an iron-carbon alloy for which the mass fractions of total ferrite and pearlite are 0.860 and 0.969, respectively? Why or why not?

Multiple-Choice Questions

1. First material known to be used by man

(a) Cotton (b) Bronze
(c) Iron (d) Rock

2. First metal known to be used by man

(a) Iron (b) Bronze
(c) Silver (d) Aluminium

3. Which one of the following is not a basic component of Materials Science?

(a) Cost (b) Properties
(c) Structure (d) Performance

4. Figure out the odd statement about ceramics in the following:

(a) Good insulators of heat and electricity
(b) Usually less desired than metals
(c) Ductile in nature
(d) Contains both metallic and nonmetallic elements

5. Pick a composite from the list

(a) Wood (b) Steel
(c) Nylon (d) Mica

6. Particles that most affects material properties are

(a) Neutrons (b) Protons
(c) Electrons (d) Valence electrons

7. Mean distance between atoms in the range of
   (a) 25 nm (b) 2.5 nm
   (c) 0.25 nm (d) 0.025 nm
8. Which one of the following is not a strong bond?
   (a) van der Waals bond (b) Covalent bond
   (c) Metallic bond (d) Ionic bond
9. Bond strength of secondary bonds is in the range of
   (a) 1 kJ/mol (b) 10 kJ/mol
   (c) 100 kJ/mol (d) 1000 kJ/mol
10. Electron sea exists in
    (a) Polar bonds (b) Ionic bond
    (c) Covalent bond (d) Metallic bond
11. A family of directions is represented by
    (a) (*hkl*) (b) <*uvw*>
    (c) {*hkl*} (d) [*uvw*]
12. Miller indices for octahedral plane in cubic crystal
    (a) (100) (b) (110)
    (c) (111) (d) None
13. The plane (1–11) is parallel to
    (a) (–11–1) (b) (–1–11)
    (c) (111) (d) (1–11)
14. The angle between [111] and [11–2] directions in a cubic crystal is (in degrees)
    (a) 0 (b) 45
    (c) 90 (d) 180
15. Miller indices of the line of intersection of (–1–11) and (110) are
    (a) [110] (b) [101]
    (c) [10–1] (d) [–110]
16. Theoretical strength is about __________ times to average real strength of a material.
    (a) 1 (b) 10
    (c) 100 (d) 1000

17. Hooke's law
    (a) Elastic range, strain is proportional to stress
    (b) Plastic range, strain is proportional to stress
    (c) In both elastic and plastic range, strain is proportional to stress
    (d) None
18. Following is not the 2-dimensional imperfection
    (a) Twin boundary (b) Dislocation
    (c) Surface (d) Grain boundary
19. Figure out the odd one in the following:
    (a) Frenkel defect (b) Tilt boundary
    (c) Twist boundary (d) Stacking fault
20. Thermodynamically stable defects are
    (a) Point defects (b) Line defects
    (c) Surface defects (d) Volume defects
21. Negative screw dislocation is represented by
    (a) $\perp$ (b) Counterclockwise rotation
    (c) Clockwise rotation (d) $\top$
22. Average frequency of atomic vibrations in a solid (in Hz) is
    (a) 10^{-12} (b) 10^{-13}
    (c) 10^{12} (d) 101^{13}
23. Requirement for cross-slip movement of dislocation is
    (a) Preferred slip plane (b) Preferred slip direction
    (c) Not preferred slip plane (d) Not preferred slip direction
24. Beneficial property of foreign particles
    (a) Reduces density (b) Acts as stress raisers
    (c) Obstructs dislocation motion (d) None
25. Stacking fault energies are in the range of
    (a) 0.01-0.1 J/m^2 (b) 0.01-0.1 J/cm^2
    (c) 0.1-10 J/m^2 (d) 0.1-10 J/m^2
26. Not a Hume-Ruthery condition:
    (a) Crystal structure of each element of solid solution must be the same.
    (b) Size of atoms of each two elements must not differ by more than 15%.

(c) Elements should form compounds with each other.

(d) Elements should have the same valence.

27. Pick the odd one in the following:

(a) Isomorphous alloy
(b) Terminal solid solution
(c) Intermediate solid solution
(d) Compound

28. The boundary line between (liquid) and (liquid+solid) regions must be part of __________.

(a) Solvus
(b) Solidus
(c) Liquidus
(d) Tie-line

29. The boundary line between (liquid+solid) and (solid) regions must be a part of __________.

(a) Solvus
(b) Solidus
(c) Liquidus
(d) Tie-line

30. The boundary line between (alpha) and (alpha+beta) regions must be a part of __________.

(a) Solvus
(b) Solidus
(c) Liquidus
(d) Tie-line

31. On heating, one solid phase results in another solid phase plus on liquid phase during __________ reaction.

(a) Eutectic
(b) Peritectic
(c) Monotectic
(d) Syntectic

32. A solid phase results in a solid plus another solid phase up on cooling during __________ reaction.

(a) Eutectoid
(b) Peritectoid
(c) Eutectic
(d) Peritectic

33. A solid phase results in a solid plus another solid phase up on heating during __________ reaction.

(a) Eutectoid
(b) Peritectoid
(c) Monotectoid
(d) None

34. A liquid phase produces two solid phases during __________ reaction up on cooling.

(a) Eutectic
(b) Eutectoid
(c) Peritectic
(d) Peritectoid

35. Liquid phase is involved in the following reaction:
    (a) Eutectoid (b) Peritectoid
    (c) Monotectoid (d) None
36. Overall transformation rate changes with temperature as follows:
    (a) Monotonically decreases with temperature
    (b) First increases, then decreases
    (c) Initially it is slow, and then picks up
    (d) Monotonically increases with temperature
37. wt.% of carbon in mild steels
    (a) <0.008 (b) 0.008-0.3
    (c) 03-0.8 (d) 0.8-2.11
38. Eutectic product in Fe-C system is called
    (a) Pearlite (b) Bainite
    (c) Ledeburite (d) Spheroidite
39. Eutectoid product in Fe-C system is called
    (a) Pearlite (b) Bainite
    (c) Ledeburite (d) Spheroidite
40. Phases that exist on the left side of an invariant reaction line are called
    (a) Prophase (b) Hypophase
    (c) Hyperphase (d) None
41. Diffusion can occur in __________ materials.
    (a) Solid (b) Liquid
    (c) Gaseous (d) All
42. Probably the fastest diffusing species in Fe is
    (a) H (b) Ni
    (c) W (d) C
43. The value of error function for '∞' is
    (a) $-\infty$ (b) -1
    (c) 1 (d) ∞
44. The value of error function for '0' is
    (a) -1 (b) 0
    (c) 1 (d) None

45. The units for diffusivity, *D*, are
   (a) m^2sec^1 (b) m^2sec^{-1}
   (c) $m^{-2}sec^{-1}$ (d) $m^{-2}sec^1$
46. Macro-segregation can be removed by
   (a) Annealing (b) Hot working
   (c) Both (d) None
47. Both the elements of 1st period contains valence electrons in
   (a) M shell (b) N shell
   (c) K shell (d) S shell
48. In periodic table, helium is placed at
   (a) top left corner (b) bottom right corner
   (c) bottom left corner (d) top right corner
49. Across period atomic size decreases due to
   (a) Shielding effect
   (b) Photoelectric effect
   (c) Increase in nuclear force of attraction
   (d) Decrease in nuclear force of attraction
50. First three periods are
   (a) Long Periods (b) Short Periods
   (c) Moderate Periods (d) None
51. Not an example for actuator
   (a) Optical fibre
   (b) Shape memory alloys
   (c) Magneto-strictive materials
   (d) Electro-/Magneto-rheological fluids
52. Strong and ductile materials
   (a) Polymers (b) Ceramics
   (c) Metals (d) Semiconductors
53. Presently most used metal in the world is
   (a) Aluminium (b) Gold
   (c) Steel (d) Silver

54. Detrimental property of a material for shock load applications
    (a) High density (b) Low toughness
    (c) High strength (d) Low hardness
55. Democratic material
    (a) Diamond (b) Titanium
    (c) Iron (d) Gold
56. Repeatable entity of a crystal structure is known as
    (a) Crystal (b) Lattice
    (c) Unit cell (d) Miller indices
57. Coordination number for closest packed crystal structure is
    (a) 16 (b) 12
    (c) 8 (d) 4
58. Atomic packing factor is
    (a) Distance between two adjacent atoms
    (b) Projected area fraction of atoms on a plane
    (c) Volume fraction of atoms in cell
    (d) None
59. Coordination number in simple cubic crystal structure
    (a) 1 (b) 2
    (c) 3 (d) 4
60. The atomic diameter of an BCC crystal (if a is lattice parameter) is
    (a) a (b) $a/2$
    (c) $a/(4/\sqrt{3})$ (d) $a/(4/\sqrt{2})$
61. Repeatable unit of polymers
    (a) Isomer (b) Copolymer
    (c) Homopolymer (d) Mer
62. Pick the thermoplast from the following
    (a) Vinyls (b) Epoxies
    (c) Resins (d) Vulcanized rubber
63. For c coordination number of four, anion sits at the centre of ________ where corners are occupied by cations
    (a) Cube (b) Tetrahedron
    (c) Triangle (d) Octahedron

64. Layered silicate structures in clays consists of the following group
    (a) SiO_4^{4-} (b) $Si_2O_5^{2-}$
    (c) $Si_2O_7^{6-}$ (d) SiO_4^{4-}
65. *Schottky defect* in ceramic material is
    (a) Interstitial impurity
    (b) Vacancy-interstitial pair of cations
    (c) Pair of nearby cation and anion vacancies
    (d) Substitutional impurity
66. Taylor dislocation cannot move by the following way
    (a) Slip (b) Climb
    (c) Cross-slip (d) All
67. Conservative movement of dislocations
    (a) Slip (b) Climb
    (c) Both slip and climb (d) None
68. Typical density of dislocations in a solid
    (a) 10^8-10^{10} μm^{-2} (b) 10^8-10^{10} mm^{-2}
    (c) 10^8-10^{10} cm^{-2} (d) 10^8-10^{10} m^{-2}
69. Burger's vector changes with
    (a) Kind of dislocation (b) Length of dislocation
    (c) Both kind and length of dislocation (d) None
70. Which of the following is false?
    (a) Line defects are thermodynamically stable
    (b) Dislocation can end inside a crystal without forming loop
    (c) ABC ABC ABC…is stacking sequence for HCP crystal
    (d) All
71. Gibbs phase rule for general system:
    (a) $P + F = C - 1$ (b) $P + F = C + 1$
    (c) $P + F = C - 2$ (d) $P + F = C + 2$
72. In a single-component condensed system, if degree of freedom is zero, maximum number of phases that can coexist is ________.
    (a) 0 (b) 1
    (c) 2 (d) 3

73. The degree of freedom at triple point in unary diagram for water is ________.

(a) 0
(b) 1
(c) 2
(d) 3

74. Above the following line, liquid phase exists for all compositions in a phase diagram.

(a) Tie-line
(b) Solvus
(c) Solidus
(d) Liquidus

75. Following is wrong about a phase diagram.

(a) It gives information on transformation rates.
(b) Relative amount of different phases can be found under the given equilibrium conditions.
(c) It indicates the temperature at which different phases start to melt.
(d) Solid solubility limits are depicted by it.

76. Horizontal arrest in a cooling curve represents:

(a) Continuous cooling
(b) Invariant reaction
(c) Both
(d) None

77. Relative amounts of phases in a region can be deduced using

(a) Phase rule
(b) Lever rule
(c) Either
(d) None

78. An invariant reaction that produces a solid up on cooling two liquids:

(a) Eutectic
(b) Peritectic
(c) Monotectic
(d) Syntectic

79. A solid + a liquid result in a liquid up on heating during ________ reaction.

(a) Eutectic
(b) Peritectic
(c) Monotectic
(d) Syntectic

80. A solid + a liquid result in a solid up on cooling during ________ reaction.

(a) Eutectic
(b) Peritectic
(c) Monotectic
(d) Syntectic

81. Not a basic step of precipitate strengthening

(a) Solutionizing
(b) Mixing and compacting
(c) Quenching
(d) Aging

82. Both nucleation and growth require change in free energy to be ________.

(a) –ve (b) zero

(c) +ve (d) Any

83. During homogeneous nucleation, critical size of a particle ________ with increase in under-cooling.

(a) Increases (b) Decreases

(c) Won't change (d) Not related

84. Not a typical site for nucleation during solid state transformation

(a) Container wall (b) Grain boundaries

(c) Stacking faults (d) Dislocations

85. Growth occurs by

(a) Diffusion controlled individual movement of atoms

(b) Diffusion-less collective movement of atoms

(c) Both

(d) None

86. Alloying element that decreases eutectoid temperature in Fe-C system is

(a) Mo (b) Si

(c) Ti (d) Ni

87. Nose of a C-curve represents

(a) Shortest time required for specified fraction of transformation

(b) Longest time required for specified fraction of transformation

(c) Average time required for specified fraction of transformation

(d) No information regarding time required for specified fraction of transformation

88. Phase formed of diffusion-less reaction:

(a) Pearlite (b) Lower Bainite

(c) Upper bainite (d) Martensite

89. Ms for Fe-C system is round ________ °C.

(a) 725 (b) 550

(c) 450 (d) 210

90. Impurity not responsible for temper embrittlement is

(a) Sn (b) Sb

(c) Si (d) As

91. The following mechanism contributes very little diffusivity
    (a) Vacancy (b) Interstitial
    (c) Substitutional (d) Self-interstitial
92. Not an example for short-circuit path
    (a) Vacancy (b) Dislocations
    (c) Grain boundaries (d) External surfaces
93. Example for steady-state diffusion
    (a) Hydrogen purification by palladium sheet
    (b) Doping semiconductors
    (c) Corrosion resistance of duralumin
    (d) Decarburization of steel
94. The most influencing factor of diffusivity
    (a) Diffusing species (b) Temperature
    (c) Lattice structure (d) Presence of defects
95. The following phenomena are useful in zone-refining process
    (a) Coring (b) Segregation
    (c) Both (d) None
96. The following enhances the coring
    (a) Minute difference between liquidus and solidus
    (b) Marked difference between liquids and solids
    (c) Either
    (d) None
97. On basis of electronic configuration group and period of $^{9}_{5}B$ is
    (a) 2 and IIIA (b) 3 and IIA
    (c) 4 and VIA (d) 5 and VIIA
98. Chemical properties depend on
    (a) Electronic configuration
    (b) Valence shell electronic configuration
    (c) Atomic Mass
    (d) None

99. Nobel gases are present in

(a) VIIA (b) VIA

(c) VA (d) VIIIA

100. Energy required to remove an electron from the outermost shell is called

(a) Potential energy (b) Kinetic energy

(c) Mechanical energy (d) Ionization energy

Answers

1. (d)	**2.** (b)	**3.** (a)	**4.** (c)	**5.** (a)
6. (d)	**7.** (c)	**8.** (a)	**9.** (b)	**10.** (d)
11. (b)	**12.** (c)	**13.** (a)	**14.** (c)	**15.** (d)
16. (c)	**17.** (a)	**18.** (b)	**19.** (a)	**20.** (a)
21. (b)	**22.** (d)	**23.** (c)	**24.** (c)	**25.** (a)
26. (c)	**27.** (a)	**28.** (c)	**29.** (b)	**30.** (a)
31. (b)	**32.** (a)	**33.** (b)	**34.** (a)	**35.** (d)
36. (b)	**37.** (b)	**38.** (c)	**39.** (a)	**40.** (c)
41. (d)	**42.** (a)	**43.** (c)	**44.** (b)	**45.** (b)
46. (d)	**47.** (c)	**48.** (d)	**49.** (c)	**50.** (b)
51. (a)	**52.** (c)	**53.** (c)	**54.** (b)	**55.** (c)
56. (c)	**57.** (b)	**58.** (c)	**59.** (b)	**60.** (c)
61. (d)	**62.** (a)	**63.** (b)	**64.** (b)	**65.** (c)
66. (b)	**67.** (a)	**68.** (d)	**69.** (d)	**70.** (d)
71. (d)	**72.** (c)	**73.** (a)	**74.** (c)	**75.** (a)
76. (b)	**77.** (b)	**78.** (d)	**79.** (c)	**80.** (b)
81. (b)	**82.** (a)	**83.** (b)	**84.** (a)	**85.** (c)
86. (d)	**87.** (a)	**88.** (d)	**89.** (d)	**90.** (c)
91. (d)	**92.** (a)	**93.** (a)	**94.** (b)	**95.** (c)
96. (b)	**97.** (a)	**98.** (b)	**99.** (d)	**100.** (d)

SECTION II

Properties and Strengthening of Engineering Materials

5

Mechanical Behaviour of Solids and Properties of Materials

Learning Objectives

After going through this chapter the reader will be able to:

- Understand engineering stress and engineering strain.
- State Hooke's law and note its validation.
- Gather idea on elastic and plastic behaviours of materials.
- List the mechanical properties of polymers, ceramics and glasses.
- Cite the electrical, thermal, magnetic and optical properties of materials.

5.1 INTRODUCTION

It is incumbent on engineers to understand how the various mechanical properties are measured and what these properties represent. They may be called upon to design structures/components using predetermined materials such that unacceptable levels of deformation and/or failure will not occur. Numerous materials, when in administration, are subjected to force or loads; illustrations incorporate the aluminium combination from which a plane wing is developed and the steel in a vehicle pivot. In such circumstances it is important to know the qualities of the material and to plan the part from which it is made with the end goal that any subsequent twisting won't be unnecessary and break won't happen. The mechanical conduct of a material mirrors the relationship between its reaction or twisting to a connected load or drive. Important mechanical properties are strength, hardness, ductility, and stiffness.

The mechanical properties of materials are ascertained by performing carefully designed laboratory experiments that replicate as nearly as possible the service conditions. Elements to be considered incorporate the way of the connected load and its span, and additionally the ecological conditions. It is possible for the load to be tensile, compressive, or shear, and its magnitude may be constant with time, or may fluctuate continuously. Application time might be just a small amount of a second, or it might reach out over a time of numerous years. Service temperature might be a critical variable.

In electrical engineering, for example, the packaging of microelectronic chips has raised a host of issues related to mechanical reliability and failure of substrates, metal solder joints, polymer boards, contacts, and connectors. Although some components may be minuscule in size compared with structures dealt with by civil or mechanical engineers, they are subject to the same limitations imposed on elastic and plastic phenomena. This chapter deals with the issues and facets related to the mechanical behaviour of solids. At one extreme the subject matter is concerned with **elasticity** or deformation that is recoverable. In this regime materials deform (e.g., elongate or shorten) *linearly* with applied load. Elastic behaviour forms the basis for virtually all structural and machine design. The reason we can be so confident of the mechanical integrity of our engineering structures and components is that elastic deformation phenomena are predictable to a high degree of accuracy. Design loads are such as to keep dimensional changes small, and upon unloading, the material springs back and regains its original shape without any apparent damage.

Materials and metallurgical engineers, on the other hand, are concerned with producing and fabricating materials to meet service requirements as predicted by these stress analyses. This necessarily involves an understanding of the relationships between the microstructure (i.e., internal features) of materials and their mechanical properties.

5.2 ELASTIC BEHAVIOUR

5.2.1 Concepts of Stress and Strains

If a load is static or changes relatively slowly with time and is applied uniformly over a cross section or surface of a member, the mechanical behaviour may be ascertained by a simple stress-strain test; these are most commonly conducted for metals at room temperature. There are three principal ways in which a load may be applied: namely, tension, compression, and shear (Figures 5.1a, b, c). In engineering practice many loads are torsional rather than pure shear; this type of loading is illustrated in Figure 5.1d.

Tension Test: One of the most common mechanical stress-strain tests is performed in *tension*. As will be seen, the tension test can be used to ascertain several mechanical properties of materials that are important in design. A specimen is deformed, usually to fracture, with a gradually increasing tensile load that is applied uniaxially along the long axis of a specimen. A standard tensile specimen is shown in Figure 6.2. Normally, the cross section is circular, but rectangular specimens are also used. This "dog-bone" specimen configuration was chosen so that, during testing, deformation is confined to the narrow centre region (which has a uniform cross section along its length), and, also,

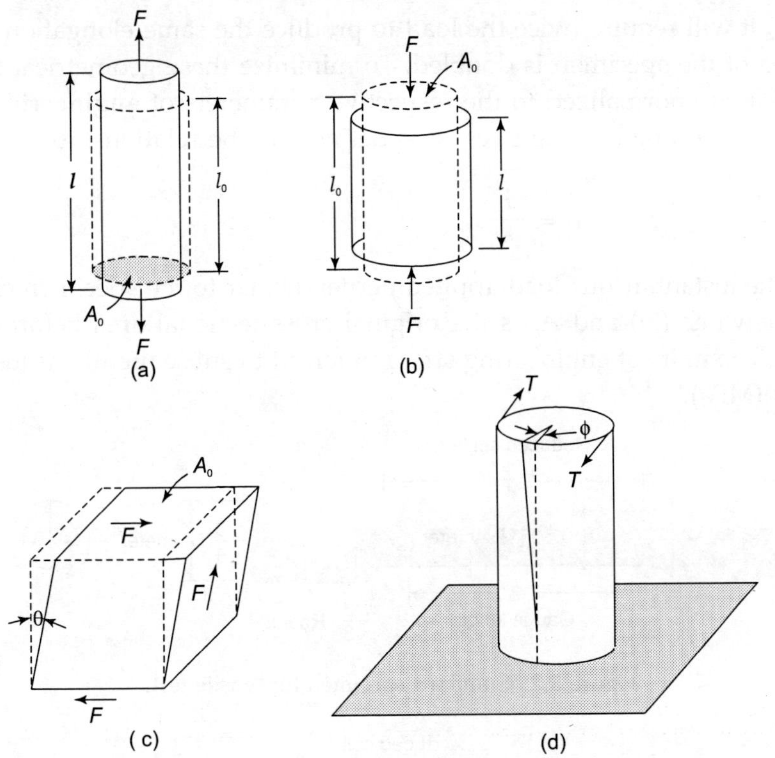

Figure 5.1 (a) Diagrammatic representation of how a tensile load produces an elongation and positive linear strain. Dashed lines refer to the shape before deformation; solid lines, after deformation. (b) Schematic representation of how a compressive load produces contraction and a negative linear strain. (c) Schematic illustration of shear strain, where $\gamma = \tan \theta$ (d) Schematic illustration of torsional deformation.

to reduce the likelihood of fracture at the ends of the specimen. The standard diameter is approximately 12.8 mm (0.5 in), whereas the reduced section length should be at least four times this diameter; 60 mm is common. Gauge length is used in ductility computations, as discussed in Section 6.6; the standard value is 50 mm (2.0 in). The specimen is mounted by its ends into the holding grips of the testing apparatus (Figure 5.3). The tensile testing machine is designed to elongate the specimen at a constant rate, and to continuously and simultaneously measure the instantaneous applied load (with a load cell) and the resulting elongations (using an extensometer). A stress-strain test typically takes several minutes to perform and is destructive; that is, the test specimen is permanently deformed and usually fractured.

The output of such a tensile test is recorded (usually on a computer) as load or force versus elongation. These load deformation characteristics depend on the specimen size.

For example, it will require twice the load to produce the same elongation if the cross-sectional area of the specimen is doubled. To minimize these geometrical factors, load and elongation are normalized to the respective parameters of **engineering stress** and **engineering strain.** Engineering stress σ is defined by the relationship

$$\sigma = \frac{F}{A_o}$$

where, F is the instantaneous load applied perpendicular to the specimen cross section, in units of newtons (N) and A_o is the original cross-sectional area before any load is applied (m^2). The units of engineering stress (referred to subsequently as just stress) are megapascals (MPa).

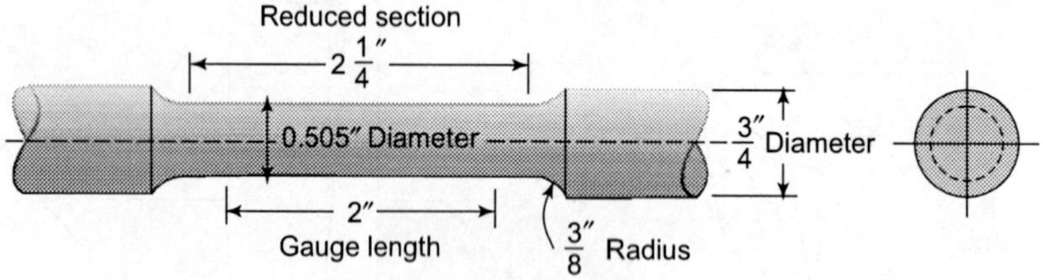

Figure 5.2 Standard specimen for tensile test.

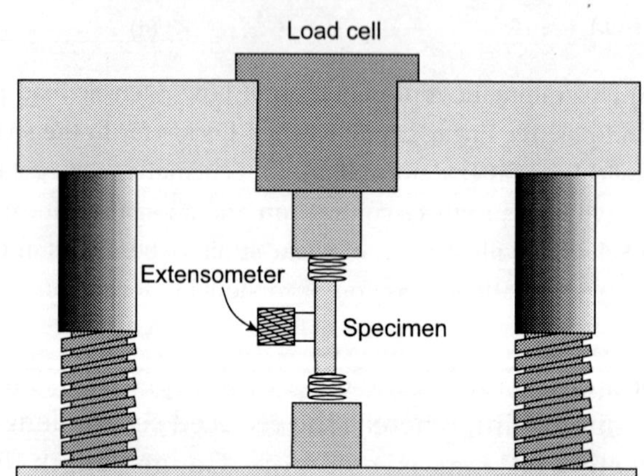

Figure 5.3 Schematic illustration of the setup used to conduct tensile stress-strain tests. The specimen gets elongated by the moving crosshead; load cell and extensometer measure, respectively, the magnitude of the applied load and the elongation.

Engineering strain is defined according to

$$\varepsilon = \frac{l_i - l_o}{l_o} = \frac{\Delta l}{l_o}$$

where, l_o is the original length before any load is applied, and is the instantaneous length. Sometimes the quantity $l_i - l_o$ is denoted as Δl and is the deformation elongation or change in length at some instant, as referenced to the original length. Engineering strain (subsequently called just strain) does not have a unit, but meters per meter or inches per inch are often used.

Compression Test: Compression stress-strain tests may be conducted if in-service forces are of compressive type. A compression test is conducted in a manner similar to the tensile test, except that the force is compressive and the specimen contracts along the direction of the stress. Basic equation are utilized to compute compressive stress and strain, respectively. By convention, a compressive force is taken to be negative, which yields a negative stress.

Shear and Torsional Tests: For tests performed using a pure shear force as shown in Figure 5.1c, the shear stress is computed according to

$$\tau = \frac{F}{A_o}$$

where, F is the load or force imposed parallel to the upper and lower faces, each of which has an area A_o. The shear strain γ is defined as the tangent of the strain angle, as indicated in the figure. The units for shear stress and strain are the same as for their tensile counterparts.

Torsion is a variation of pure shear, wherein a structural member is twisted in the manner of Figure 5.1d; torsional forces produce a rotational motion about the longitudinal axis of one end of the member relative to the other end. Examples of torsion are found for machine axles and drive shafts, and also for twist drills. Torsional tests are normally performed on cylindrical solid shafts or tubes. A shear stress τ is a function of the applied torque T, whereas shear strain γ is related to the angle of twist ϕ, in Figure 5.1d.

5.2.2 Stress-Strain Behaviour

The degree to which a structure deforms or strains depends on the magnitude of an imposed stress. For most metals that are stressed in tension and at relatively low levels, stress and strain are proportional to each other through the relationship

$$\sigma = E\varepsilon$$

This is known as Hooke's law, and the constant of proportionality E (GPa) is the **modulus of elasticity,** or Young's modulus. For most typical metals the magnitude of this modulus ranges between 45 GPa, for magnesium, and 407 GPa, for tungsten. Modulus of elasticity values for several metals at room temperature are presented in Table 5.1.

Table 5.1 Room-temperature elastic and shear moduli, and Poisson's ratio for various metal alloys

Metal Alloy	Modulus of Elasticity		Shear Modulus		Poisson's Ratio
	GPa	10^6 psi	GPa	10^6 psi	
Aluminium	69	10	25	3.6	0.33
Brass	97	14	37	5.4	0.34
Copper	110	16	46	6.7	0.34
Magnesium	45	6.5	17	2.5	0.29
Nickel	207	30	76	11.0	0.31
Steel	207	30	83	12.0	0.30
Titanium	107	15.5	45	6.5	0.34
Tungsten	407	59	160	23.2	0.28

Deformation in which stress and strain are proportional is called elastic deformation; a plot of stress (ordinate) versus strain (abscissa) results in a linear relationship, as shown in Figure 5.4. The slope of this linear segment corresponds to the modulus of elasticity *E*. This modulus may be thought of as stiffness, or a material's resistance to elastic deformation. The greater the modulus is, the stiffer the material, or the smaller the elastic strain that results from the application of a given stress. The modulus is an important design parameter used for computing elastic deflections. Elastic deformation is non-permanent, which means that when the applied load is released, the piece returns to its original shape. As shown in the stress-strain plot (Figure 5.4), application of the load corresponds to moving from the origin up and along the straight line. Upon release of the load, the line is traversed in the opposite direction, back to the origin.

There are some materials (e.g., gray cast iron, concrete, and many polymers) for which this elastic portion of the stress-strain curve is not linear (Figure 5.5); hence, it is not possible to determine the modulus of elasticity as described above. For this nonlinear behaviour, either *tangent* or *secant modulus* is normally used. Tangent modulus is taken as the slope of the stress-strain curve at some specified level of stress, while secant modulus represents the slope of a secant drawn from the origin to some given point of the σ-ε curve. The determination of these moduli is illustrated in Figure 5.5.

As would be expected, the imposition of compressive, shear, or torsional stresses also evokes elastic behaviour. The stress-strain characteristics at low stress levels are virtually the same for both tensile and compressive situations, to include the magnitude of the modulus of elasticity. Shear stress and strain are proportional to each other through the expression

$$\tau = G\gamma$$

where, *G* is the shear modulus, the slope of the linear elastic region of the shear stress-strain curve. Table 5.1 also gives the shear moduli for a number of the common metals.

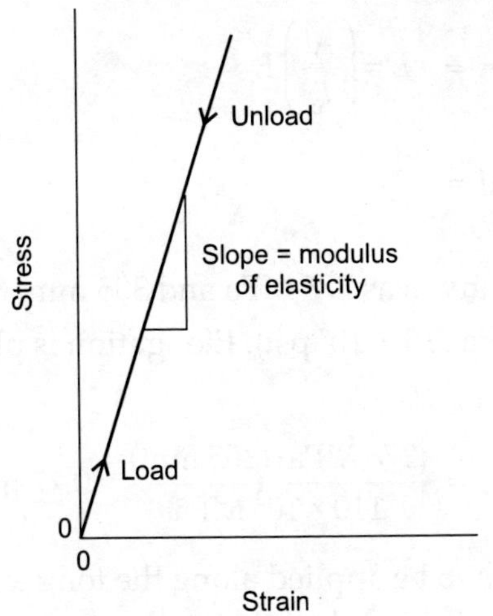

Figure 5.4 Schematic stress-strain diagram showing linear elastic deformation for loading and unloading cycles.

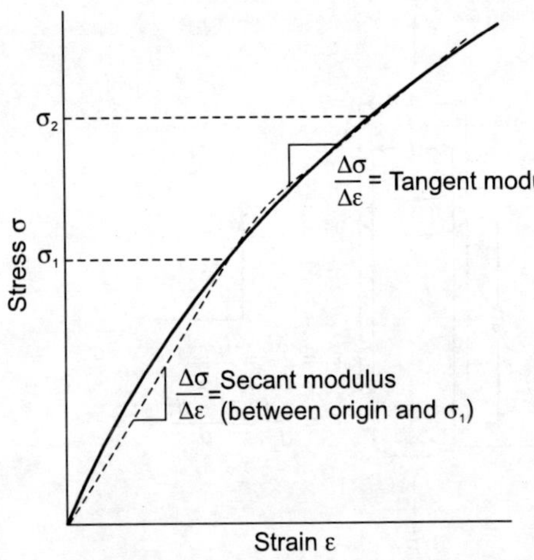

Figure 5.5 Schematic stress-strain diagram showing nonlinear elastic behaviour, and how secant and tangent moduli are determined.

Example 5.1 A piece of copper originally 305 mm (12 in) long is pulled in tension with a stress of 276 MPa (40,000 psi). If the deformation is entirely elastic, what will be the resultant elongation?

Solution: Since the deformation is elastic, strain depends on stress according to equation given below. Furthermore, the elongation Δl is related to the original length l_0 through equation given below. Combining these two expressions and solving for Δl yields

$$\sigma = \varepsilon E = \left(\frac{\Delta l}{l_0}\right) E$$

$$\Delta l = \frac{\sigma l_0}{E}$$

The values of σ and l_0 are given as 276 MPa and 305 mm, respectively, and the magnitude of E for copper is 110 GPa (16×10^6 psi). Elongation is obtained by substitution into the expression above as

$$\Delta l = \frac{(276\ \text{MPa})\ (305\ \text{mm})}{110 \times 10^3\ \text{MPa}} = 0.77\ \text{mm}\ (0.03\ \text{in.})$$

Example 5.2 A tensile stress is to be applied along the long axis of a cylindrical brass rod that has a diameter of 10 mm (0.4 in.). Determine the magnitude of the load required to produce a 2.5×10^{-3} mm (in) change in diameter if the deformation is entirely elastic.

Solution: This deformation situation is represented in the accompanying drawing.

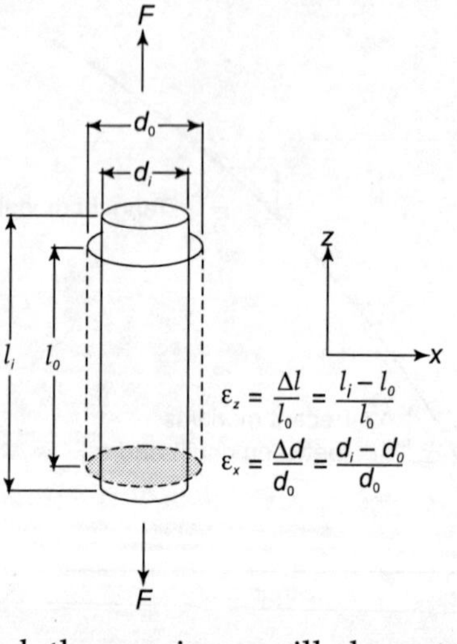

When the force F is applied, the specimen will elongate in the z direction and at the same time experience a reduction in diameter, Δd, of 2.5×10^{-3} mm in the x direction. For the strain in the x direction,

$$\varepsilon_x = \frac{\Delta d}{d_0} = \frac{-2.5 \times 10^{-3}\ \text{mm}}{10\ \text{mm}} = -2.5 \times 10^4$$

which is negative, since the diameter is reduced.

It becomes necessary to calculate the strain in the z direction using equation given below. The value for Poisson's ratio for brass is 0.34 and thus

$$\varepsilon_z = \frac{\varepsilon_x}{v} - \frac{(-2.5 \times 10^{-4})}{0.34} = 7.35 \times 10^{-4}$$

The applied stress may now be computed using equation given below and the modulus of elasticity, given in Materials Data Table as 97 GPa (14 × 10^6 psi), as

$$\sigma = \varepsilon_z E = (7.35 \times 10^{-4})\ (97 \times 10^3\ \text{MPa}) = 71.3\ \text{MPa}$$

Finally, from basic equations, the applied force may be determined as

$$F = \sigma A_o = \sigma\left(\frac{d_0}{2}\right)^2 \pi$$

$$= (71.3 \times 10^6\ \text{N/m}^2)\left(\frac{10 \times 10^{-3}\ \text{m}}{2}\right)^2 \pi = 5600\ \text{N}\ (1293\ \text{lb}_f)$$

5.3 PLASTIC DEFORMATION OF METALS

Up to this point, it has been assumed that elastic deformation is time independent, that is, an applied stress produces an instantaneous elastic strain that remains constant over the period of time the stress is maintained. It has also been assumed that upon release of the load the strain is totally recovered, that is, the strain immediately returns to zero. In most engineering materials, however, there will also exist a time-dependent elastic strain component. That is, elastic deformation will continue after the stress application, and upon load release some finite time is required for complete recovery. This time-dependent elastic behaviour is known as **anelasticity,** and it is due to time-dependent microscopic and atomistic processes that are attendant to the deformation. For metals the anelastic component is normally small and is often neglected. However, for some polymeric materials its magnitude is significant; in this case it is termed *viscoelastic behaviour*.

For most metallic materials, elastic deformation persists only to strains of about 0.005. As the material is deformed beyond this point, the stress is no longer proportional to strain (Hooke's law), and permanent, non-recoverable, or plastic deformation occurs. Figure 5.6a plots schematically the tensile stress-strain behaviour into the plastic region for a typical metal. The transition from elastic to plastic is a gradual one for most metals; some curvature results at the onset of plastic deformation, which increases more rapidly with rising stress.

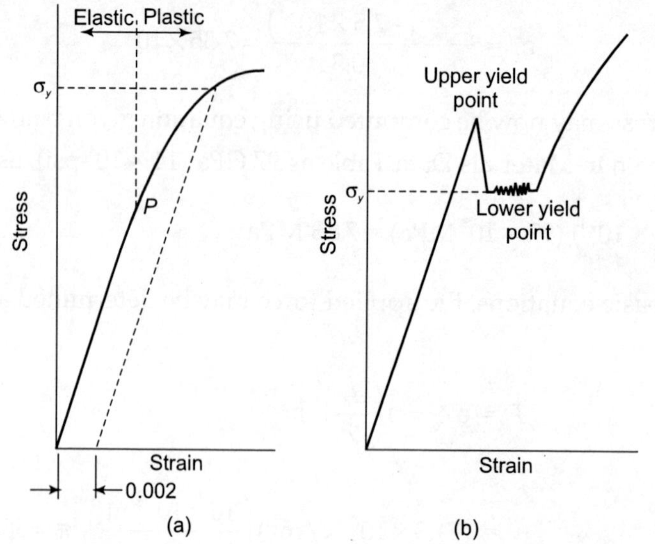

Figure 5.6 (a) Typical stress-strain behaviour for a metal showing elastic and plastic deformations, the proportional limit *P*, and the yield strength σ_y as determined using the 0.002 strain offset method. (b) Representative stress-strain behaviour for some steels demonstrating the yield point phenomenon.

From an atomic perspective, plastic deformation corresponds to the breaking of bonds with original atom neighbours and then reforming bonds with new neighbours as large number of atoms or molecules move relative to one another; upon removal of the stress they do not return to their original positions. The mechanism of this deformation is different for crystalline and amorphous materials.

Example 5.3 A cylindrical specimen of steel having an original diameter of 12.8 mm (0.505 in) is tensile tested to fracture and found to have an engineering fracture strength of 460 MPa (67,000 psi). If its cross-sectional diameter at fracture is 10.7 mm (0.422 in.), determine:

(a) The ductility in terms of percent reduction in area

(b) The true stress at fracture

Solution:

(a) Ductility is computed as

$$\% RA = \frac{\left(\dfrac{12.8\text{ mm}}{2}\right)^2 \pi - \left(\dfrac{10.7\text{ mm}}{2}\right)^2 \pi}{\left(\dfrac{12.8\text{ mm}}{2}\right)^2 \pi} \times 100$$

$$= \frac{128.7\ \text{mm}^2 - 89.9\ \text{mm}^2}{128.7\ \text{mm}^2} \times 100 = 30\%$$

(b) True stress is defined where in this case the area is taken as the fraction area A_f. However, the load at fracture must first be computed from the fracture strength as

$$F = \sigma_f A_0 = (460 \times 10^6\ \text{N/m}^2)(128.7\ \text{mm}^2)\left(\frac{1\ \text{m}^2}{10^6\ \text{mm}^2}\right) = 59{,}200\ \text{N}$$

Thus, the true stress is calculated as

$$\sigma_T = \frac{F}{A_f} = \frac{59{,}200\ \text{N}}{(89.9\ \text{mm}^2)\left(\frac{1\ \text{m}^2}{10^6\ \text{mm}^2}\right)}$$

$$= 6.6 \times 10^8\ \text{N/m}^2 = 660\ \text{MPa (95,700 psi)}$$

Example 5.4 Compute the strain-hardening exponent n for an alloy in which a true stress of 415 MPa (60,000 psi) produces a true strain of 0.10; assume the value of 1035 MPa (150,000 psi) for K.

Solution: This requires some algebraic manipulation so that n becomes the dependent parameter. This is accomplished by taking logarithms and rearranging. Solving for n yields

$$n = \frac{\log \sigma_T - \log K}{\log \varepsilon_T}$$

$$= \frac{\log(415\ \text{MPa}) - \log(1035\ \text{MPa})}{\log(0.1)} = 0.40$$

5.4 MECHANICAL BEHAVIOUR OF POLYMERS, CERAMICS AND GLASSES

5.4.1 Polymers

Young's Modulus: Understanding Young's modulus (£) is a key to appreciating the mechanical properties of polymers. Unlike the case for metals and ceramics, E for polymers is often dependent so that it is, in effect, a viscoelastic modulus. After load is applied, polymers not only instantaneously stretch, but also viscously elongate further at a slower rate as polymer molecules find new avenues for relaxing or uncoiling. A more or less constant stress coupled with a time-dependent, but recoverable strain means a time-dependent modulus

$$E(t) = \sigma/\varepsilon(t)$$

In addition, we are reasonably confident that E for metals and ceramics will vary only by a very small and predictable amount over a several hundred degree range about room temperature. The same is not true of polymers. This is immediately apparent in Figure 5.7, where E for a typical linear polymer is plotted against temperature; as much as a huge 10^6-fold variation in E may occur over a relatively small temperature range (e.g., ~200°C). As before, T_G and T_M are the glass transition and melting temperatures. Glassy and viscous regimes of polymer mechanical behaviour exist at the temperature extremes, with leathery and rubbery behaviour in between. Note that very large changes in modulus can also occur by increasing the extent of either crystallinity or cross-linking.

Glassy Regime: For values of $T/T_G < 1$ the polymer behaves like a relatively brittle glass. The free volume in the structure has just about been eliminated and the polymer molecules are either densely packed amorphous tangles or collections of mis-oriented crystals permeated by an amorphous matrix. Applied stresses (σ) stretch the strong covalent bonds of the chain spine, as well as the weaker secondary bonds between chains. The resultant strain, ε, reflects the differential stiffness of both kinds of bonds weighted by their respective concentrations in the polymer.

Viscous Regime: Above about 1.5 T_G all of the secondary bonds break and the modulus suffers another large decline. Linear polymers become viscous liquids and undergo so-called Newtonian flow. Applying a shear stress (τ) to such fluids causes them to shear or flow at a strain rate of γ (or *dy/dt)* that is determined by the coefficient of viscosity (η) such that

$$\tau = \eta\gamma$$

Viscosity has units of poise P ($1P = 10^{-1}$ Pa-s) and *drops* with temperature in a thermally activated manner.

Rubbery Plateau: The glassy region gives way to the rubbery plateau above T_G. A thousandfold drop in E occurs, a consequence of the partial breaking of secondary bonds. Molecular chains do not yet slither in an unrestricted way as they do in the melt. Rather they are now loosely entangled with one another; under loading, chains extend but then wriggle back when the load is removed. If there is a dilute concentration of covalent cross-links that do not break, the polymer becomes a true elastomer. It can now elastically extend a few hundred percent and snap back when the load is withdrawn. But, if there is a high concentration of cross-links, the covalent bonds associated with them will create a three-dimensional network of relatively short linear molecular segments. No longer are there long linear molecules to stretch large amounts in a unidirectional way.

Surprisingly, in elastomers, unlike other materials, E increases slightly as the temperature rises. This is caused by the increased random thermal motion of polymer segments, which makes it more difficult to orient coiled chains. The higher uncoiling load required means a higher elastic modulus. Table 5.2 depicts the mechanical behaviour and properties of polymers.

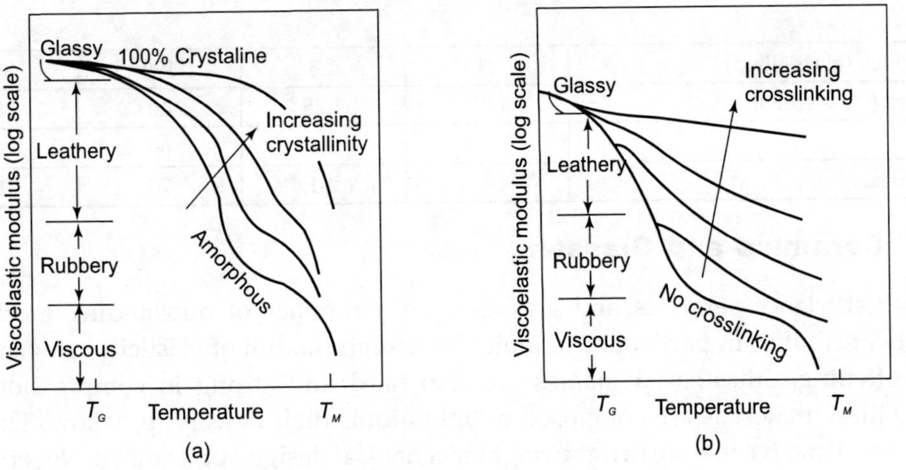

Figure 5.7 Typical temperature-dependent changes in the viscoelastic modulus of long-chain polymers. (a) Effect of degree of crystallinity. (b) Effect of extent of cross-linking.

Table 5.2 Mechanical properties of polymers

Polymer	E (GN/m^2)	ρ (Mg/m^3)	σ_0^c (MN/m^2)	T_G (°C)
Thermoplastics				
Polyethylene (PE)				
High-density	0.56	0.96	30	
Low-density	0.18	0.91	11	– 20
Polyvinyl chloride (PVC)	2.0	1.2	25	80
Polypropylene (PP)	1.3	0.9	35	0
Styrene				
ABS[b]	2.5	1.2	50	80
Polystyrene (PS)	2.7	1.1	50	100
Polycarbonate (PC)	2.5	1.2-1.3	68	150
Polyethylene terephthalate (PET)	8	0.94	135	67
Polymethylene methacrylate (PMMA)	2.8	1.2	70	100
Polyesters	1.3-4.5	1.1-1.4	65	67
Polyamide	2.8	1.1-1.2	70	60
Polytetrafluoroethylene (PTFE)	0.4	2.3	25	120
Thermosets				
Epoxies	2.1-5.5	1.2-1.4	60	(107-200)
Phenolics	18	1.5	80	(200-300)
Polyesters (glass-filled)	14	1.1-1.5	70	200

Contd...

Contd...

Polyimides (glass-filled)	21	1.3	190	350
Silicones (glass-filled)	8	1.25	40	300
Ureas	7	1.3	60	80
Urethanes	7	1.2-1.4	70	100

5.4.2 Ceramics and Glasses

At the first glance ceramics and glasses have a number of outstanding mechanical property attributes. In particular, ceramics have high moduli of elasticity, are hard, and resist softening. Silica-based glasses are also hard and strong in compression. And because these materials are composed of light atoms their density (ρ) is low. Thus, *E/p* is large, an attractive feature in lightweight mechanical design applications. Nevertheless, the use of these materials is severely restricted in structural applications because of their fatal tendency to fracture in a brittle manner under tension. There are *intrinsic* as well as *extrinsic* reasons for the brittle behaviour of these materials. It has already been mentioned that passage of dislocations through crystalline ceramics means rupturing very strong inter-atomic bonds.

Therefore, dislocations are not prevalent and their absence generally limits plastic deformation. Factors that adversely affect their mechanical behaviour stem from the surface and interior cracks and notches that are unavoidable in processing or that develop in use. Because they cannot be easily melted, ceramics are formed into shape by pressing and sintering powders. In consolidating powders, sharp crevices at internal pores can then behave as incipient cracks. As the mechanical behaviour of these materials is severely limited by macroscopic cracks and pores, much of the subsequent discussion is devoted to the implications of stressing matrices containing such defects.

Cracks and Stress Concentration: Surface and interior cracks in objects under load can greatly alter local stress distributions relative to the case of flaw-free material. This intuitive result is illustrated in Figure 5.8, where a body (plate) containing elliptical-shaped surface and interior holes or cracks is pulled by uniaxial forces, *F*. The undisturbed force lines in the defect-free portion of the plate mean that the stress distribution is uniform there at a level σ_a. But, the force lines are concentrated in the vicinity of cracks; stress is magnified there because there are effectively more force lines per unit area. A stress concentration factor k_σ, can be defined as the ratio of amplified local to background stress, that is given by $k_\sigma = \sigma/\sigma_a$. For flat elliptical surface cracks the maximum value of k_σ is at the crack tip and given by

$$k_\sigma = 2\left(\frac{c}{\rho}\right)^{\frac{1}{2}}$$

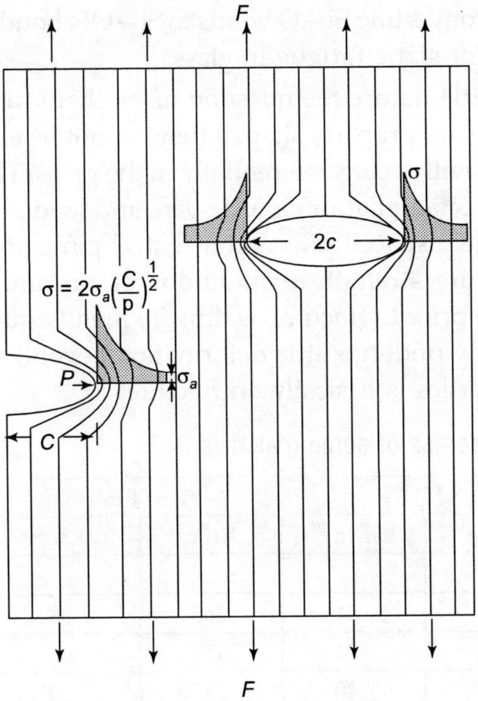

Figure 5.8 Stress concentration around internal and surface cracks in a uniformly stretched solid. Parallel lines denote the presence of uniform stress.

Here c is half the length of the major axis and p is the radius of curvature at the crack tip. It is apparent that longer and sharper cracks rise (k_σ). Although cracks can exist within the bulk, they are more commonly found at the surface. Furthermore, they pose a much greater hazard in brittle ceramics and glasses than they do in ductile metals. In metals, the elevated stress ahead of the crack causes local plastic deformation, which strengthens the matrix. It is this effect that blunts the advance of the crack. But no such mechanism to arrest crack growth exists in brittle materials. The simultaneous magnification of stress at a crack tip coupled with a lack of dislocations and plastic deformation to absorb or dissipate the strain energy build up is a disastrous combination. Quite simply put, metals are tough and ceramics and glasses are not: the former plastically yield before fracture; the reverse is true for the latter.

On this basis we can understand why both ceramics and glasses respond differently in tension and compression. Young's modulus measured either way is virtually identical, but the fracture strength in compression is typically 5 to 10 or more times that in tension. Flat cracks are closed shut in compression but tend to be pulled open in tension. Glasses can be melted and made dense, but atmospheric attack can enlarge natural surface flaws into cracks in the presence of tensile stress. Water can penetrate these surface cracks and weaken the glass by converting Si – O bonds to Si – OH bonds. This is the mechanism that appears to account for **static fatigue** in glass.

In view of their brittle nature the question of mechanical testing of ceramics and glasses arises. Unless they are properly shaped they cannot be easily gripped and pulled in tension because failure will occur preferentially in the grips. Therefore, these materials are often fashioned into rectangular or circular bars and loaded in three-point bending. The bars rest on two widely spaced pins and a central pin is driven into the surface of the bar. **Transverse rupture strength** (or **modulus of rupture)**, the maximum tensile stress at the beam surface prior to fracture, is thus frequently quoted for these materials. As brittle materials usually undergo little deformation beyond the elastic regime, focus in their mechanical properties is basically on fracture.

Table 5.3 Mechanical properties of some materials

Material	K_{IC}		Yield Stress		Impact Energy (Charpy)	
	MPa-m$^{1/2}$	ksi-in.$^{1/2}$	MPa	ksi	J	ft-lb
Metals						
Aluminium	45	41	260	38	27	~ 20
2024, hardened	26	24	455	66	14	~ 10
2024, annealed	44	40	345	50		
7075, hardened	26	22	495	72		
Ti-6A1-4V	55	50	1035	150	54	~ 40
Mild steel	140	127	220	32	203	~ 150
4340, hardened	60.4	55	1515	220	18	13
4340, annealed	99	90	860	125	106	78
Maraging steel	~ 133	~ 120	1700	247	54	~ 40
Pressure vessel steel	200	182	500	72		
			Modulus of ruputure			
			MPa	**ksi**		
Ceramics						
Soda glass	0.7	70	10			
Al_2O_3	3.7	270	40			
SiC	3.1	170	25			
Si_3N_4	4					
ZrO_2, stabilized	12					
Concrete	0.2					

						Izod (ft-lb)[a]
Polymers						
Polyethylene (high density)	~ 2					0.4-14
PMMA	~ 1.3					
Polystyrene	~ 1					0.2-0.5
Polystyrene (high impact)						0.5-4
Polycarbonate	2					12-16
PVC	~ 0.5					0.4-20
Epoxy	~ 40					0.2-3.0
Epoxy (fibreglass reinforced)						2-30

5.5 ELECTRICAL PROPERTIES OF MATERIALS

The prime objective of this section is to explore the electrical properties of materials, that is, their responses to an applied electric field. Consideration of the electrical properties of materials is often important when materials selection and processing decisions are made during the design of a component or structure. The electrical behaviours of the various materials are diverse. Some need to be highly electrically conductive (e.g., connecting wires), whereas electrical insulation is required of others (e.g., the protective package encapsulation) Engineering materials are essential in regular day-to-day existence as a result of their adaptable basic properties. Other than these properties, they do assume an essential part due to their physical properties. Prime physical properties of materials include: electrical properties, thermal properties, magnetic properties, and optical properties. The electrical practices of designing materials differ, as are their uses in electrical applications. Couple of cases of electrical applications: copper and aluminium wires used in power transmission; amorphous silicon in solar cells; LEDs; gallium-arsenide chips used in modern-day computers, cell phones, digital assistants, etc.

5.5.1 Electrical Conduction

Electrical conductivity of a material is defined in terms of ease with which a material transmits an electrical current. Electrical current (I) is the flow of electrons, and driving force for the flow of electrons is called voltage (V). Ohm's law relates these parameters as follows:

$$V \alpha I$$

$$V = IR$$

where, R is the materials resistance to flow of electrons through it. V, I, and R respectively have units as volts, amperes, and ohms (Ω).

Electrical resistance of a material is influenced by its geometric configuration; hence a new parameter called *electrical resistivity* (ρ) is defined such as it is independent of the geometry..

$$\rho = \frac{RA}{l}$$

where, A is cross-sectional area perpendicular to the direction of the current, and l is the distance between points between which the voltage is applied. Units for ρ are ohm-meters (Ω-m).

Reciprocal of the electrical resistivity, known as *electrical conductivity* (σ), is used to express the electrical behaviour of a material, which is indicative of the ease with which a material allows flow of electrons.

$$\sigma = \frac{1}{\rho} = \frac{l}{RA}$$

Electrical conductivity has the following units: $(\Omega\text{-m})^{-1}$ or mho/metre.

Resistivity or *conductivity* does not depend on the dimensions of the material. These properties are microstructure-sensitive instead, like many other intrinsic properties, for example, yield strength. Thus, either of it (usually *conductivity*) allows us to compare different materials. Solid engineering materials exhibit very wide range of electrical conductivity (about 27 orders of variation). Hence, the materials for electrical applications are classified according to their electrical conductivity as: *conductors, semiconductors* and *insulators/dielectrics*. Metals are conductors having conductivities in the range of 10^7 $(\Omega\text{-m})^{-1}$, while semiconductors have conductivities in range from 10^{-6} to 10^4 $(\Omega\text{-m})^{-1}$, and materials with conductivity lower than 10^{-10} $(\Omega\text{-m})^{-1}$ are termed insulators.

In components designed to conduct electrical energy, minimizing losses is important to conserve energy as well as minimize heating. The electric power (P) lost when a current flows through a material is given by:

$$P = VI = I^2R$$

Electrical losses are also known as Joule heating losses.

Second form of Ohm's law can be obtained by combining the equations above,

$$\frac{I}{A} = \sigma\frac{V}{l}$$

$$J = \sigma E$$

where, J is current density (A/cm^2), and E is electric field strength (V/cm).

In all conductors, semiconductors, and many insulating materials, only electronic conduction exists, i.e., electrons are the only charge carriers. Ionic conduction results because of net motion of charged ions. Movement of different particles in various materials depends on more than one parameter. These include: atomic bonding, imperfections, microstructure, ionic compounds, diffusion rates and temperature. Because of these, electrical conductivity of materials varies tremendously.

In metals not all electrons accelerate in the presence of an electric field. Instead the number of charge carriers is dependent on the arrangement of electron states/ levels with respect to energy, and the manner in which these states are occupied by electrons. Electrons in metals are arranged into shells and sub-shells in accordance with Pauli exclusion principle. Each of these shells/sub-shells is distinguished from every other by their energy state/level. Basic existence of metals is dependent on metallic bond where large groups of atoms come together and share valence electrons in what is called *metallic bond*. Thus, there are N atoms in the lattice with N atomic orbitals and N molecular orbitals, many of which are of same energy. This leads to bands of electrons. The electrons not involved in bonding remain in what is called the core band, whereas the valence electrons that form the electron gas enter into the valence band. The remaining un-filled orbitals form higher-energy bands, called conduction band. It is the conduction band that gives metals and alloys the ability to freely conduct electrons.

Four different types of band structures are possible at 0 K: (1) Where one outermost band is only partially filled with electrons. The energy corresponding to the highest filled state at 0 K is called the Fermi energy, E_f. (2). There is overlap of an empty band and a filled band. (3) and (4). One filled band (valence band) and an empty band (conduction band) are separated by an energy band gap which lies between them. These are schematically shown in Figure 5.9.

Band structures (1) and (2) are found in metals/conductors. Band structures (3) and (4) are distinguished by the size of energy band gap. Narrower energy band gap, i.e., size < 2 eV, is found in semiconductors, while the broader energy band gap, i.e., size > 4 eV, is found in insulators. Fermi energy for these two band structures lies within the band gap near its centre.

Metals have high conductivities because of the large number of free electrons that have been excited into empty states above the Fermi energy. If the metal crystal lattices were perfect and there are no lattice vibrations, the electrons would pass through the lattice unscattered, encountering no resistance. However, lattice vibrations and phonon scattering play a role in disrupting the mean free path of electrons. In addition, crystalline defects and impurity atoms affect the conductivity. These scattering mechanisms act independently on one another.

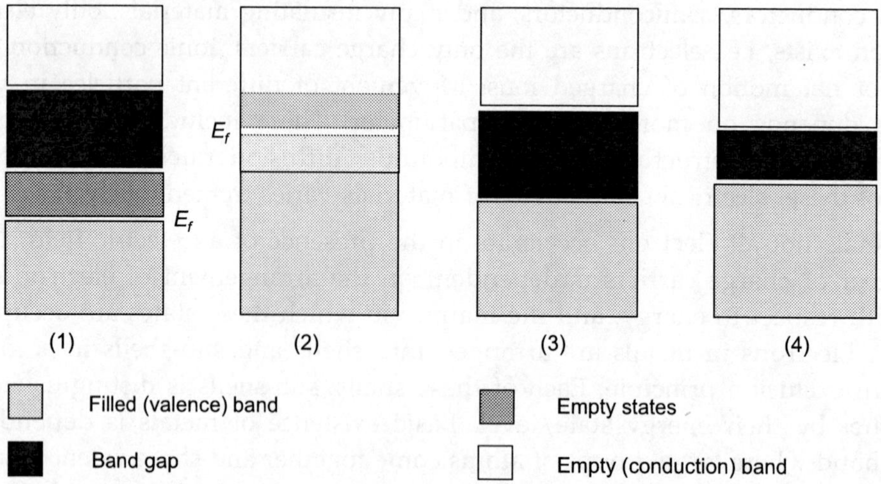

Figure 5.9 Possible electron band structures.

5.5.2 Semiconductivity

Conductivity of semiconductors is not as high as metals/channels. Notwithstanding, their electrical properties are one of a kind, as in their electrical properties are to a great degree touchy to even moment convergences of debasements. For characteristic semiconductors, their electrical conduct depends on inalienable electronic structure of the unadulterated material. Then again, if the electrical properties are dominated by impurities, they are called extrinsic semiconductors.

Intrinsic semiconduction: Energy band gap between the valence and conduction bands can be used to classify materials as conductors, insulators, and semiconductors. The magnitude of the band gap is characteristic of the lattice alone, and varies widely for different crystals. In semiconductors, the valence and conduction bands do not overlap as in metals, but they possess enough electrons in the valence band those can be promoted to the conduction band at a certain temperature. Promotion of electrons leaves behind positively charged holes in the valence band which maintain the charge neutrality. These holes can be regarded as moving positive charges through the crystal with a charge magnitude as that of an electron. Conduction that arises from thermally/optically excited electrons is called intrinsic semiconduction. It usually takes place at elevated temperatures since sufficient thermal agitation is necessary to transfer a reasonable number of electrons from the valence band to the conduction band. At still higher temperatures, the concentration of thermally excited electrons in the conduction band becomes so high that the semiconductor behaves more like a metal. The elements capable of intrinsic conduction are limited, and include: Si, Ge, Sn, Pb, etc.

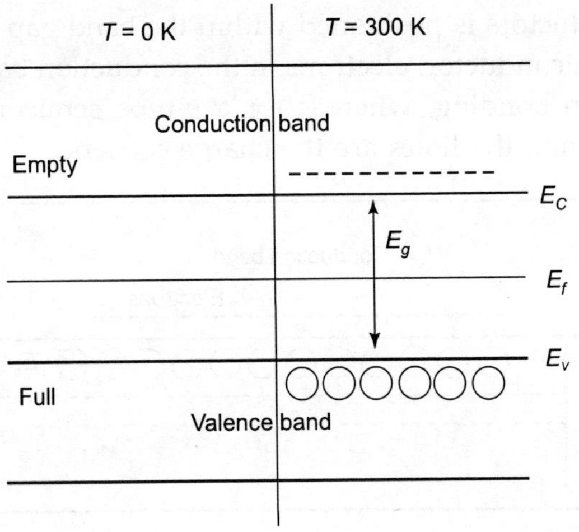

Figure 5.10 Band structure and intrinsic conductivity.

Extrinsic Semiconduction: The charge carrier density can also be increased by adding impurities of either higher or lower valence. This addition of impurities is known as doping, and impure atoms in the element are called donor atoms.

When a pentavalent substitutional atoms like P, As, or Sb are placed into a covalently bonded tetravalent materials such as Si or Ge, only four of their five valence electrons are required to participate in covalent bonding, while the fifth electron remains weakly bound to the donor atom. However, this binding energy is in order of 0.01 eV, much less than that of a covalently bonded electron. The energy state of this electron is indicated as E_d. Energy gap between the donor and conducting bands is much smaller than that between the valence band and conduction band, E_g. Thermal agitation, even at room temperature, is good enough to excite this electron to the conduction band, leaving behind a positively charged hole in the donor band. Conductivity is due to motion of negatively charged electrons in the conduction band, and so the material is called *n-type semiconductor*. Fermi level (E_f) of *n*-type extrinsic semiconductor is higher than that of intrinsic semiconductor, i.e., it shifts upward in the band gap, within the vicinity of the donor state; and its exact position is a function of both temperature and donor concentration.

Other possibility of susbstituion involves impurity with lower valence. For example, trivalent elements like B, Al, Ga, In, Th doped to Si or Ge. Since the donor atoms are deficient in bonding electrons, one of their bonding orbitals will contain a hole that is capable of accepting an electron from elsewhere in the crystal. Binding energy is small, and promotion of an electron from the valence band to the acceptor band, E_a, leaves a positively charged hole in the valence band that can act as a charge carrier. Materials with this *p*-type extrinsic semiconduction are called *p-type semiconductors*. Fermi level

(E_f) for p-type semiconductors is positioned within the band gap and near the acceptor level. For an n-type semiconductor, electrons in the conduction band are charge carriers and holes participate in bonding, whereas for a p-type semiconductor, the electrons participate in bonding, and the holes are the charge carriers.

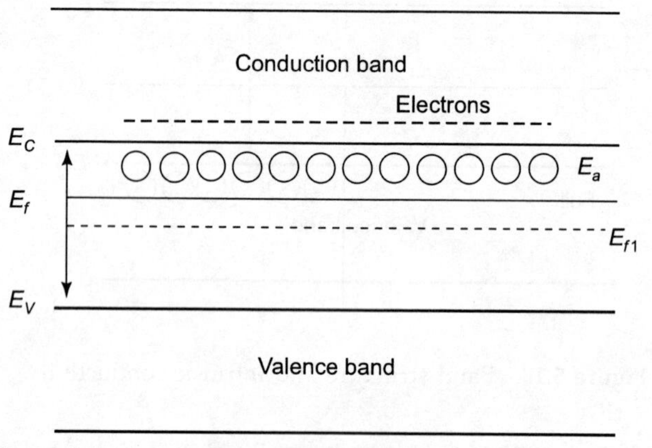

Figure 5.11 Band structure and n-type semiconductivity.

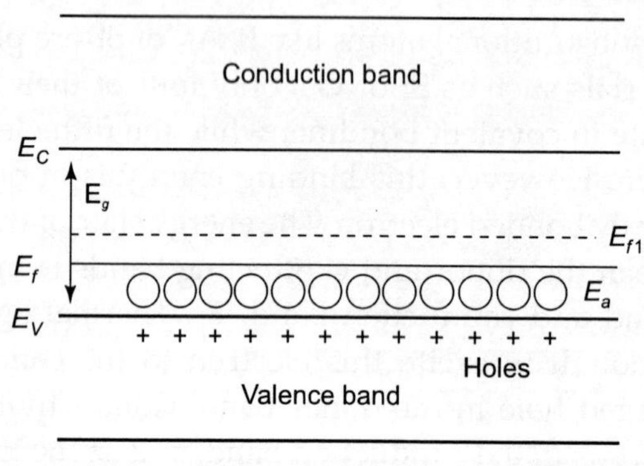

Figure 5.12 Band structure and p-type semiconductivity.

In both types of extrinsic semiconductors, the doping elements are chosen so that both acceptor and donor levels are located closer to the corresponding energy bands and only a small energy gap in involved when exciting electrons. Intrinsic semiconductors are pure to the order greater than 10^7, while extrinsic semiconductors are made with impurity levels in order of 1-1000 ppm.

In *intrinsic semiconductors*, the conductivity is dominated by the exponential relation between temperature and band gap. However, the conductivity of extrinsic semiconductors is governed by competing forces: charge carrier density and charge carrier mobility. At low temperatures, the number of charge carriers initially increases with temperature. At intermediate temperatures, for n-type extrinsic semiconductor, most of the donor electrons have been promoted and the number of charge carriers is nearly independent of temperature in what is known as the exhaustion range. At higher temperatures, the number of valence electrons excited to the conduction band greatly exceeds the total number of electrons from the donor atoms, and extrinsic semiconductor behaves like an intrinsic semiconductor. That is, at low temperatures the conductivity varies with temperature as the charge carrier concentration, at higher temperatures the charge carrier mobility dominates and the conductivity decreases with temperature, and at still higher temperatures the extrinsic semiconductor behaves like an intrinsic semiconductor.

5.5.3 Electron Mobility

When an electric field is applied, a force is brought to bear on the free electrons; as a consequence, they all experience an acceleration in a direction opposite to that of the field, by virtue of their negative charge. According to quantum mechanics, there is no interaction between an accelerating electron and atoms in a perfect crystal lattice. Under such circumstances all the free electrons should accelerate as long as the electric field is applied, which would give rise to an electric current that is continuously increasing with time. However, we know that a current reaches a constant value the instant that a field is applied, indicating that there exist what might be termed "frictional forces," which counter this acceleration from the external field. These frictional forces result from the scattering of electrons by imperfections in the crystal lattice, including impurity atoms, vacancies, interstitial atoms, dislocations, and even the thermal vibrations of the atoms themselves. Each scattering event causes an electron to lose kinetic energy and to change its direction of motion. There is, however, some net electron motion in the direction opposite to the field, and this flow of charge is the electric current.

The scattering phenomenon is manifested as a resistance to the passage of an electric current. Several parameters are used to describe the extent of this scattering; these include the *drift velocity* v_d and the **mobility** of an electron. The drift velocity represents the average electron velocity in the direction of the force imposed by the applied field. It is directly proportional to the electric field.

5.6 THERMAL PROPERTIES OF MATERIALS

Engineering materials are essential in regular daily existence in light of their flexible basic properties. Other than these properties, they do assume an imperative part in light of their physical properties. Prime physical properties of materials include: electrical properties, thermal properties, magnetic properties, and optical properties. Determination of materials for use at elevated temperatures or potentially temperature changes require a specialist to know and comprehend their thermal properties. This section manages the thermal properties of materials. Physical property of a strong body identified with utilization of thermal properties is characterized as a thermal property.

This area might portray thermal properties like heat capacity, thermal expansion, thermal conductivity, and thermal stresses. It is vital to know and comprehend the idea of concept of thermal expansion which is the root cause for thermal stresses. Thermal stresses are stresses leading to failure of engineering structures at elevated temperatures.

By "thermal property" is meant the response of a material to the application of heat. As a solid absorbs energy in the form of heat its temperature rises and its dimensions increase. The energy may be transported to cooler regions of the specimen if temperature gradients exist, and ultimately, the specimen may melt. Heat capacity, thermal expansion, and thermal conductivity are properties that are often critical in the practical utilization of solids.

5.6.1 Heat Capacity

Many engineering solids when exposed to heat experiences an increase in temperature, i.e., it absorbs heat energy. This property of a material, i.e., material's ability to absorb heat energy is called its *heat capacity*, C. It is defined as the energy required for change a material's temperature by one degree. Mathematically, it is expressed as:

$$C = \frac{dQ}{dT}, \text{ J/mol-K} \quad \text{or} \quad \text{Cal/mol-K}$$

where, dQ is the energy required to produce a temperature change equal to dT.

Heat capacity is not an intrinsic property, i.e., total heat a material can absorb depends on its volume/mass. Hence, another parameter called *specific heat, c*, it defined as heat capacity per unit mass (J/kg-K, Cal/kg-K).

With increase of internal energy, geometrical changes may occur. Accordingly, heat capacity is measured either at constant volume, C_v, or at constant external pressure, C_p. The magnitude of C_p is always greater than C_v but only marginally.

Heat energy absorption of a (solid, liquid or gaseous) material exists in the mode of thermal energy vibrations of constituent atoms or molecules, apart from other mechanisms of heat absorption such as electronic contribution. With increase of energy, atoms vibrate at higher frequencies. However, the vibrations of adjacent atoms are coupled through atomic bonding, which may lead to movement of lattices. This may be represented as elastic waves (phonon) or sound waves. Vibrational contribution of heat capacity of solids varies with temperature according to the following relation at low temperatures:

$$C_v = AT^3$$

C_v is equal to zero at 0 K, but increases rapidly with temperature. It represents increased ability of atomic vibrations or enhanced energy of lattice waves with ascending temperature. Above a temperature called Debye temperature, θ_d, dependence of volumetric heat capacity value reaches saturation. This saturation value if approximately equal to 3R (≈ 6 cal/mol-K), R being the universal gas constant. For many solids, value of θ_d is below the room temperature.

In most solids the principal mode of thermal energy assimilation is by the increase in vibrational energy of the atoms. Again, atoms in solid materials are constantly vibrating at very high frequencies and with relatively small amplitudes. Rather than being independent of one another, the vibrations of adjacent atoms are coupled by virtue of the atomic bonding. These vibrations are coordinated in such a way that travelling lattice waves are produced. These may be thought of as elastic waves or simply sound waves, having short wavelengths and very high frequencies, which propagate through the crystal at the velocity of sound. The vibrational thermal energy for a material consists of a series of these elastic waves, which have a range of distributions and frequencies. Only certain energy values are allowed (the energy is said to be quantized), and a single quantum of vibrational energy is called a **phonon.** (A phonon is analogous to the quantum of electromagnetic radiation, the **photon.**) On occasion, the vibrational waves themselves are termed phonons.

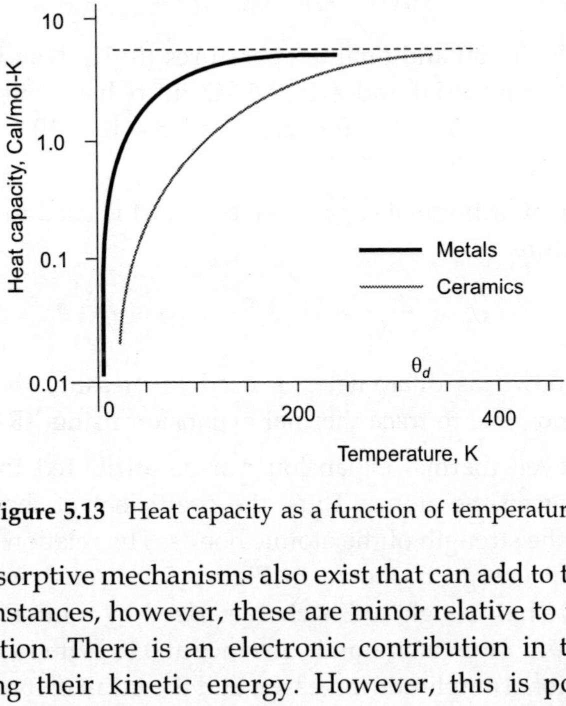

Figure 5.13 Heat capacity as a function of temperature.

Other energy-absorptive mechanisms also exist that can add to the total heat capacity of a solid. In most instances, however, these are minor relative to the magnitude of the vibrational contribution. There is an electronic contribution in that electrons absorb energy by increasing their kinetic energy. However, this is possible only for free electrons—those that have been excited from filled states to empty states above the Fermi energy. In metals, only electrons at states near the Fermi energy are capable of such transitions, and these represent only a very small fraction of the total number. An even smaller proportion of electrons experiences excitations in insulating and semiconducting materials. Hence, this electronic contribution is ordinarily insignificant, except at temperatures near 0 K. Furthermore, in some materials other energy-absorptive processes occur at specific temperatures, for example, the randomization of electron spins in a ferromagnetic material as it is heated through its Curie temperature. A large spike is produced on the heat capacity-versus-temperature curve at the temperature of this transformation.

5.6.2 Thermal Expansion

After heat absorption, vibrating atoms behave as though they have larger atomic radius, which leads to increase in materials dimensions. The phenomenon is called thermal expansion. It is quantified in terms of thermal expansion coefficient. Linear coefficient of thermal expansion (α) defined as the change in the dimensions of the material per unit length, and is expressed as:

$$\alpha = \frac{l_f - l_o}{l_o(T_f - T_o)} = \frac{\Delta l}{l_o \Delta T} = \frac{\varepsilon}{\Delta T}$$

where T_0 and T_f are the initial and final temperatures (in K), l_0 and l_f are the initial and final dimensions of the material and ε is the strain. α has units as $(°C)^{-1}$. For most metals α is the range of $5 - 25 \times 10^{-6}$, for ceramics $0.5 - 15 \times 10^{-6}$, and for polymers $50 - 400 \times 10^{-6}$.

A volume coefficient of thermal expansion, α_v (= 3α) is used to describe the volume change with temperature.

$$\alpha_v = \frac{\Delta v}{v_o \Delta T}$$

An instrument known as dilatometer is used to measure the thermal expansion coefficient. It is also possible to trace thermal expansion using XRD.

At microscopic level, thermal expansion can be attributed to the increase in the average distance between the atoms. Thus, the coefficient of thermal expansion of a material is related to the strength of the atomic bonds. The relation between interatomic distance and potential energy is shown in Figure 5.14. As shown in the figure, the potential energy curve has trough and is characterized by a minimum. This minimum corresponds to the equilibrium interatomic distance at 0 K. With increase in temperature of the material, vibrational amplitude increases, and the mean of it represents the average inter-atomic distance. Because of asymetric nature of the potential energy trough, rather than the increase in atomic vibration amplitude, the average inter-atomic distance increases with the temperature of a material.

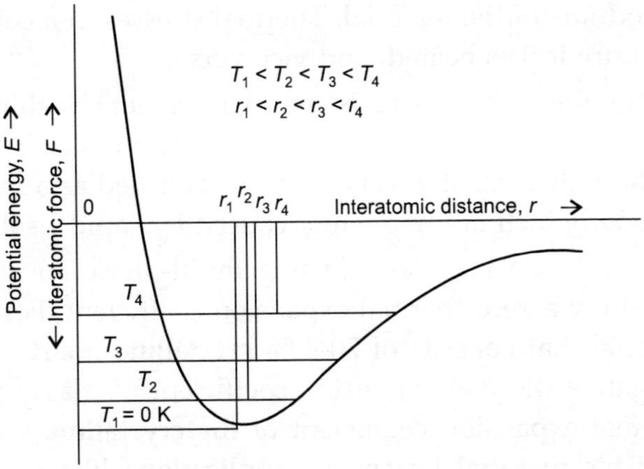

Figure 15.14 Change in interatomic distance with temperature.

Thermal expansion is due to the asymmetric curvature of this potential energy trough, rather than the increased atomic vibrational amplitudes with rising temperature. If the potential energy curve were symmetric there would be no net change in interatomic separation and, consequently, no thermal expansion.

Relatively strong interatomic bonding forces are found in many ceramic materials as reflected in comparatively low coefficients of thermal expansion; values typically range between about 0.5×10^{-6} and 15×10^{-6} (°C^{-1}). For noncrystalline ceramics and also those having cubic crystal structures, it is isotropic. Otherwise, it is anisotropic; and, in fact, some ceramic materials, upon heating, contract in some crystallographic directions while expanding in others. For inorganic glasses, the coefficient of expansion is dependent on composition. Fused silica (high-purity SiO_2 glass) has a small expansion coefficient, 0.4×10^{-6} (°C^{-1}). This is explained by a low atomic packing density such that interatomic expansion produces relatively small macroscopic dimensional changes.

Some polymeric materials experience very large thermal expansions upon heating as indicated by coefficients that range from approximately 50×10^{-6} to 400×10^{-6} (°C^{-1}) the highest values are found in linear and branched polymers because the secondary intermolecular bonds are weak, and there is a minimum of cross-linking. With increased cross-linking, the magnitude of the expansion coefficient diminishes; the lowest coefficients are found in the thermosetting network polymers such as phenol-formaldehyde, in which the bonding is almost entirely covalent.

5.6.3 Thermal Stresses

Another of problem exists with thermal expansion of a material where there is no scope of dimensional changes. Thus, due to temperature changes, material may experience thermal stresses ($\sigma_{thermal}$).

$$\sigma_{thermal} = \alpha E \Delta T$$

where, E is elastic modulus of the material. Thermal stresses in a constrained body will be of compressive nature if it is heated, and vice versa.

Another source for thermal stresses is thermal gradient within the body when a solid body is heated or cooled. It is because temperature distribution will depend on its size and shape. These thermal stresses may be established as a result of temperature gradients across a body, which are frequently caused by rapid heating or cooling.

Engineering materials can be tailored using multi-phase constituents so that the overall material can show a zero thermal expansion coefficient. For example, Zerodur, a glass-ceramic material that consists of 70-80% crystalline quartz, and the remaining as glassy phase. Negative thermal expansion coefficient of glassy phase compensates for the positive thermal expansion coefficient of the crystalline base, leading to zero thermal expansion. This material has many applications like parts of telescopes in astronomy. Many ceramic materials for critical thermal applications are developed based on sodium-zirconium-phosphate (NZP) that have a near-zero thermal expansion coefficient.

When a solid body is heated or cooled, the internal temperature distribution depends on its size and shape, the thermal conductivity of the material, and the rate of temperature change. Thermal stresses may be established as a result of temperature gradients across a body, which are frequently caused by rapid heating or cooling, in that the outside changes temperature more rapidly than the interior; differential dimensional changes serve to restrain the free expansion or contraction of adjacent volume elements within the piece. For example, upon heating, the exterior of a specimen is hotter and, therefore, will have expanded more than the interior regions. Hence, compressive surface stresses are induced and are balanced by tensile interior stresses. The interior-exterior stress conditions are reversed for rapid cooling such that the surface is put into a state of tension.

5.7 MAGNETIC PROPERTIES OF MATERIALS

Engineering materials are important in everyday life because of their versatile structural properties. Other than these properties, they do play an important role because of their physical properties. Prime physical properties of materials include: electrical properties, thermal properties, magnetic properties, and optical properties. Magnetic properties play an important role in daily life. Magnetic materials are used in electric motors, generators, transformers. Modern-day devices use data storage technology that is based on magnetic particles. Magnetic materials are used in devices like telephones, televisions, supercomputers, etc. they are also used in medical technology, for example, DNA sequencing.

This section shall describe magnetic properties like dia-, para-, and ferro-magnetism along with antiferro- and ferrimagnetism. It also includes discussion about temperature effects on magnetic behaviour, followed by magnetic domains and hysteresis.

Magnetism is a phenomenon by which a material exerts either attractive or repulsive force on another material. Basic source of magnetic force is movement of electrically charged particles. Magnetic behaviour of a material can be traced to the structure of atoms. Electrons in atoms have a planetary motion in that they go around the nucleus. This orbital motion and its own spin cause separate magnetic moments, which contribute to the magnetic behaviour of materials. Thus, every material can respond to a magnetic field. However, the manner in which a material responds depend much on its atomic structure, and determines whether a material will be strongly or weakly magnetic.

Magnetic moment due to spin of an electron is known as *Bohr magneton*, M_B, is the most fundamental magnetic moment.

$$M_B = \frac{qh}{4\pi m_e} = 9.274 \times 10^{-24} \text{ A.m}^2$$

where, h is Planck's constant, q is the charge on the electron, m_e is mass of electron.

This moment is directed along the axis of electron spin. Magnetic moment resulted from a magnetic moment of electrons could sum up, then the world would be a magnetic place. Fortunately, two reasons are found to explain this phenomenon: First: according to *Pauli exclusion rule*, two electrons with same energy level must have opposite spins – thus so are their magnetic moments, which cancel out each other. Second: orbital moments of electrons also cancel out each other – thus no net magnetic moments if there is no unpaired electron(s). Some elements such as transition elements, lanthanides, and actinides have a net magnetic moment since some of their energy levels have an unpaired electron.

Magnetic dipoles are found to exist in magnetic materials, analogous to electric dipoles. A magnetic dipole is a small magnet composed of north and south poles instead of positive and negative charges. Within a magnetic field, the force of field exerts a torque that tends to orient the dipoles with the filed. Magnetic forces are generated by moving electrically charged particles. These forces are in addition to any electrostatic forces that may already exist. It is convenient to think magnetic forces in terms of distributed field, which is represented by imaginary lines. These lines also indicate the direction of the force. If a magnetic field is generated by passing current I through a coil of length l and number of turns n, then the magnetic field strength is given by:

$$H = \frac{nI}{l}$$

Units for magnetic field strength, thus, are A/m.

Magnetic flux density (induction) is the measure of lines within a medium. It has units as weber (Wb) /m^2 or tesla.

$$B = \mu H$$

where, μ is permeability. It is a specific property of the medium, and has units as Wb/A.m or henries (H) /m.

Flux density, B, is determined by the manner in which induced and permanent dipoles interact with the applied field. If the magnetic moments reinforce the applied field, i.e., greater number of lines of flux are created, and the field is magnified. This is represented as

$$\mu > \mu_o$$

where, μ_o is magnetic permeability of vacuum.

Several parameters may be used to describe magnetic properties of solids. One of them is relative permeability (μ_r). It is a measure of the degree to which the material can be magnetized.

$$\mu_r = \frac{\mu}{\mu_o}$$

Another field quantity called magnetization, M, is defined as

$$B = \mu_o H + \mu_o M = \mu_o \mu_r H$$

$$M = \chi_m H$$

$$\chi_m = \mu_r - 1$$

χ_m is called the magnetic susceptibility.

There are numerous ways a material can be polarized, i.e., numerous sorts of attraction. Three fundamental attractions are: dia-attraction, para-attraction and ferro-magnetisms. Against ferromagnetisms and ferri-attractions are considered subclasses of ferromagnetisms. A material displays one of these magnetisms.

5.7.1 Dia-, Para-, and Ferromagnetism

Diamagnetism is very weak form of magnetism. It exists only when an external field is applied, and is non-permanent. The applied external field acts on atoms of a material, slightly unbalancing their orbiting electrons, and creates small magnetic dipoles within atoms which oppose the applied field. This action produces a negative magnetic effect known as diamagnetism. The induced magnetic moment is small, and the magnetization (M) direction is opposite to the direction of applied field (H). Thus, the relative permeability is less than unity, i.e., magnetic susceptibility is negative, and is in order of -10^{-5}. Diamagnetism is virtually found in all materials; however, it is observable only in the absence of other magnetisms. This form of magnetism is of no practical importance. Materials such as Cu, Ag, Si, Ag and alumina are diamagnetic at room temperature. Superconductors are perfect diamagnets ($\chi_m = -1$); they lose their superconductivity at higher temperatures or in the presence of a magnetic field.

Materials which exhibit a small positive magnetic susceptibility in the presence of a magnetic field are called paramagnetic, and the effect is termed *paramagnetism.* When materials have unpaired electrons, a net magnetic moment due to electron spin is associated with each atom. In the absence of an external field, the orientations of these atomic magnetic moments are random leading to no net magnetization. When an external field is applied dipoles line-up with the field, resulting in a positive magnetization. However, because the dipoles do not interact, extremely large magnetic fields are required to align all of the dipoles. In addition, the effect is lost as soon as the magnetic field is removed. Since thermal agitation randomizes the directions of the magnetic dipoles, an increase in temperature decreases the paramagnetic effect. Paramagnetism is produced in many materials like aluminium, calcium, titanium, alloys of copper. Magnetic susceptibility of these materials is slightly positive, and lies in the range 10^{-5} to 10^{-2}.

Both dia- and paramagnetic materials are considered non-magnetic because they exhibit magnetization only in the presence of an external field. Certain materials possess permanent magnetic moments even in the absence of an external field. This is the result of permanent unpaired dipoles formed from unfilled energy levels. These dipoles can easily line up with the imposed magnetic field due to the exchange interaction or mutual reinforcement of the dipoles. These are characteristics of *ferromagnetism.* Materials with ferromagnetism (Examples: Fe, Co, Ni, Gd) possess magnetic susceptibilities approaching 10^6. Consequently,

$$H << M$$

$$B \cong \mu_o M$$

The mutual spin alignment in ferromagnetic materials exists over relatively large volume regions of the crystal called *domains.* The susceptibility of ferromagnetic materials depends on the intensity of the applied field. Above the Curie temperature, ferromagnetic materials behave as paramagnetic materials and their susceptibility is given by the Curie-Weiss law, defined as follows:

$$\chi_m = \frac{C}{T - T_c}$$

where C is material constant, T is temperature, T_c is Curie temperature.

In some materials such as Mn, Cr, MnO, NiO, CoO, $MnCl_2$ the magnetic moments produced in neighbouring dipoles line up in opposition to one another in the magnetic field, even though the strength of each dipole is very high. This results in zero magnetization, and the effect is called *anti-ferromagnetism.* Exchange interaction which is responsible for parallel alignment of spins is extremely sensitive to inter-atomic spacing and to the atomic positions. This sensitivity causes antiparallel alignment of spins. When the strength of antiparallel spin magnetic moments is equal, no net spin moment exists, and resulting susceptibilities are quite small. One noticeable characteristic of anti-ferromagnets is that they attain maximum susceptibility at a critical temperature called *Neel temperature*. At temperatures above this, anti-ferromagnets become paramagnetic.

On the other hand, some ceramic materials exhibit net magnetization. This is because of either the strength or number of opposing dipoles is not equal. In a magnetic field, the dipoles of a cation may line up with the field, while dipoles of other cation may not. These ceramics are called ferrites, and the effect is known as *ferrimagnetism*. Ferrimagnetism is similar to anti-ferromagnetism in that the spins of different atoms or ions line up antiparallel. However, the spins do not cancel each other out, and a net spin moment exists. Below the Neel temperature, therefore, ferromagnetic materials behave very much like ferromagnetic materials and are paramagnetic above the Neel temperature. These materials exhibit a large but field dependent magnetic susceptibility similar to ferromagnets. They also show Curie-Weiss behaviour. As these ceramics are good insulators, electrical losses are minimal, and hence ferrites have lot of applications in devices such as high frequency transformers. Examples: Fe_3O_4, $NiFe_2O_4$, (Mn.Mg) Fe_2O_4, $PbFe_{12}O_{19}$, Ba $Fe_{12}O_{19}$, YIG – yttrium iron garnet $Y_3Fe_5O_{12}$.

Table 5.4 compares different magnetisms.

Table 5.4 Various types of magnetisms

Magnetism	Magnetic Susceptibility		Examples
	Sign	Magnitude	
Dia	–	Small, Constant	Organic materials, superconducting materials, metals like Bi
Para	+	Small, Constant	Alkali and transition metals, rate earth elements
Ferro	+	Large, Function of *H*	Transition metals (Fe, Ni, Co), rate earth elements (Gd)
Anti-Ferro	+	Small, Constant	Salts of transition elements (MnO)
Ferri	+	Large, Function of *H*	Ferrites ($MnFe_2O_4$, $ZnFe_2O_4$) and chromites

5.7.2 Influence of Temperature on Magnetic Behaviour

Temperature influences the attractive qualities of materials, as it impacts electrical properties. With rising temperature, extent of the particle heat vibrations increments. This may prompt to more randomization of nuclear attractive minutes as they are allowed to pivot.

Usually, atomic thermal vibrations counteract forces between the adjacent atomic dipole moments, resulting in dipole misalignment up to some extent both in the presence and absence of external field. As a consequence of it, saturation magnetization initially decreases gradually, then suddenly drops to zero at a temperature called *Curie temperature*, T_c. The magnitude of the Curie temperature depends on the material. For example: for cobalt – 1120°C, for nickel – 335°C, for iron – 768°C, and for Fe_3O_4 – 585°C.

5.7.3 Domains and Hysteresis

In addition to susceptibility differences, the different types of magnetism can be distinguished by the structure of the magnetic dipoles in regions called domains. Each domain consists of magnetic moments that are aligned, giving rise to a permanent net magnetic moment per domain. Each of these domains is separated from the rest by domain boundaries/domain walls. Boundaries, also called *Bolch walls*, are narrow zones in which the direction of the magnetic moment gradually and continuously changes from that of one domain to that of the next. The domains are typically very small about 50 μm or less, while the Bloch walls are about 100 nm thick. For a polycrystalline specimen, each grain may have more than one microscopic sized domain.

Domains exist even without an outer field. In a material that has never been presented to an attractive field, the individual spaces have an arbitrary introduction. This sort of plan speaks to the most reduced free vitality. The area structure of material is dictated by numerous sorts of energies, with most stable structure being achieved when the general potential energy of the material is a minimum.

Total magnetic energy of a material is the sum of the contributions of the following: exchange energy, magneto-static energy, magneto-crystalline anisotropy energy, domain wall energy and magnetostrictive energy. When the bulk material is un-magnetized, the net magnetization of these domains is zero, because adjacent domains may be orientated randomly in any number of directions, effectively canceling each other out.

The normal attractive enlistment of a ferro-attractive material is personally identified with the space structure. At the point when an attractive field is forced on the material, areas that are almost agreed with the field develop to the detriment of unaligned spaces. This procedure proceeds until just the most positively arranged areas remain. Altogether for the areas to develop, the Bloch dividers must move, the outer field gives the compel required to this moment.

When the domain growth is completed, a further increase in the magnetic field causes the domains to rotate and align parallel to the applied field. At this instant the material reaches saturation magnetization and no further increase takes place on increasing the strength of the external field. Saturation magnetization is the greatest amount of magnetization that the material can obtain. Under these conditions the permeability of these materials becomes quite small. Figure 5.15 in the following presents the relation between the applied field strength and magnetization of the material.

As shown in Figure 5.15, magnetization increases with applied field, and reaches a saturation value in a sufficiently stronger field. Once magnetic saturation has been achieved, a decrease in the applied field back to zero results in a macroscopically permanent or residual magnetization, known as *remanance*, M_r. The corresponding induction, B_r is called *retentivity* or *remanent induction* of the magnetic material. This effect of retardation by material is called *hysteresis*. The material acts as a permanent magnet, even at zero applied field. At this point, spin orientations within domains have readily rotated back to their favourable position, but the original random domain arrangement is not achieved. This is on account of the resistance offered by the space dividers averts redevelopment of the areas, i.e., area development process is not by any stretch of the imagination reversible, and space divider movement is constrained. The attractive field quality expected to convey the instigated charge to zero is named as coercivity, *Hc*. This must be connected hostile to parallel to the original field.

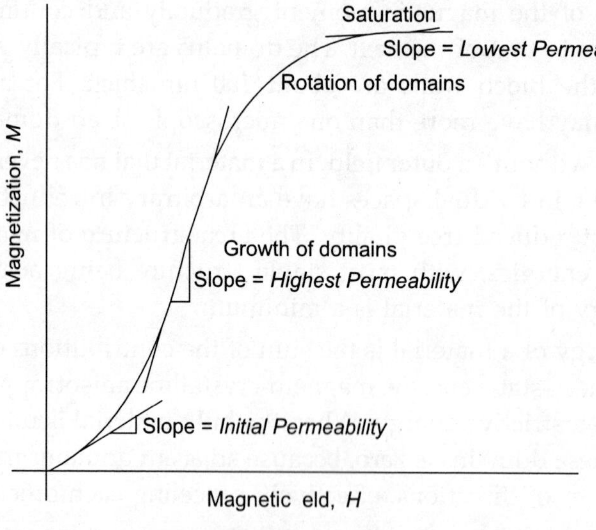

Figure 5.15 Magnetization saturation of a material with applied field.

A further increment in the field the other way brings about the most extreme acceptance, i.e., immersion charge, however the other way. The field can by and by be turned around, and the field-magnetization loop can be shut, thus the comparing field-induction loop.

This loop is known as *hysteresis loop* or *B-H plot* or *M- H plot*. The region inside the hysteresis loop speaks to the vitality misfortune per unit volume of material for one cycle. The B-versus H bend in Figure 5.16 speaks to a hysteresis loop taken to immersion. Notwithstanding, it is not important to build the field quality to immersion to create the loop.

The coercivity of a material is a microstructure sensitive property. This dependence is known as *magnetic shape anisotropy*. The coercivity of recording materials needs to be smaller than that for others since data written onto a data storage medium should be erasable. On the other hand, the coercivity values should be higher since the data need to be retained. Thus, such materials are called magnetically semi-hard. Examples: Hard ferrites based on Ba, CrO_2, γ-Fe_2O_3; alloys based on Co-Pt-Ta-Cr, Fe-Pt and Fe-Pd, etc.

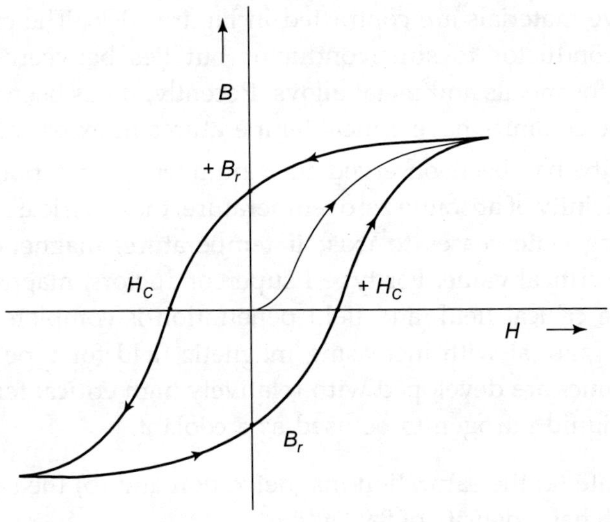

Figure 5.16 Schematic presentation of a typical hysteresis loop.

Soft magnets are characterized by low coercive forces and high magnetic permeabilities; and are easily magnetized and demagnetized. They generally exhibit small hysteresis losses. Application of soft magnets include: cores for electromagnets, electric motors, transformers, generators, and other electrical equipment. Examples: ingot iron, low-carbon steel, Silicon iron, superalloy (80% Ni-5% Mo-Fe), 45 Permalloy (55%Fe-45%Ni), 2-79 Permalloy (79% Ni-4% Mo-Fe), MnZn ferrite / Ferroxcube A (48% $MnFe_2O_4$-52%$ZnFe_2O_4$), NiZn ferrite / Ferroxcube B (36% $NiFe_2O_4$-64% $ZnFe_2O_4$), etc.

Hard magnets are characterized by high remanent inductions and high coercivities. These are also called *permanent magnets* or *hard magnets*. They generally exhibit large hysteresis losses. Examples: Co-steel, Tungsten steel, $SmCo_5$, $Nd_2Fe_{14}B$, ferrite Bao.6Fe_2O_3, Cunife (60% Cu 20% Ni-20% Fe), Alnico (alloy of Al, Ni, Co and Fe), etc. Applications include fractional horse-power motors, automobiles, audio- and video-recorders, earphones, computer peripherals, and clocks.

Superconductivity is basically an electrical phenomenon; however, its discussion has been deferred to this point because there are magnetic implications relative to the superconducting state, and, in addition, superconducting materials are used primarily in magnets capable of generating high fields. As most high-purity metals are cooled down to temperatures nearing 0K, the electrical resistivity decreases gradually, approaching some small yet finite value that is characteristic of the particular metal. There are a few materials, however, for which the resistivity, at a very low temperature, abruptly plunges from a finite value to one that is virtually zero and remains thereupon further cooling. Materials that display this latter behaviour are called *superconductors,* and the temperature at which they attain **superconductivity** is called the critical temperature, T_c. Resistivity– temperature behaviours for superconductive and non-superconductive materials are contrasted in Figure 20.26. The critical temperature varies from superconductor to superconductor but lies between less than 1K and approximately 20K for metals and metal alloys. Recently, it has been demonstrated that some complex oxide ceramics have critical temperatures in excess of 100K.

Superconductivity has been observed in a number of materials, in which, upon cooling and in the vicinity of absolute zero temperature, the electrical resistivity vanishes. The superconducting state ceases to exist if temperature, magnetic field, or current density exceeds the critical value. For type I superconductors, magnetic field exclusion is complete below a critical field, and field penetration is complete once is exceeded. This penetration is gradual with increasing magnetic field for type II materials. New complex oxide ceramics are developed with relatively high critical temperatures, which allow inexpensive liquid nitrogen to be used as a coolant.

Example 5.5 Calculate (a) the saturation magnetization and (b) the saturation flux density for nickel, which has a density of 8.90 g/cm^3.

Solution:

(a) The saturation magnetization is just the product of the number of Bohr magnetons per atom (0.60 as given above), the magnitude of the Bohr magneton μ_B, and the number N of atoms per cubic metre or

$$M_s = 0.60\ \mu_B N$$

Now, the number of atoms per cubic metre is related to the density ρ, the atomic weight A_{Ni}, and Avogadro's number N_A, as follows:

$$N = \frac{\rho N_A}{A_{Ni}}$$

$$= \frac{(8.90 \times 10^6\ \text{g/m}^3)(6.023 \times 10^{23}\ \text{atoms/mol})}{58.71\ \text{g/mol}}$$

$$= 9.13 \times 10^{28}\ \text{atoms/m}^3$$

Finally,

$$M_s = \left(\frac{0.60\ \text{Bohr magneton}}{\text{atom}}\right)\left(\frac{9.27 \times 10^{-24}\ \text{A-m}^2}{\text{Bohr magneton}}\right)\left(\frac{9.13 \times 10^{28}\ \text{atoms}}{\text{m}^3}\right)$$

$$= 5.1 \times 10^5\ \text{A/m}$$

(b) The saturation flux density is given by

$$B_s = \mu_0 M_s$$

$$= \left(\frac{4\pi \times 10^{-7}\ \text{H}}{\text{m}}\right)\left(\frac{5.1 \times 10^{5}\ \text{A}}{\text{m}}\right)$$

$$= 0.64 \text{ tesla}$$

Example 5.6 Calculate the saturation magnetization for Fe_3O_4 given that each cubic unit cell contains 8 Fe^{2+} and Fe^{3+}16 ions, and that the unit cell edge length is 0.839 nm.

Solution: The saturation magnetization will be equal to the product of the number N' of Bohr magneton per cubic metre of Fe_3O_4, and the magnetic moment per Bohr magneton μ_B

$$M_s = N' \mu_B$$

Now , N' is just the number of Bohr magnetons per unit cell n_B divided by the unit cell volume V_C, or

$$N' = \frac{n_B}{V_C}$$

Again, the net magnetization results from the Fe^{2+} ions only. Since there are 8 Fe^{2+} ions per unit cell and 4 Bohr magnetons per Fe^{2+} ion, n_B is 32. Furthermore, the unit cell is a cube, and $V_C = a^3$, a being the unit cell edge length. Therefore,

$$M_s = \frac{n_B \mu_B}{a^3}$$

$$= \frac{(32 \text{ Bohr magnetons/ unit cell})\,(9.27 \times 10^{-24} \text{ A-m}^2/\text{Bohr magneton})}{(0.839 \times 10^{-9} \text{ m})^3 / \text{unit cell}}$$

$$= 5.0 \times 10^5 \text{ A/m}$$

5.8 OPTICAL PROPERTIES OF MATERIALS

Designing materials are critical in regular day-to-day existence on account of their adaptable basic properties. Other than these properties, they do assume a vital part in light of their physical properties. Prime physical properties of materials include: electrical properties, thermal properties, attractive properties, and optical properties. Optical properties assume an essential part in day-by-day life. They have significantly affected the improvement of the interchanges foundation and the data innovation. They are likewise helpful in fields like medicine, manufacturing, astronomy, etc. The objective of this segment is to present essential ideas about optical properties, optical properties of metals and non-metals, lastly uses of optical wonders.

Optical property of a material is related to the interaction of it with electromagnetic radiation. This radiation may have characteristics that fall in the visible light spectrum, or may be even out of it. Electromagnetic spectrum of radiation spans the wide range from γ-rays with wavelength as 10^{-12} m, through x-rays, ultraviolet, visible, infrared, and finally radiowaves with wavelengths as along as 10^5 m. Visible light is one form of electromagnetic radiation with wavelengths ranging from 0.39 to 0.77 μm. It contains colour bands from violet through red, as shown in Figure 5.17. White light is simply a mixture of all colours. The ultraviolet region covers the range from about 0.01 to about 0.40 μm, and the infrared region extends from about 0.75 to 1000 μm.

The genuine way of the light will most likely never be known. In any case, light can be considered having waves and comprising of particles called photons. The imperative attributes of the photons are connected by the accompanying condition. This condition permits us to consider the photon as a molecule of vitality, or as a wave with a characteristic wavelength and frequency.

$$E = h\nu = \frac{hc_o}{\lambda}$$

where, E is energy, h is Planck's constant (6.62×10^{-34} J.sec), ν is frequency, c_0 is speed of light in vacuum (3×10^8 m/sec), and λ is wavelength.

All materials communicate somehow with light. Communication of photons with the electronic or gem structure of a material prompts to various phenomenon. The photons may give their vitality to the material (assimilation); photons give their vitality, however, photons of indistinguishable vitality are promptly discharged by the material (reflection); photons may not cooperate with the material structure (transmission); or amid transmission photons are changes in velocity (refraction).

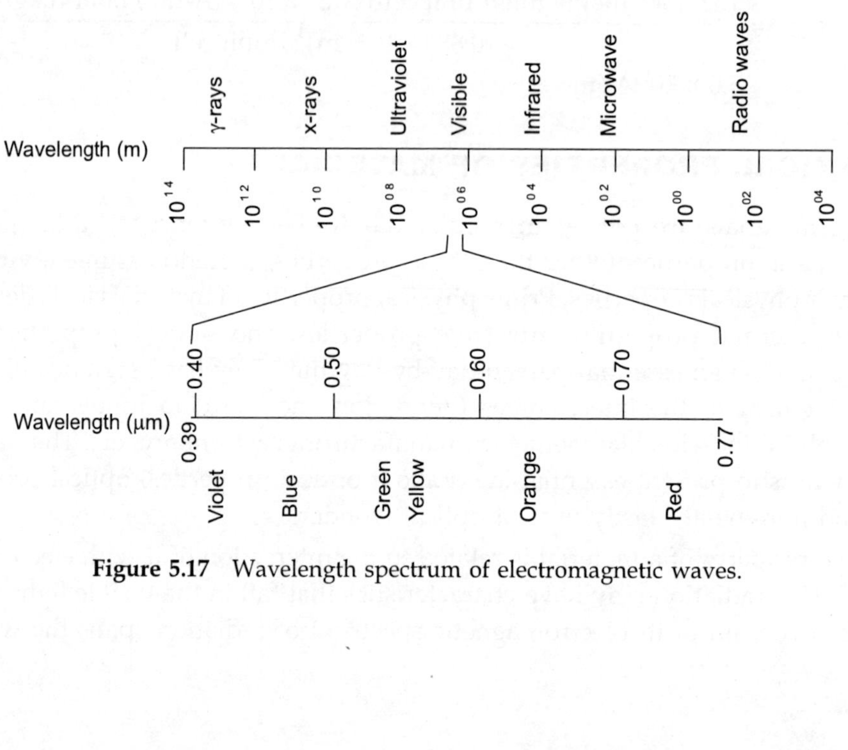

Figure 5.17 Wavelength spectrum of electromagnetic waves.

At any instance of light interaction with a material, the total intensity of the incident light striking a surface is equal to sum of the absorbed, reflected, and transmitted intensities, i.e.,

$$I_o = I_A + I_R + I_T$$

The intensity is defined as the number of photons impinging on a surface per unit area per unit time. Materials that are capable of transmitting light with relatively little absorption and reflection are called *transparent materials*, i.e., we can see through them. *Translucent materials* are those through which light is transmitted diffusely, i.e., objects are not clearly distinguishable when viewed through. Those materials that are impervious to the transmission of visible light are termed *opaque materials.*

5.8.1 Optical Properties of Metals and Alloys

A typical characteristic of metals as for crystal structure is that they have a high-energy band that is just in part loaded with electrons. At the point when noticeable light in coordinated on a metal surface, the vitality is utilized to energize electrons into abandoned vitality states over the Fermi level, subsequently making metals carry on as hazy materials, i.e., light is absorbed.

Except of thin sections, metals strongly reflect and/or absorb incident radiation for long wavelengths to the middle of the ultraviolet range, i.e., metals are opaque to all electromagnetic radiation on the low end of the frequency spectrum, from radiowaves, through infrared, visible, into middle of the ultraviolet radiation. However, metals are transparent to high end frequencies, e.g., x-ray and γ-ray radiation. Total absorption by metals is within a very thin outer layer, usually less than 0.1 μm; thus only metallic films thinner than 0.1 μm are capable of transmitting visible light.

The vast majority of the retained radiation is transmitted from the metallic surface as unmistakable light of an indistinguishable wavelength from the reflected light. The reflectivity of metals is around 0.95, while whatever is left of encroached vitality is disseminated as heat. The measure of vitality consumed by metals relies on the electronic structure of every specific metal. For instance, with copper and gold there is more noteworthy assimilation of the short wavelength hues, for example, green and blue and a more prominent impression of yellow, orange and red wavelengths. Different metals, for example, silver and aluminum emphatically mirror all parts of the unmistakable range and demonstrate a white silvery shading.

5.8.2 Optical Properties of Non-metallic Materials

By excellence of their electron structure with characteristic energy band structures, non-metallic materials might be transparent to visible lights. Along these lines, every one of the four optical phenomena, for example, absorption, reflection, transmission and refraction are vital for these materials.

Refraction is at the point when light photons are transmitted through a material, they cause polarization of the electrons in the material and by connecting with the polarized materials, photons lose some of their energy. As a consequence of this, the speed of light is diminished and the light emission alters course.

The relative velocity of light passing through a medium is expressed by the optical property called the index of refraction (*n*), and is defined as:

$$n = \frac{c_0}{c}$$

where, c_0 is speed of light in vacuum, c is speed of light in the concerned material. If the angle of incidence from a normal to the surface is θ_i, and the angle of refraction is θ_r, the refractive index of the medium, n, is given by

$$n = \frac{\sin \theta_i}{\sin \theta_r}$$

provided that the incident light is coming from a phase of low refractive index such as vacuum or air. Reflection and refraction of light is shown in Figure 5.18.

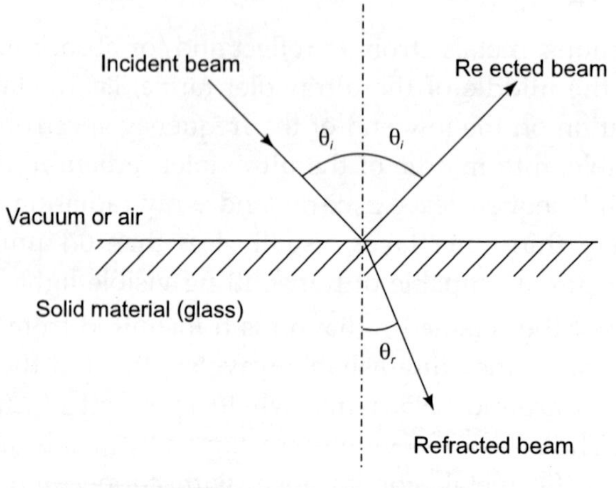

Figure 5.18 Reflection and refraction of light.

5.8.3 Optical Phenomenon

As said in before, light cooperates with a material from numerous points of view. Contingent upon the material, its precious stone/miniaturized scale structure, furthermore on the qualities of episode light, there are numerous impossible to miss marvels happens, which are known as optical wonders. These include: luminescence, lasers, thermal emission, photo-conductivity, and optical fibres. All these find quite many applications in technology for every-day life.

Luminescence: It is characterized as the procedure in which a material assimilates vitality and after that quickly emanates unmistakable or close obvious radiation. This happens as a consequence of excitation of electrons of a material from the valence band into the conduction band. The source of input energy may be high energy electrons or light photons. During luminescence, the excited electrons drop back to lower energy levels. If the emission takes place within 10^{-8} s after excitation, the luminescence is called *fluorescence*, and if it takes longer than 10^{-8} s, it is known as *phosphorescence*.

Luminescence happens in outer valence and conduction groups, while x-rays are created amid electron moves in the internal vitality levels of a molecule. Radiance does not happen in metals. In specific pottery and semi-directors, notwithstanding, the vitality crevice between the valence and conduction groups is to such an extent that an electron dropping through this hole delivers a photon in the visible range.

Ordinarily, pure materials do not display this phenomenon. Special materials called *phosphors* have the capability of absorbing high-energy radiation and spontaneously emitting lower-energy radiation. Ex.: some sulfides, oxides, tungstates, and few organic materials. The emission spectra of these are controlled by added impurities referred as *activators* which provide discrete energy levels in the gap between valence and conduction bands. The excited electrons first drop to donor level, and get trapped. When the source is removed, they gradually escape the trap and emit light over some additional period of time. The intensity of the luminescence is given by

$$I = I_o \exp\left(-\frac{t}{\tau}\right)$$

where, I_0 is initial intensity of luminescence, I is fraction of luminescence after time, t, τ is relaxation time, constant for a material.

Luminescence process is classified based on the energy source for electron excitation as *photo-luminescence, cathode-luminescence,* and *electro-luminescence*. Photo-luminescence occurs in fluorescent lamps. Here ultraviolet radiation from low-pressure mercury arc is converted to visible light by calcium halo-phosphate phosphor ($Ca_{10}F_2P_6O_{24}$). In commercial lamps, about 20% of F^- ions are replaced with Cl^- ions. Antimony, Sb^{3+}, ions provide a blue emission while manganese, Mn^{2+}, ions provide an orange-red emission band.

Cathode-luminescence is produced by an energized cathode which generates a beam of high-energy bombarding electrons. Applications of these include: electron microscope; cathode-ray oscilloscope; colour television screens. Here, however, relaxation time must not be too long which makes the picture blur or they may even overlap. The modern televisions have very narrow, about 0.25 mm wide, vertical stripes of red-, green-, and blue-emitting phosphors deposited on the inner surface of the screens. With help of steel shadow mask with small elongated holes, the incoming signal is scanned over the entire screen at 30 times per second. The small and large number of phosphors consecutively exposed in the rapid scan of 15750 horizontal lines per second. The persistence of human eye makes possible a clear visible picture with good resolution. Commercial phosphors for different colours are: red – yttrium oxy-sulphide (Y_2O_2S) with 3% europium (Eu); green – (Zn,Cd)S with a Cu^+ acceptor and Al^{3+} donor; blue – zinc sulphide (ZnS) with Ag^+ acceptor and Cl^- donor.

Electroluminescence happens in gadgets with p-n amending intersections which are engaged by a remotely associated voltage. Exactly when a forward uneven voltage is associated over gadget, electrons and gaps recombine at the intersection and radiate photons in the noticeable range (mono-chromatic light, i.e., burn shading). These diodes are called light-emitting diodes (LEDs). The trademark shade of a LED relies on the specific semi-leading material that is utilized. GaAs, GaP, GaAlAs, and GaAsP are common materials for LEDs. LEDs discharge light of numerous hues, from red to violet, contingent upon the piece of the semiconductor material utilized. Some even emanate light outside of the obvious range, i.e., infrared and bright. Table 5.5 records materials utilized as a part of various shaded LEDs.

Table 5.5 Materials for coloured LEDs

Wavelength (nm)	Colour	Material
–	Infra-red	GaAs
660	Red	$GaP_{0.40}As_{0.60}$ or $Al_{0.25}Ga_{0.75}As$
635	Orange	$GaP_{0.65}As_{0.35}$
578	Yellow	$GaP_{0.85}As_{0.15}$
556	Green	GaP ($GaP_{1.00}As_{0.00}$)
–	Blue	$Ga_{0.94}NIn_{0.06}$

Lasers: Laser is an acronym for *light amplification by stimulated emission of radiation.* It is in fact special application of luminescence. Unlike most radiation processes, such as luminescence, which produces incoherent light, the light produced by laser emission is coherent, i.e., light waves are all in phase with each other. Consequently, laser light waves are not spread out, i.e., parallel, directional, and monochromatic, i.e., entirely of one wavelength.

In specific materials, electrons energized by a boost create photons which thus energize extra photons of indistinguishable wavelength. Consequently, a substantial enhancement of the photons transmitted in the material happens. By selecting stimulant and material legitimately, laser bar can be in the obvious range.

Lasers are valuable in numerous applications, for example, welding, metal cutting, heat treatment, surgery, mapping, perusing minimized plates, and so on. An assortment of materials is utilized to deliver lasers. Ex.: Ruby, single precious stone of Al_2O_3 doped with little measure of Cr_2O_3; yttrium aluminium garnet ($Y_3A_{15}O_{12}$ - YAG) doped with neodymium, Nd; CO_2 gas; He-Ne gas; some semiconductors like GaAs and InGaAsP.

Gas lasers for the most part bring down forces and powers, however are more appropriate for consistent operation since strong state lasers create apparent measures of heat.

Laser operation: When the laser material is exposed to stimulant, for example, flash lamp, electrons that initially fill the lowest-energy levels get excited into higher energy-levels. These electrons can decay back by two paths: one in which they fall directly back – associated photon emissions are not part of the laser beam; others decay into a intermediate meta-stable state where they reside for about 3 ms before spontaneous emission. This initial spontaneous emission acts as stimulus and triggers an avalanche of emissions from the remaining electrons in the metastable state. The photons are of the same energy and are in phase. The beam is collimated through the use of a tube with silvered mirrors at each end. As stimulated emission occurs, only those photons travelling nearly parallel to the log axis of the material are reflected. These reflected photons stimulate the emission of more photons. Reflected photons travelling up and down the length of the crystal produce ever-increasing number of stimulated photons. Finally, high-energy, highly-collimated, monochromatic beam of coherent light is emitted from the laser device.

Thermal emission: When a material is heated, electrons are excited to higher energy levels, particularly in the outer energy levels where the electrons are less strongly bound to the nucleus. These excited electrons, upon dropping back to the ground state, release photons in process what is called *thermal emission*.

During thermal emission a consistent range of radiation is transmitted with a base wavelength and the force dissemination is reliant on the temperature. Higher the temperature, more extensive will be the scope of wavelengths radiated. By measuring the force of a limited band of the transmitted wavelengths with a pyrometer, material's temperature can be evaluated.

Photoconductivity: As mentioned in earlier section upon absorption of photons at surface, electron may be released from its atom nucleus. Thus, electricity can be generated from the surface of a metal when it is bombarded with photons. Similarly, bombardment of semiconductors by photons, with energy equal to or greater than the band gap, may result in creation of electron-hole pairs that can be used to generate current. This process is called photoconductivity, and is different from photoelectric effect in the sense that an electron-hole pair is generated whose energy is related to the band gap energy instead of free electron alone whose energy is related to the Fermi level. The current produced in photoconductivity is directly related to the incident light intensity.

This phenomenon is used in photographic light metres. Cadmium sulphide (CdS) is ordinarily utilized for the recognition of noticeable light, as in light meters. Photoconductivity is additionally the basic rule of the photograph voltaic cell, referred to regular man as sun powered cell, utilized for transformation of sun based vitality into power.

Optical Fibres: These are recently the buzz word in the communications sector is the optical fibre, using in place of metallic copper wires. Signal transmission through a metallic wire conductor is electronic, whereas in fibres it is photonic, i.e., by photons. This enables faster transmission at higher densities to longer distances with reduction in error rate. These systems consist of transmitter (a semiconductor laser) to convert electrical signals to light signals, optical fibre to transmit the light signals, and a photodiode to convert light signals back to electrical signals.

Optical fibre is the heart of the correspondence system. It must have to a great degree low loss of light, must have the capacity to direct the light heartbeats over long separations without huge misfortune and additionally mutilation. It principally comprises core, cladding and coating. The centre transmits the signs, while the cladding compels the light bar to the centre; external covering shields the centre and cladding from the outer environment.

Optical fibre operates on the principle of total internal reflectance. Typically, both the core and cladding are made of special types of glass with carefully controlled indices of refraction. The indices of refraction are selected such that

$$n_{\text{cladding}} < n_{\text{core}}$$

Once the light enters the centre from the source, it is reflected inside and spreads along the length of the fibre. Interior reflection is refined by shifting the list of refraction of the centre and cladding glass materials. Normally, two plans are utilized in such manner. In step-index optical fibre, there is a sharp change in refractive index between the centre and cladding. In this outline output pulse will be more extensive than the info one. It is on the grounds that light beams going in various directions have an assortment of wavelengths.

It is possible to avoid pulse broadening by using graded-index fibre. This results in a helical path for the light rays, as opposed to zig-zag path in a step-index fibre. Here impurities such as boron oxide (B_2O_3) or germanium dioxide /GeO_2) are added to the silica glass such that the index of refraction varies gradually in parabolic manner across the cross section. This enables light to travel faster while close to the periphery than at the centre. This avoids pulse broadening, i.e., light rays arrive at the output at approximately same time. Both step- and graded-index fibres are termed multi-mode fibres. Third type of optical fibre is called single-mode fibre in which light travels largely parallel to the fibre axis with little distortion of the digital light pulse. These are used for long transmission lines.

Core and cladding materials are selected not only on the basis of their refractive indices, but also on the basis of ease of manufacturability, light loss, mechanical strength properties and dispersion properties. However, density (ρ) and refractive index (n) are critical. These two parameters are related approximately as

$$n = \frac{\rho + 10.4}{8.6}$$

High-purity silica-based glasses are used as fibre material, with fibre diameter ranging from 5 to 100 μm. The fibres are carefully fabricated to be virtually free from flaws, as a result, are extremely strong and flexible. Uniformity of fibre cross-sectional dimensions and core roundness are critical; allowable tolerances of these parameters are within a micrometre over 1 km of length.

5.9 SUMMARY

This chapter initially deals with engineering properties and then the extensive study of several properties of materials are considered. The mechanical behaviour of solids is broadly divided into the elastic and plastic regimes. Elasticity theory and engineering simplifications of it form the basis of the engineering design of structures and machine parts. The tensile, compressive, and shear strains are always small in the elastic regime. Elastic tensile stresses exert a pull on atomic bonds, and the resultant atomic separation is manifested macroscopically as strain; both stress and strain vanish when the load is removed. The reason we can be so confident about stress analysis in the elastic range is the validity of Hooke's law or the linear connection between stress and strain. (Ohm's law linearly connecting voltage and current plays a similar singular role in the design and analysis of electrical circuits.) Young's modulus and Poisson's ratio are the only material constants of concern in isotropic elasticity, but there are more moduli to contend with when single crystals are involved. With computers it is now possible to determine accurately the stress and strain distribution throughout arbitrarily shaped bodies loaded elastically in almost any given way. Later in this chapter the electrical, thermal, magnetic and the optical properties of the materials are studied with a primary objective of learning the properties of materials.

This chapter discussed heat absorption, thermal expansion, and thermal conduction—three important thermal phenomena. Heat capacity represents the quantity of heat required to produce a unit rise in temperature for one mole of a substance; on a per unit mass basis, it is termed specific heat. Most of the energy assimilated by many solid materials is associated with increasing the vibrational energy of the atoms; contributions to the total heat capacity by other energy-absorptive mechanisms (i.e., increased free-electron kinetic energies) are normally insignificant.

Free electrons being acted on by an electric field are scattered by imperfections in the crystal lattice. The magnitude of electron mobility is indicative of the frequency of these scattering events. In many materials, the electrical conductivity is proportional to the product of the electron concentration and the mobility. For metallic materials, electrical resistivity increases with temperature, impurity content, and plastic deformation. The contribution of each to the total resistivity is additive.

Solid materials expand when heated and contract when cooled. The fractional change in length is proportional to the temperature change, the constant of proportionality being the coefficient of thermal expansion. Thermal expansion is reflected by an increase in the average interatomic separation, which is a consequence of the asymmetric nature of the potential energy versus inter-atomic spacing curve trough. The larger the interatomic bonding energy, the lower is the coefficient of thermal expansion.

The transport of thermal energy from high- to low-temperature regions of a material is termed thermal conduction. For steady-state heat transport, the flux is proportional to the temperature gradient along the direction of flow; the proportionality constant is the thermal conductivity.

Diamagnetism results from changes in electron orbital motion that are induced by an external field. The effect is extremely small and in opposition to the applied field. All materials are diamagnetic. Paramagnetic materials are those having permanent atomic dipoles, which are acted on individually and are aligned in the direction of an external field. Since the magnetizations are relatively small and persist only while an applied field is present, diamagnetic and paramagnetic materials are considered to be nonmagnetic. Large and permanent magnetizations may be established within the ferromagnetic metals (Fe, Co, Ni). Atomic magnetic dipole moments are of spin origin, which are coupled and mutually aligned with moments of adjacent atoms.

Antiparallel coupling of adjacent cation spin moments is found for some ionic materials. Those in which there is total cancellation of spin moments are termed antiferromagnetic. With ferrimagnetism, permanent magnetization is possible because spin moment cancellation is incomplete. For cubic ferrites, the net magnetization results from the divalent ions that reside on octahedral lattice sites, the spin moments of which are all mutually aligned.

For soft magnetic materials, domain wall movement is easy during magnetization and demagnetization. Consequently, they have small hysteresis loops and low energy losses. Domain wall motion is much more difficult for the hard magnetic materials, which results in larger hysteresis loops; because greater fields are required to demagnetize these materials, the magnetization is more permanent.

Metals appear opaque is a result of the absorption and then re-emission of light radiation within a thin outer surface layer. Absorption occurs via the excitation of electrons from occupied energy states to unoccupied ones above the Fermi energy level. Re-emission takes place by decay electron transitions in the reverse direction. The perceived colour of a metal is determined by the spectral composition of the reflected light.

The optical behaviour of a solid material is a function of its interactions with electromagnetic radiation having wavelengths within the visible region of the spectrum. Possible interactive phenomena include refraction, reflection, absorption, and transmission of incident light. When light passes from one transparent medium to another having a different index of refraction, some of it is reflected at the interface. The degree of the reflectance depends on the indices of refraction of both media, as well as the angle of incidence.

Both crystalline and noncrystalline states are possible for ceramics. The crystal structures of those materials for which the atomic bonding is predominantly ionic are determined by the charge magnitude and the radius of each kind of ion. Some of the simpler crystal structures are described in terms of unit cells; several of these were discussed (rock salt, cesium chloride, zinc blende, diamond cubic, graphite, fluorite, perovskite, and spinel structures).

Diamond is a gem stone and, because of its hardness, is used to cut and grind softer materials. Furthermore, it is now being produced and utilized in thin films. The layered structure of graphite gives rise to its excellent lubricative properties and a relatively high electrical conductivity. Graphite is also known for its high strength and chemical stability at elevated temperatures and in nonoxidizing atmospheres. Fullerenes exist as hollow and spherical molecules composed of 60 carbon atoms. The recently discovered carbon nanotubes have extraordinary properties: high stiffnesses and strengths, low densities, and unusual electrical characteristics. Structurally, a nanotube is a graphite cylinder with fullerene hemispheres on its ends.

Any plastic deformation of crystalline ceramics is a result of dislocation motion; the brittleness of these materials is explained, in part, by the limited number of operable slip systems. The mode of plastic deformation for noncrystalline materials is by viscous flow; a material's resistance to deformation is expressed as viscosity. At room temperature, the viscosities of many noncrystalline ceramics are extremely high.

Many ceramic bodies contain residual porosity, which is deleterious to both their moduli of elasticity and fracture strengths. In addition to their inherent brittleness, ceramic materials are distinctively hard. Also, since these materials are frequently utilized at elevated temperatures and under applied loads, creep characteristics are important.

EXERCISE

1. A specimen of copper having a rectangular cross section 15.2 mm 19.1 mm (0.60 in 0.75 in) is pulled in tension with 44,500 N (10,000 lb) force, producing only elastic deformation. Calculate the resulting strain.
2. A cylindrical specimen of a nickel alloy having an elastic modulus of 207 GPa (psi) and an original diameter of 10.2 mm (0.40 in) experiences only elastic deformation when a tensile load of 8900 N (2000 lb) is applied. Compute the maximum length of the specimen before deformation if the maximum allowable elongation is 0.25 mm (0.010 in).
3. Wire used by orthodontists to straighten teeth should ideally have a low modulus of elasticity and a high yield stress. Why?

4. For a hardened alloy steel the true stress-true strain behaviour is $\sigma = 190\varepsilon^{0.3}$ ksi.
   (a) What is the meaning of true strain at necking?
   (b) What is the engineering strain at necking?
5. A bar is pulled in tension to a certain plastic stress level and the machine is halted at a fixed crosshead separation. What will happen to the recorded load as a function of time?
6. The elastic modulus determined from the slope of the stress-strain curve in the elastic region is more accurate at a lower testing temperature. Why?
7. What is the difference between electronic and ionic conduction?
8. How does the electron structure of an isolated atom differ from that of a solid material?
9. In terms of electron energy band structure, discuss reasons for the difference in electrical conductivity between metals, semiconductors, and insulators?
10. Briefly explain thermal expansion using the potential energy versus inter-atomic spacing curve.
11. A bimetallic strip is constructed from strips of two different metals that are bonded along their lengths. Explain how such a device may be used in a thermostat to regulate temperature.
12. We might think of a porous material as being a composite wherein one of the phases is a pore phase. Estimate the upper and lower limits for the room-temperature thermal conductivity of an aluminium oxide material having a volume fraction of 0.25 of pores that are filled with still air.
13. Cite the differences between hard and soft magnetic materials in terms of both hysteresis behaviour and typical applications.
14. Briefly explain why the magnitude of the saturation magnetization decreases with increasing temperature for ferromagnetic materials, and why ferromagnetic behaviour ceases above the Curie temperature.
15. Briefly describe the phenomenon of magnetic hysteresis, and why does it occur for ferromagnetic and ferrimagnetic materials.
16. Distinguish between materials that are opaque, translucent, and transparent in terms of their appearance and light transmittance.
17. Briefly explain why metals are opaque to electromagnetic radiation having photon energies within the visible region of the spectrum.
18. Briefly describe the phenomenon of dispersion in a transparent medium.
19. Briefly describe the phenomenon of photoconductivity.
20. Describe briefly the phenomenon of luminescence.
21. Explain dia, para, and ferromagnetism in detail.
22. Explain the concept of crack and stress concentration.

23. Discuss the phenomenon of plastic deformation.
24. Summarize the tension test with a neat sketch.
25. What are the advantages of compression and torsion test to be carried out in engineering materials?
26. An aluminium bar 125 mm (5.0 in) long and having a square cross section 16.5 mm (0.65 in) on an edge is pulled in tension with a load of 66,700 N (15,000 lb), and experiences an elongation of 0.43 mm. Assuming that the deformation is entirely elastic, calculate the modulus of elasticity of the aluminium.
27. Consider a cylindrical nickel wire 2.0 mm (0.08 in.) in diameter and 3×10^4 mm (1200 in) long. Calculate its elongation when a load of 300 N (67 lb) is applied. Assume that the deformation is totally elastic.
28. A cylindrical rod of steel (E = 270 GPa, 30×10^6 psi) having a yield strength of 310 MPa (45,000 psi) is to be subjected to a load of 11,100 N (2500 lb). If the length of the rod is 500 mm (20.0 in.), what must be the diameter to allow an elongation of 0.38 mm (0.015 in)?
29. A cylindrical specimen of some metal alloy 10 mm (0.4 in) in diameter is stressed elastically in tension. A force of 15,000 N (3370 lb_f) produces a reduction in specimen diameter of 7×10^3 mm. Compute Poisson's ratio for this material if its elastic modulus is 100 GPa (14.5×10^6 psi).
30. A cylindrical specimen of a hypothetical metal alloy is stressed in compression. If its original and final diameters are 30.00 and 30.04 mm, respectively, and its final length is 105.20 mm, compute its original length if the deformation is totally elastic. The elastic and shear moduli for this alloy are 65.5 and 25.4 GPa, respectively.

6

Strengthening of Engineering Materials

Learning Objectives

After going through this chapter the reader will be able to:

- Understand strengthening of engineering materials and heat treatment of steel.
- List the applications of ferrous and nonferrous alloys.
- Gather idea on recrystallization and strengthening of nonferrous metals.
- Understand the properties of composites and basics of engineering composites.
- Know about strengthening and toughening of ceramics.

6.1 INTRODUCTION

In this chapter we will consider materials that are employed in mechanically functional engineering applications. Those materials that achieve commercial status have necessarily passed a tough development ordeal whose hurdles include (1) fulfilment of stringent mechanical property specifications (e.g., tensile strength, elongation, toughness); (2) development of reproducible processing and fabrication methods (e.g., casting, mechanical forming, joining); (3) extensive testing to ensure reliable operation in components and structures; and (4) competitive pricing. The surviving materials are typically but a fraction of potential candidates that for one reason or another were discarded and did not see commercialization.

Historically, metals have survived longest and been the materials of choice because they have represented the best compromise in terms of a basket of engineering properties that include stiffness (Young's modulus), yield strength, toughness, melting temperature, and cost. This can be seen in the bar graphs of Figure 6.1, where the properties of metals, ceramics, and polymers are compared. If materials are scored 3 (best), 2, and 1 (worst) based on the maximum property value attained by any representative material, then the cumulative grades for the four properties considered are, roughly, 9 for metals, 10 for ceramics, and 5 for polymers. Surprisingly, ceramics score highest on this limited,

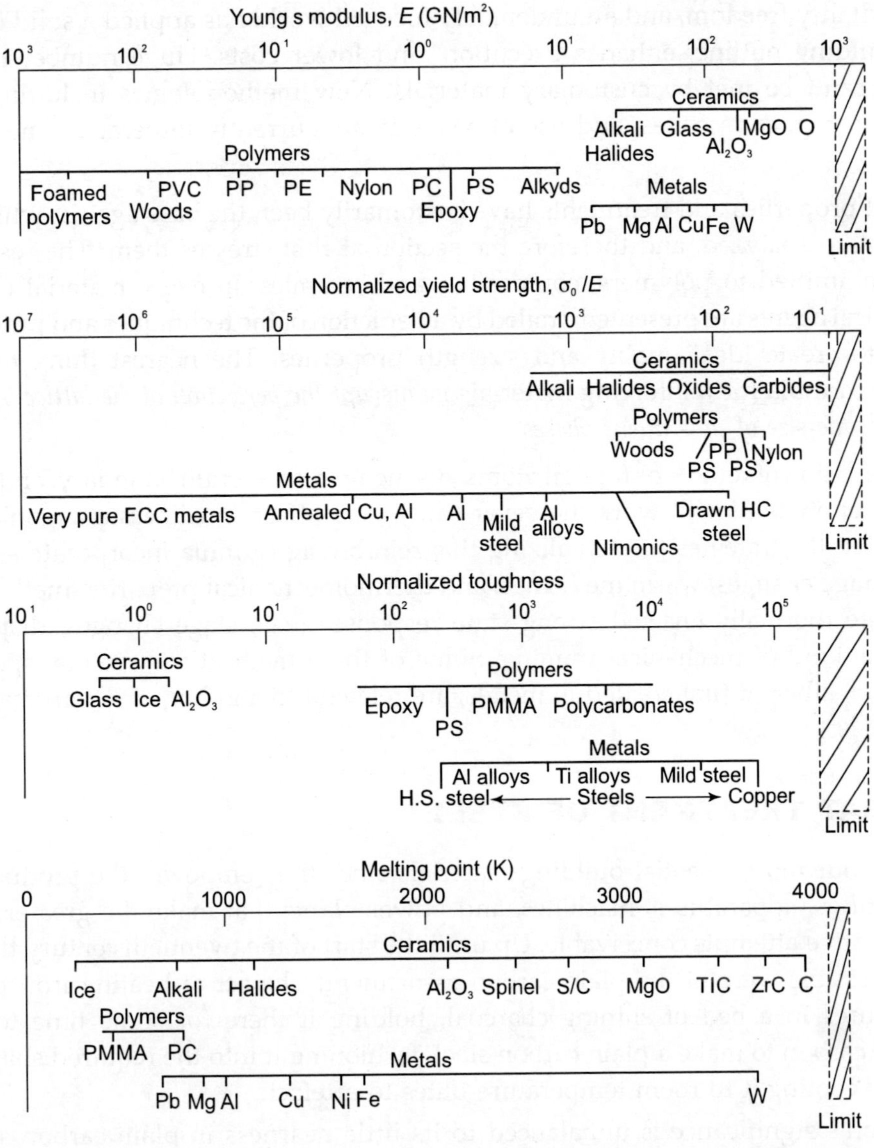

Figure 6.1 Property value ranges for metals, ceramics, and polymers.

and perhaps biased, test. They are the stiffest, strongest, and most heat-resistant materials. Their Achilles heel is toughness, which always raises the frightening specter of brittle fracture. Although metals occasionally suffer brittle fracture, their outstanding attribute is toughness. If weight considerations are critical, as in aerospace applications, polymers rise to the top when strength to density comparisons are made. The last is a case of the factors that must be considered in the inexorably complex field of materials determination and use. Worries of restricted accessibility of vital materials, the push toward vitality freedom, and an undeniably focused world has applied a solid impulse to fix building outline, enhance execution, and lower costs. Huge number of these requests can't be met by customary materials. New methodologies including novel thermal curing, composites, and use of coatings are currently imperative and will be examined.

Metal properties and treatments have customarily been the base against which new materials are analyzed, and therefore the section at first stresses them. The rest of the part is committed to polymers, composites, and ceramics. In every material class the more critical agents are presented, trailed by a depiction of the techniques and procedures utilized to create ideal quality and strength properties. The nearest thing to an all inclusive formula for reinforcing materials is: *disrupt the perfection of the lattice or matrix and reduce the size of constituent phases.*

Presentation of remote or foreign atoms, disengagements, grain boundary, and second stage disseminations will work, however sadly may affect sturdiness antagonistically. Down to earth strategies for actualizing this reinforcing formula incorporate structure control, stage changes, warm medicines, and thermomechanical preparing methods that consolidate thermally enacted strong state responses (e.g., stage changes, dispersion) with some kind of mechanical framing. Some of these medications are non-specific in nature and, albeit at first created in metals, are relevant to and honed in various classes of materials.

6.2 HEAT TREATMENT OF STEEL

Steels are our most essential building materials since they empower the production of the structures, apparatuses, machines, and power plants that make the greater part of our innovative attempts conceivable. Up until the start of the twentieth century the main steels accessible were of the plain carbon assortment. The act of heating iron to lifted temperatures in a bed of shining charcoal, holding it there for some time to retain adequate carbon to make a plain carbon steel, fashioning it into the required shape, and gradually cooling it to room temperature dates to artefact.

Carbon's significance is unbalanced to its little nearness in plain carbon steel, as well as different evaluations of steel also. A decent reason for seeing plain carbon steels and a portion of the microstructures emerging from basic warm medications is given by the Fe-C (Fe-Fe_3C) stage graph. Evaluating the balance properties of the Fe-C system now is, along these lines, advantageous. It ought to, notwithstanding, be recollected that lone balance changes, achieved by moderate cooling of steel from lifted temperatures, were considered beforehand. A sensational change in the hardness and quality of the steel comes about because of a basic warmth treatment. Rather than being gradually cooled after it is fashioned, the piece must be again warmed to a raised temperature, *rapidly quenched* (in water or oil), and afterward tempered or warmed to a somewhat low temperature preceding use. By these methods hardnesses and tensile strengths can frequently be dramatically increased, an enormous increment when one talks about mechanical properties. Furthermore, vitally, parts can be solidified after they have been framed to the last shape without apparently mutilating them amid heat treatment.

6.2.1 Isothermal Transformation

Eutectoid Steel: Let us reconsider the TTT bend for eutectoid steel that is duplicated again in Figure 6.2 for accommodation. Review that the C state of the TTT bend underscores consolidated nucleation and development impacts that describe the isothermal austenite-pearlite change. Assume stable austenite (An), over the eutectoid basic temperature of 723°C, is quickly cooled underneath it to temperature T_1 = 475°C and left to deteriorate. With time, the now insecure austenite changes isothermally (at consistent temperature) by means of strong state response to ferrite (F) and iron carbide (C) along the even way as appeared. For whatever length of time that the A + F + C field has not been entered, just A will be available. After around 1 second the limit amongst An and A + F + C is crossed, creating pearlite (F + C) to nucleate and develop with time. For roughly the following 18 seconds the change of A to F + C continues gradually at first and after that all the more quickly; at long last, it tails off towards culmination in the sigmoidal design.

Once the change enters the F + C field there is no more austenite left and the steel won't apparently change if taken to any lower temperature. This same kind of isothermal change will happen at any temperature between – 220 and 723°C, however, with some huge contrasts. Imperatively, the particular isothermal change in temperature has a noteworthy bearing on both the microstructure and hardness (quality) of the pearlite structures. At high temperatures the few F + C fixes that nucleate develop quickly, prompting to a coarse pearlitic structure. The huge interlammelar dividing between the α-Fe and Fe_3C periods of pearlite results in a generally delicate, low-quality steel. Separations experience moderately couple of hindrances to movement in such coarse structures. Later in the part it will be clear that mechanical quality is upgraded if stages in a two-stage structure are little in size, for then they all the more successfully limit disengagement movement. Impressively harder and more strong steels are created if the change happens at progressively bring down temperatures.

The numerous nuclei that can't promptly develop cause advancement of better pearlitic microstructures having higher hardness and strength. If the transformation temperature is decreased well underneath the nose of the bend (Figure 6.2), austenite isothermally breaks down into bainite. This structure is named after E. C. Bain, an early examiner of this sort of change. Indeed, even today the way of the bainite change is not by any means caught on. Like pearlite, bainite comprises ferrite and Fe_3C, however, not at all like pearlite the microstructure is nonlamellar (Figure 6.4a). In some ways bainite shows attributes of martensite, which are examined in the following segment. Qualifications in bainitic structures happen contingent upon the temperature of development. As may be normal, the better lower bainites delivered beneath ~350°C are harder than the upper bainites framed between – 350 and 450°C. Bring down bainites contain fine accelerates of carbides dispersed all through the ferrite plates, yielding a hard and extreme structure.

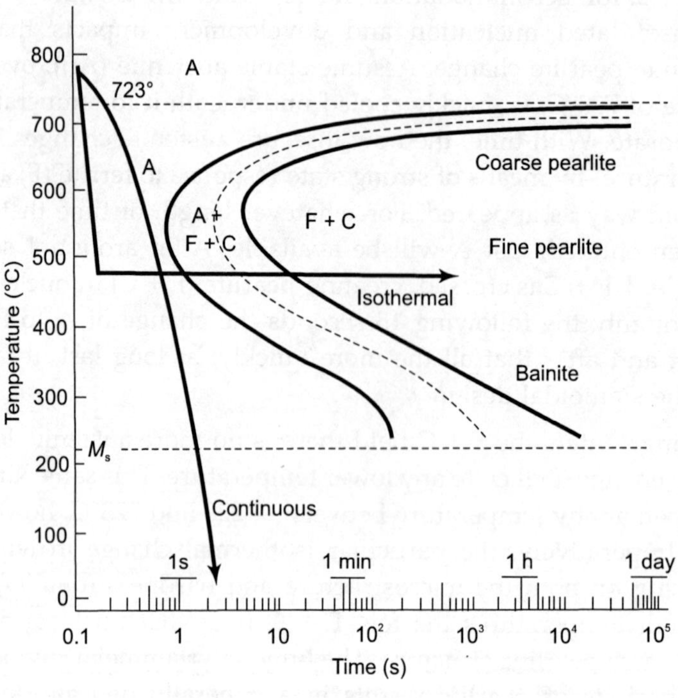

Figure 6.2 TTT curve for eutectoid steel with superimposed paths of isothermal and continuous cooling transformation of austenite.

Non-Eutectoid Steel: At first austenitized non-eutectoid carbon steels change over to either ferrite or Fe_3C when isothermally changed above 723°G, contingent on whether the included steel is hypo-eutectoid or hypereutectoid in piece, separately. This records for the individual A + F and A + C fields in the TTT graphs for these steels as shown in Figure 6.3. Beneath the basic eutectoid temperature the change of A to A + F or A + C continues initially, trailed by a consequent A to F + C change, as sometime recently, for any austenite that remaining parts.

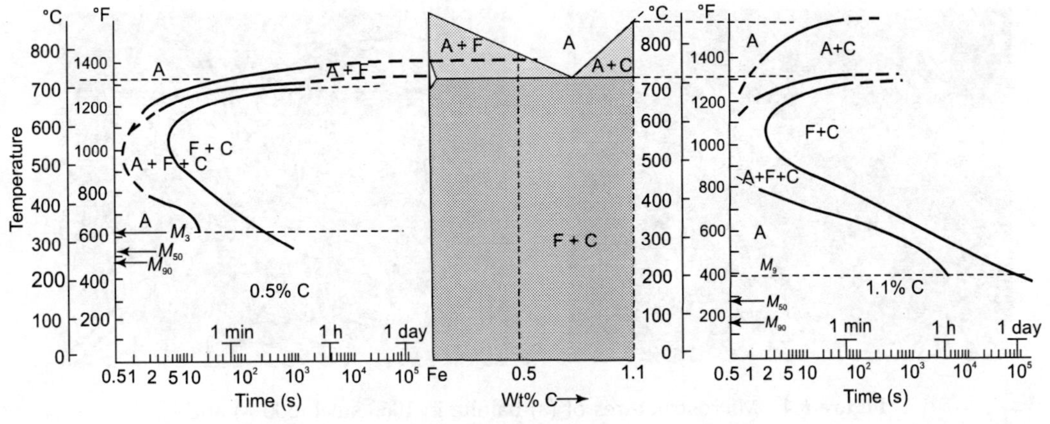

Figure 6.3 TTT curves and their relation to the Fe-Fe_3C phase diagram. Left: 1050 hypo-eutectoid steel (0.50 wt% C, 0.91 wt% Mn] Right 10110 hypereutectoid steel (1.10 wt% C, 0.30 wt% Mn).

6.2.2 Transformation of Austenite to Martensite

Life is too short to sit tight for steel to change isothermally. Assume that while eutectoid steel was isothermally changing, the procedures were abruptly hindered by extinguishing of the steel or quick bringing down of its temperature. We would find that any austenite that did not frame pearlite would rather change to martensite, yielding a blended microstructure comprising of both martensite and pearlite. On extinguishing, the metastable FCC austenite endeavours to dismiss the 0.8 wt% C held in strong arrangement since stable BCC α-Fe can just break down 0.01 wt% C. In any case, the temperature drops so quickly that carbon just has no opportunity to get away; rather it is caught inside the austenite, straining it to the point where a diffusionless or martensitic change happens. The martensite change is crystallographically fascinating.

Encased inside the two FCC y-Fe unit cells of cross section has a fixed body-centred tetragonal cell, outlined in Figure 6.5, whose grid parameters are a' and c. In the midst of complex cross section shearing, including nuclear removals of not exactly a grid dividing, austenite contorts into an extended body-focused tetragonal martensitic structure. This disentangled model of the change outlines how easily the (111) plane in FCC Fe turns into the (011) plane in BCT Fe. Expanding carbon substance really broadens the c and abbreviate the a' measurements. The microstructure of martensite duplicated in Figure 6.4b uncovers that it has a needle-like acicular morphology.

The important thing to remember is that the highly strained martensite is extremely hard and responsible for strengthening the steel. It is harder and stronger than the finest pearlitic structures described earlier. Carbon strongly influences the formation and hardness of martensite. Harder martensites are produced with higher carbon contents. And like many hard materials, quenched martensite is brittle, especially because it

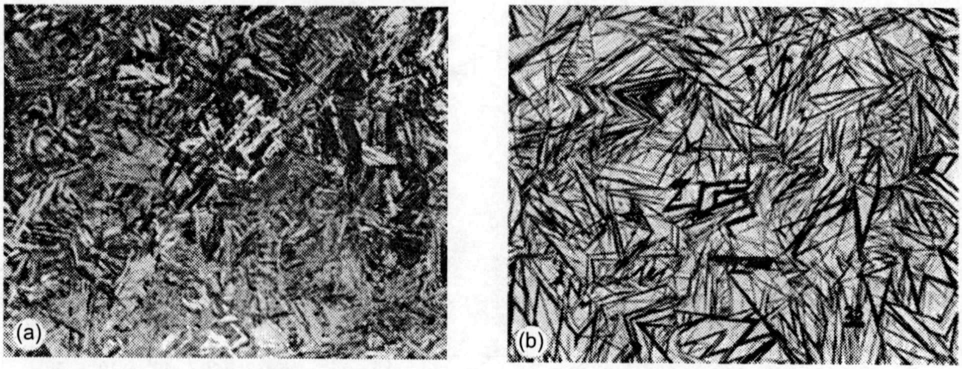

Figure 6.4 Microstructures of (a) bainite in 1060 steel (500 ×) and (b) martensite in a steel containing 1.8 wt% C (200 ×).

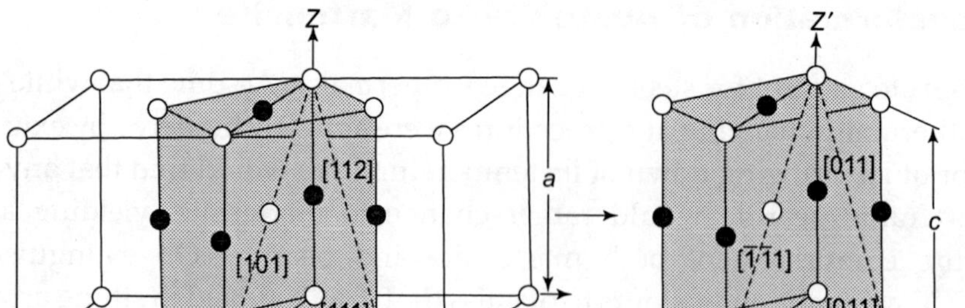

Figure 6.5 Simplified model for the crystallography of the martensite transformation. The shaded BCT cell that becomes martensite in the $X'Y'Z'$ coordinate system stems from the prior FCC cells in the *XYZ* coordinate system. Interstitial carbon sites are drawn as dark circles.

contains high levels of internal or residual stress. To relieve the latter it is recommended that freshly formed martensite be immediately annealed or **tempered.** Horror stories exist of quench-hardened dies loudly cracking, ruining the lunch of startled workers who planned to temper them after eating. Relief of the carbon supersaturation and stress occurs through out-diffusion of carbon to form carbide precipitates. The tempering process is accompanied by a slight, but tolerable reduction in hardness. Importantly, tempering toughens the steel considerably. Tempering temperatures of less than 500°C are typically employed for times on the order of minutes to a few hours depending on the steel and part size.

The martensite transformation has a number of other features relevant to its role in hardening the steel. First, because it occurs by elastic shearing, the transformation essentially proceeds with the speed of sound. Further, the martensite transformation cannot be suppressed; it is triggered when the austenite temperature drops below the M_s (martensite start) temperature. Once the first bit of martensite forms, the driving force for further transformation is reduced because the structure is stabilized. Martensite and austenite coexist as the strain energy released by the former balances the chemical driving force to convert the latter. As the temperature drops, increasing amounts of martensite form until at the M_f (martensite finish) temperature, the transformation of austenite is complete. The M_s and M_f temperatures vary with the steel. To obtain 100% martensite it is necessary first to cool the steel so as to miss the nose of the TTT curve; furthermore, the steel must be quenched below *Mf*. This eliminates the possibility that austenite, which can transform into brittle martensite if the service temperature drops sufficiently, will be retained.

6.2.3 Continuous Cooling Transformation of Austenite

Besides a couple of exceptional treatments, steel is infrequently isothermally changed. Or maybe, it is cooled persistently from austenitizing temperatures to room temperature, and changes incrementally impressively reliant on the cooling rate. The line drawn superimposed over the TTT bend of Figure 6.6 speaks to such an ordinary cooling history. We might want to outline energy of austenite disintegration along this or other slanting cooling bends as we did already along the even isotherms; be that as it may, it is a more troublesome test to create a consistent cooling transformation (CCT) bend than an isothermal transformation (IT or TTT) curve.

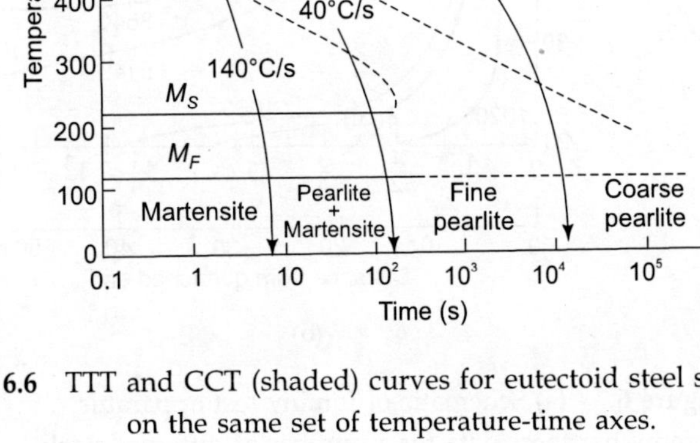

Figure 6.6 TTT and CCT (shaded) curves for eutectoid steel superimposed on the same set of temperature-time axes.

By and by, it is found that change amid ceaseless cooling is postponed. This dislodges the CCT bend down and to one side of the TTT bend as appeared in Figure 6.6 where the two practices are superimposed. We should make the critical suspicion that the more open TTT bends can be utilized to portray nonstop cooling conduct. Staying away from knee of a TTT curve guarantees change of austenite to martensite. In the event that it is fancied to solidify a section completely, then cooling rates all through must be equivalent to or more prominent than the basic one (i.e., tumble to one side of the knee). Then again, when a slower cooling-rate bend catches the upper parts of the TTT curve, change to pearlite or blends of pearlite and martensite can be normal. Albeit some judgment is required, TTT curve give a sensible comprehension of what's in store amid functional heat treatment.

6.2.4 Hardenability

Hardenability is a measure of the **ability** to **harden** a steel to a given depth. It is not the same as hardness, which measures a material's ability to withstand penetration by a hard indenter. The factors that displace TTT curves to the right, inhibiting the transformation of austenite to pearlite, also increase the hardenability of the steel. These include increased carbon content, alloying elements, and large austenitic grain size. High hardenability implies ease in producing martensite at greater depths within a part and accomplishing it with less drastic cooling rates. All of these associations have been distilled into a simple, universally accepted test, i.e., the Jominy end-quench hardenability

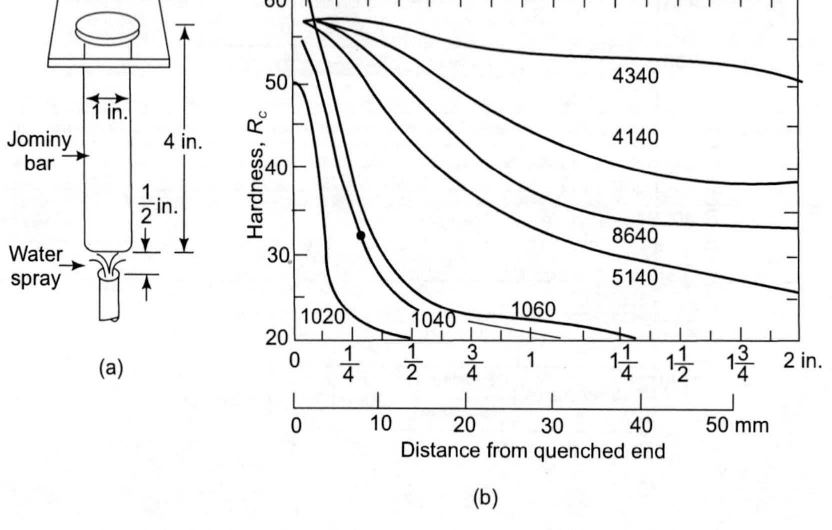

Figure 6.7 (a) Schematic of Jominy test apparatus. (b) Jominy hardness data for a number of different steels.

test, that readily rates the hardenability of steels. All that is needed is a fixture with a hole in it to support a hot test bar and a source of water placed at a fixed distance beneath it (Figure 6.7a). A round Jominy bar of standard measurements is austenized at a lifted temperature and after that immediately exchanged to the apparatus, where it is extinguished specifiedly. In the wake of cooling, a level is ground on the bar length and hardness estimations are recorded along it at consistent interims. A fundamental suspicion of the test is that all steels cool at a similar rate at relating positions along the bar. This is roughly genuine in light of the normal directional cooling and the way that heat conduction, convection, and radiation properties of all heated steel are comparative. (Stainless and other high-composite steels with low warm conductivities are special cases.) Jominy test comes about for various steels are plotted in Figure 6.7b with rough cooling rates demonstrated on the upper abscissa. Since a given cooling rate is connected with a specific microstructure having a trademark hardness, Jominy test information permits hardenability correlations among various steels in parts of different shape.

6.2.5 Additional Steel Hardening and Toughening Treatments

Various exceptional treatments for strengthening, toughening, and enhancing the properties of steel have been formulated. A portion of the fascinating ones include mixes of warm changes and plastic twisting and are known as thermomechanical treatments (TMTs). The latter are further distinguished according to whether deformation occurs: (i) *before* austenite transforms (martensite forms in work-hardened austenite); (ii) *during* austenite transformation (martensite forms in deformed metastable steels); or (iii) *after* austenite transforms (thermal aging of transformation products). They all exploit the properties of TTT diagrams and brief consideration of them is a fitting way to end our discussion on heat treatment of steel. We start with two heat treatments that require no deformation.

Austempering: In austempering, steel is first quenched to the left of and below the nose of the TTT curve to a temperature at which bainite formation is feasible. Time is then allowed for isothermal transformation to bainite to proceed to completion (Figure 6.8a). High hardness and toughness are the benefits of austempering. The process is not suited to steels that require long bainite transformation times.

Martempering (or Marquenching): Martempering involves an interrupted quench of austenitized steel to reduce the overall cooling rate. Residual stress generation in the martensite and its potential for distortion and even cracking are minimized in this way. In martempering, steel is quenched into an oil or molten salt bath held just above or slightly below the M_s Temperature. It is held there until a uniform temperature is attained, but less than the time required for bainite transformation to initiate (Figure 6.8b). Gradual cooling below the M_f followed by standard tempering completes the process.

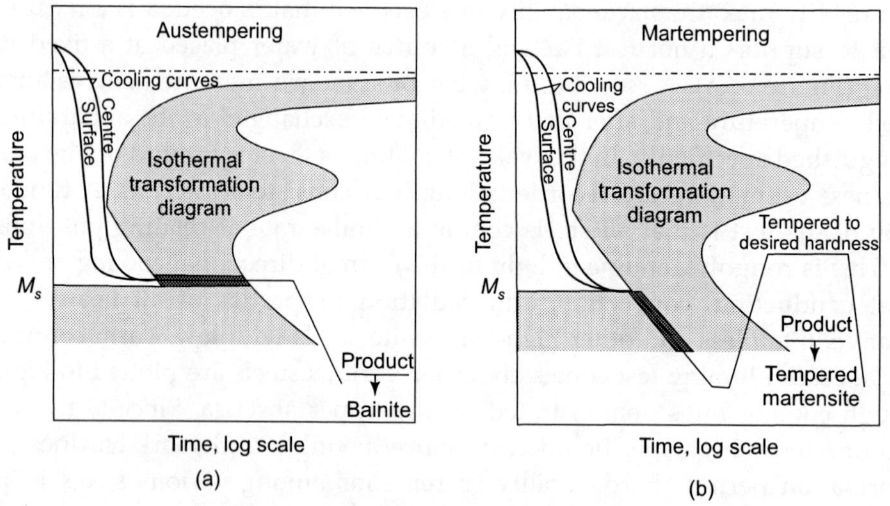

Figure 6.8 (a) Austempering and (b) martempering treatments.

Ausforming (Ausforging): An assortment of both high- and low-temperature TMT processes falls into the ausforming category and are depicted in Figure 6.9. They feature mechanical deformation of both stable and unstable austenite with subsequent transformation to either ferrite and pearlite or martensite. Low- and medium-carbon steels are strengthened, whereas tool steels are toughened and made more wear resistant by such treatments. Perhaps the most widespread of these TMT processes are carried out in **high-strength low-alloy** (HSLA) steels, also known as microalloyed steels. These contain 0.05 to 0.2 wt% C and about 0.1 wt% of elements like Nb, Ti, and V. During hot rolling the austenite transforms to a fine-grained ferrite. As we know, significant strengthening can occur through grain refinement. This plus greater corrosion resistance is the reason why these microalloyed steels are successfully supplanting mild steel in many structural applications.

Transformation-Induced Plasticity (TRIP) Steels: The strain triggers change of the martensite, which raises the strength as well as, shockingly, expands the toughness of the steel. Fortifying expands the nearby strain solidifying rate so that the neck arrangement and disappointment are hindered. Then again, if a split as of now exists, martensite change at the break tip blunts its further proliferation.

Maraging: Maraging steels are intensely alloyed (e.g., ≈ 18 wt% Ni, 8 wt% Co, 5 wt% Mo, 0.5 wt% Ti) however, contain no carbon. The absence of carbon makes for a generally malleable martensite that can be mechanically prepared without breaking a sweat. After martensite is matured or warmed to low temperatures for delayed times, accelerates shape that obviously reinforces the lattice. Maturing or precipitation solidifying medicines are normally utilized to reinforce nonferrous composites. The immense change in both toughness and strength as an after-effect of these two thermomechanical treatments is obvious in Figure 6.10.

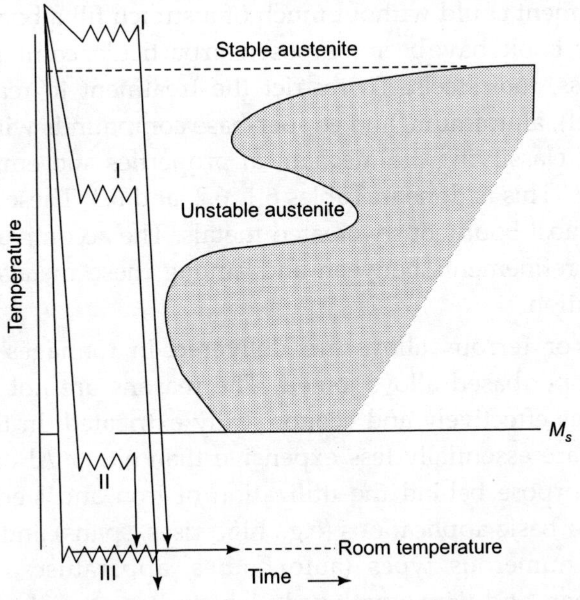

Figure 6.9 Schematic TTT curve showing classification of thermomechanical treatments during transformation of austenite. Included are TMT processes for HSLA (I), TRIP (II), and maraging steels (III).

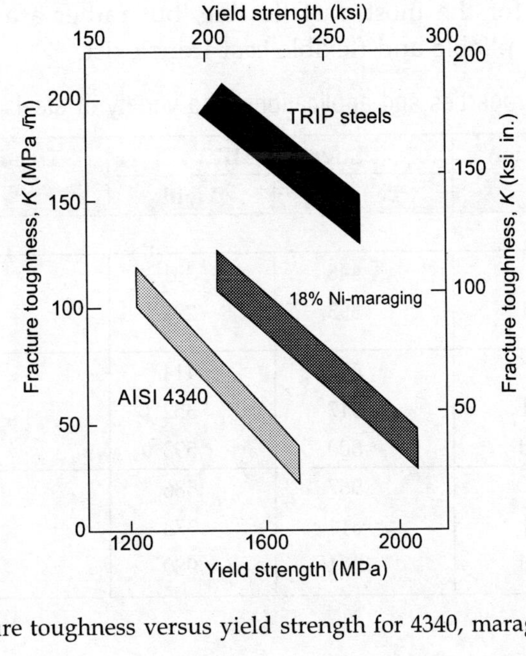

Figure 6.10 Fracture toughness versus yield strength for 4340, maraging, and TRIP steels.

6.3 PROPERTIES AND APPLICATIONS OF FERROUS AND NONFERROUS ALLOYS

The topic of this segment could without much of a stretch fill a book independent from anyone else. Actually, books have been composed on particular combination subcategories, for instance, stainless, tool steels. To restrict the treatment to reasonable extent, just illustrative iron (steel), aluminium, and copper-base compounds will be considered, and just to the degree of classifying the mechanical properties and employments of only a couple of composites. This is done in Tables 6.1, 6.2, and 6.3. These tables give a similar valuation for these most bounteously created metals. The accompanying comments are expected to elevate refinements between and among these metals and give points of view on their utilization.

Iron containing or ferrous alloys are delivered in tonnages ten times those of aluminium- and copper-based alloys joined. The reasons are not hard to see. Iron is bottomless, as well as effectively and economically extricated. In the wake of alloying and creation, steels are essentially less expensive than either Al or Cu composites. In any case, the real purpose behind the utilization of iron and steel amalgams is their prevalent quality. For basic applications (e.g., high rises, spans) and also for sheet metal walled in areas of numerous types (auto bodies, apparatuses, and so forth.) and economical instruments and wire, unalloyed plain carbon steels have a satisfactory mix of quality and pliability. For these reasons there is no need to solidify steel through alloying and extra warmth treatment. What's more, a lot of iron and steel is thrown to meet grouped requests (motor squares, machine bases, radiators, valves, and so forth.). Thrown irons are not for the most part flexible, but rather are satisfactorily solid in pressure. Furthermore, pliable and flexible cast irons exist.

Table 6.1 Mechanical properties and applications of a variety of steels

Composition (wt%)	Condition	Strength (MPa)		Ductility (% elongation)	Applications
		Tensile	Yield		
Plain Carbon					
1020 (0.20C, 0.45 Mn)	As rolled Annealed	448 393	331 297	36 36	Steel plate, structural sections
1040 (0.40C, 0.75 Mn)	As rolled Annealed Hardened	621 517 800	414 352 593	25 30 20	Shafts, studs
1080 (0.80C, 0.75 Mn)	As rolled Annealed Hardened	967 614 1304	586 373 980	12 25 1	Music wire, helical springs, chisels die blocks

Low Alloy					
4340 (0.40C, 0.90 Mn, 0.80 Cr, 0.2 Mo, 1.83 Ni)	Annealed Hardened	745 1725	469 1587	22 10	Truck, auto, and aircraft parts, landing gear
5160 (0.60C, 0.80 Cr, 0.90 Mn)	Annealed Hardened	725 2000	276 1773	17 9	Automobile coil and leaf springs
8650 (0.50C, 0.55 Ni, 0.50 Cr)	Annealed Hardened	710 1725	386 1552	22 10	Small machine axles
Stainless					
430 (17 Cr, 0.12 C max) (ferritic)		550	375	25	Decorative trim, acid tanks, restaurant equipment
304(19 Cr, 9 Ni) (austenitic)		579	290	55	Chemical and food processing equipment
Tool		**Hardness (R_c)**			**Applications**
T1 (18W, 4Cr, 1V, 0.75 C)		60 at 260°C, 50 at 600°C			Lathe and milling machine tools, drills
M2(6W, 5Mo, 4Cr, 2V, 09C)		60 at 260°C, 50 at 600°C			Taps, reamers, dies

Table 6.2 Mechanical properties and applications of a variety of aluminium alloys

Composition (wt%)	Condition	Stremgth (MPa)		Ductility (% elongation)	Application
		Tensile	Yield		
Wrougth Alloys					
1100 (99.0 min Al, 0.12 Cu)	Annealed Half-hard	89 124	24 97	25 4	Sheet metal
2024 (4.4 Cu, 1.5 Mg, 0.6 Mn)	Annealed Heat-treated (T-6)[a]	220 (max) 500	97 (max) 390	12 5	Aircraft structural part
6061 (1.0 Mg, 0.6 Si, 0.27 Cu, 0.2 Cr)	Annealed Heat-treated (T-6)	152 (max) 315	82 (max) 280	16 12	Truck and marine structures

Contd...

Contd...

7075 (5.6 Zn, 2.5 Mg, 1.6 Cu, 0.23 Cr)	Annealed Heat-treated (T-6)	276 (max) 570	145 (max) 500	10 8	Aircraft structural part
Cast Alloys					
356 (7Si, 0.3 Mg)	Sand cast Heat-treated (T-6)	207 (min) 229 (min)	138 (min) 152 (min)	3 3	Transmission and axle housings, truck wheels, pump housings
413 (12 Si, 2Fe)	Die cast	297	145	2.5	Intricate die castings

Table 6.3 Mechanical properties and applications of a variety of copper alloys

Composition (wt%)	Condition	Stremgth (MPa)		Ductility (% elongation)	Application
		Tensile	Yield		
Wrought Alloys					
OFHC[a] (99.99 Cu)	Annealed Cold worked	220 345	69 310	45 6	Bus conductors, vacuum seals, glass-to-metal seals, electrical wiring
Electrolytic Tough, Pitch (0.04O)	Annealed Cold worked	220 345	69 310	45 6	Electrical wiring, auto radiators, gutters, roofing
Brass (70 Cu, 30 Zn)	Annealed Cold worked	325 525	105 435	62 8	Radiators, lamp fixtures, musical instruments, ammunition shells, locks, electrical sockets
Beryllium copper (1.7 Bc, 0.20 Co)	Solution treated Precipitation hardened	410 1240	190 1070	60 4	Bellows, spring contacts, diaphragms, fasteners, fuse clips, switch parts, welding equipment

Casting Alloys					
Leaded bronze (5Sn, 5Pb, 5Zn)	As cast	255	117	30	Flanges, pipe fittings, plumbing goods, water pump impellers and housings, sculpture, ornamental fixtures
Aluminium bronze (4Fe, 11Al)	As cast	586	242	18	Bearings, gears, corrosion-resistant vessels

Steels are much harder than Al and Cu and, with higher melting points, are considerably more heat resistant. This reserves the utilization of steels in a wide range of forming and cutting devices for metals, as well as for polymers, wood, ceramic production, textures, etc., and for the basic assembling operations examined in the past section. Also, there are a large group of load-bearing parts (e.g., gears, shafts) in static and pivoting hardware, transportation vehicles, and earth-moving hardware that require hard wearing surfaces. For these reasons, medium- and high-alloy steels that can be hardened are crucial.

They looked for after hardness and quality are achieved, however, to the detriment of lower flexibility and some give up in strength. A correlation between plain carbon and composite steels uncovers the colossal quality advantages gave by alloying and the extinguishing and hardening heat medications they make conceivable. Among the most vigorously alloyed iron-base composites are the stainless and device steels. Stainless steels come in ferritic, austenitic, and martensitic assortments that accentuate their structure. They consolidate preparing ability with sensible quality and high malleability in applications in which consumption-free operation is vital. Device steels are basic materials since they empower the critical machining operations of turning, boring, and processing to be figured it out.

6.4 MECHANICAL WORKING AND RECRYSTALLIZATION

6.4.1 Cold Working

The first metals shaped by humans (e.g., Au, Ag, Cu) were presumably pounded frosty into the coveted shape. Paul Revere, the American silversmith, carried on the custom of pounding silver to make wonderful pieces that are saved in numerous historical centres. Beginning with a level round sheet of silver he peened it consistently in concentric circles, raising the metal into a bowl. Advance pounding would tear the now less flexible silver so he toughened it. This mollified the metal and reestablished his capacity to frosty work and shape it facilitate until the following temper was required, etc. In this way narrow-necked decanters, candlesticks, goblets, and other items were fabricated.

Lead does not solidify on the grounds that nuclear dispersion is apparent amid room temperature toughening. It is unimportant whether chilly working happens by pounding, icy moving, expulsion, fashioning, or different procedures; the microstructural change and property adjustment are comparable if the plastic strains are equivalent. Today, all the ductile metals in engineering use undergo cold working operations at one time or another to meet strength and dimensional specifications.

6.4.2 Recrystallization

Annealing cold worked metals introduces a situation, by and large known as recrystallization, that turns around the impacts of the deformation. Specifically, the earlier gentler however more bendable, unworked framework is reestablished. Such a succession of changes is taken after metallographically in a generally immaculate iron as an element of both temperature and time in Figure 9.15. Last recrystallized grain sizes may even surpass those of the undeformed grains. Obviously, temperature has a more prominent impact than time in affecting microstructural change amid recrystallization. The obvious basic reversibility amid the working/strengthening cycle gives it the presence of a sort of stage change that can be portrayed by the reaction.

$$\text{metal (soft)} \underset{\text{annealing}}{\overset{\text{cold work}}{\rightleftharpoons}} \text{metal (hard)}$$

Strain energy, consumed by the metal amid deformation, is conveyed in the push handle that encompass separations. This all the while fortifies the lattice and raises its free vitality. Diminishment of the disengagement thickness brings down the framework free vitality and this is absolutely what happens amid tempering and recrystallization. These procedures by and large require hoisted temperatures and adequate time for molecules to diffuse; at exactly that point can separations move, destroy, climb, or vanish in expansive numbers. A vital component to comprehension recrystallization conduct of various metals is the opposite relationship between liquefying point and rate of dispersion (see Section 6.3.2). This records for the way that Ag must be warmed to a few hundred degrees centigrade to toughen it, while Pb recrystallizes at room temperature.

Recovery: Also known as primary recrystallization, the recovery stage is characterized by little drop in hardness and virtually no change in grain structure. Upon closer examination a redistribution of dislocations into vertical arrays of small misorientation angle, or so-called **subgrain boundaries,** occurs. The involved dislocation climb indicates that vacancies are mobile during recovery. Nevertheless, the density of dislocations is only slightly reduced, leaving barriers to their motion largely intact. As a result, the metal still retains much of its high strength, low-ductility character. Notably, however, residual stress is largely eliminated.

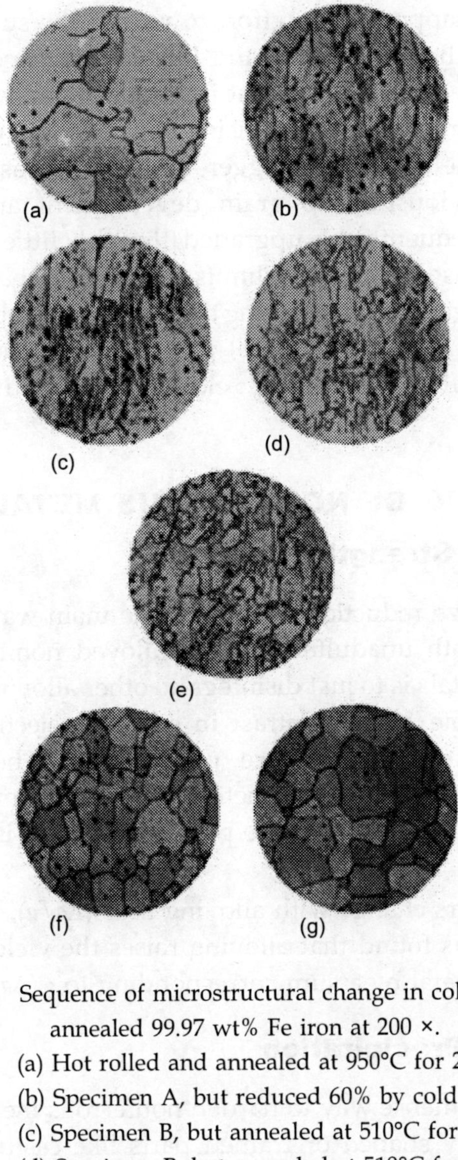

Figure 6.11 Sequence of microstructural change in cold worked and annealed 99.97 wt% Fe iron at 200 ×.

(a) Hot rolled and annealed at 950°C for 2 hours.
(b) Specimen A, but reduced 60% by cold rolling.
(c) Specimen B, but annealed at 510°C for 1 hour.
(d) Specimen B, but annealed at 510°C for 10 hours.
(e) Specimen B, but annealed at 510°C for 100 hours.
(f) Specimen B, but annealed at 675°C for 1 hour.
(g) Specimen B, but annealed at 900°C for 1 hour.

Grain Growth: After grains recrystallize they are stable, but their size can often be enlarged in a spectacular way. All that is required is to raise the temperature. As the grains grow, the matrix softens significantly and becomes even more ductile. Morphological change remarkably parallels the enlargement of soap bubble grains with time. Grains coalesce as boundaries disappear in an effort to reduce the surface tension or area of the liquid soap film. The thermodynamic and kinetic factors controlling grain growth were introduced. One aspect not treated, but important in alloy design, is the fact that foreign atoms reduce the grain size by exercising a drag on migrating grain boundaries. This implies the purer the metal, the bigger the grain measure and the lower the estimation of *TR*. Imperviousness to grain development and warm softening of immaculate metals can frequently be upgraded through little increases of particular alloying components that isolate to grain limits. The official between the alloying and host particles immobilizes grain limits, making them successful obstructions to separation accidents and obliteration forms. Therefore, raised temperature quality is presented without horribly trading off properties (e.g., electrical conductivity) touchy to alloying content.

6.5 STRENGTHENING OF NONFERROUS METALS

6.5.1 Solid Solution Strengthening

Cool working and grain size reduction have been the main ways presented up to this point for strengthening both unadulterated and alloyed nonferrous metals. Another strategy pertinent to all metals is to just disintegrate other alloying components in them. Since solute and dissolvable atoms contrast in size and electronic character, nearby cooperations with separation centres are unavoidable. The size confuse $(\Delta a/a)$ characterized as the proportion of the distinction in the two atomic radii to the span of the dissolvable atoms, makes a strain whose push field makes it harder for separations to move.

Matrix lattice parameters change with alloying thus $(\Delta a/a)$, will be a component of solute concentration, *C*. It is found that alloying raises the yield stress with respect to that of an unadulterated metal by a sum corresponding to $\sim(\Delta a/a)^{3/2}\,C$.

6.5.2 Hardening by Precipitation

Mechanical working is an intense way to harden nonferrous metals, yet in the occasion that endeavoured on viably shaped or framed parts like castings, they will twist or, perhaps, even break. Precipitation cementing solutions do have the capacity of further strengtheneing shaped parts safely and are altogether easy to realize. But not all alloy compositions are suitable candidates for these hardening heat treatments, and optimal ones must be selected. The first requirement is that the alloy system has a phase diagram (only binaries will be considered) with solid-state solubility limits that increase appreciably as the temperature is raised. Fortunately, many binary systems display this behaviour, the classic example being the Al-Cu system shown in Figure 6.12. Aluminium-rich alloys containing up to ~5 wt% Cu cross the sloped solidus phase boundary line and appear to be promising. If an alloy containing 4 wt% Cu were slowly cooled down to the ambient following high-temperature processing (forming or casting) in the α field, coarse θ ($CuAl_2$) phase precipitates would be rejected from the prior solid solution. The resulting two-phase structure is harder and stronger than pure Al, but not appreciably so.

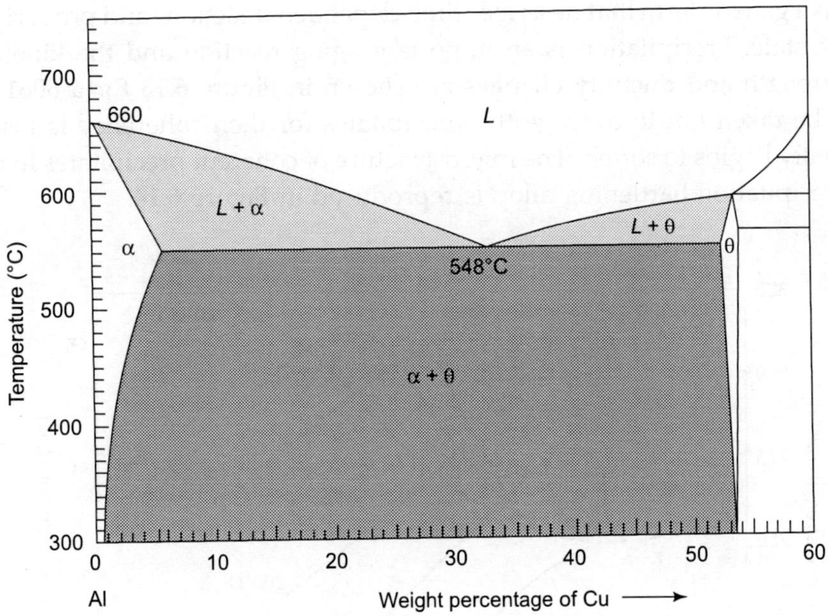

Figure 6.12 Equilibrium phase diagram of Al-rich portion of the Al-Cu system.

More significant strengthening can be achieved if, instead of slow cooling from the α field, the alloy were quenched. Now Cu atoms are frozen into a solid solution and corresponding (metastable) θ phase formation is suppressed. This is a case of simple solid solution strengthening that was discussed briefly in the preceding section. Although normally a stepping stone in precipitation hardening treatments, significant strengthening can sometimes be achieved by quenching to retain a supersaturated solid solution. In Al-Mg alloys, for example, a quadrupling of the yield stress relative to annealed Al is attainable with a 5 wt% Mg addition. Nevertheless, it is usual to liberate the trapped solute from supersaturated solution by controlling the nucleation and growth rates of the rejected θ precipitates. An optimal precipitate phase size and dispersion are required for maximum strengthening.

6.5.3 Precipitation Hardening Treatment

Prerequisites for hardening small, optimally dispersed precipitates have been suggested in the previous paragraphs. A recipe for hardening an Al-4 wt% Cu alloy therefore necessitates the following steps:

1. **Solution** heat-treat to solution temperature T_s (e.g., 540°C) in the α phase field.
2. **Quench** to a low temperature (e.g., 25°C) in the α + θ field.
3. **Anneal** or age at temperature T_p (e.g., 150°C) for the time necessary to optimally form coherent θ^{l} phase precipitates and develop maximum mechanical properties.

Aging is a generic term that describes time-dependent reactions and property changes in the solid state. Precipitation is an important aging reaction and the kinetics of the resultant strength and ductility changes are shown in Figure 6.13 for a 6061 Al alloy. Care must be taken not to overage the precipitates for then coherency is lost and the matrix actually begins to soften. The microstructure of coherent precipitates in the Al-3.8 wt% Cu precipitation hardening alloy is reproduced in Figure 6.14.

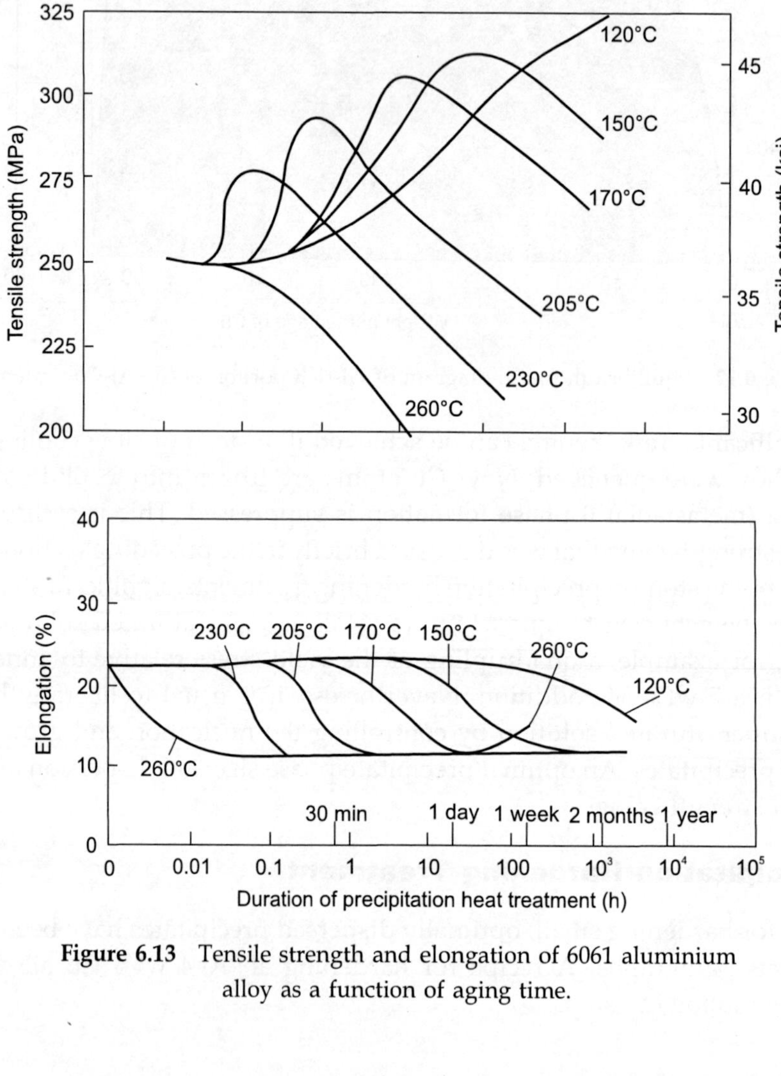

Figure 6.13 Tensile strength and elongation of 6061 aluminium alloy as a function of aging time.

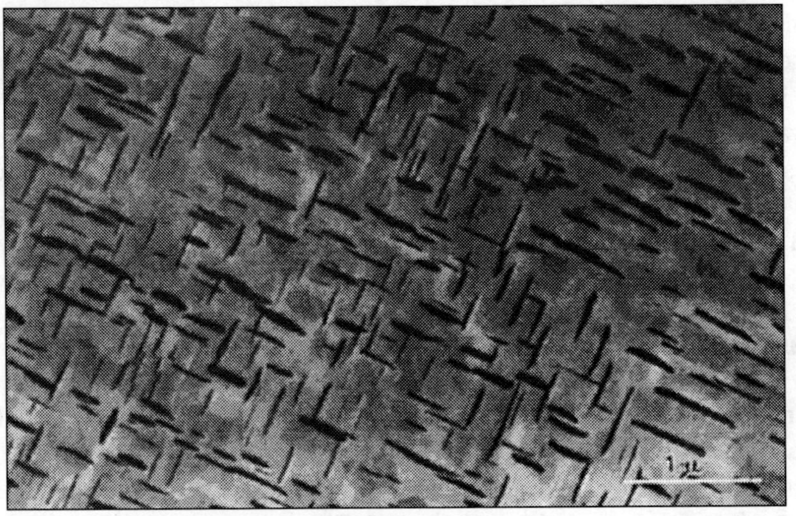

Figure 6.14 TEM image of an Al-3.8 wt% C alloy aged at 240°C for 2 hours showing θ^{l} $CuAl_2$ precipitates.

A similarity between isothermal transformation of austenite and decomposition of the extinguished Al α phase amid precipitation hardening deserves a remark. Truth be told, a TTT curve of the sort drawn above is suitable for any isothermal, nucleation, and development or growth transformation. At aging (transformation) temperature T_p, α phase is stable until θ nucleates and the $\alpha + \theta$ field is entered. As in steel, aging at a lower temperature increases the driving force for transformation. There is a larger precipitate nucleation rate leading to a finer precipitate size. A harder microstructure results, but it takes a longer time for it to evolve.

6.6 MODELLING COMPOSITE PROPERTIES

Around 30 years prior Dustin Hoffman, the legend of the film *The Graduate*, was given a single word bit of sage exhortation that characterized the future—"plastics." Today the word would be "composites." The development and wide-scale utilization of manufactured composites as of late, especially those containing filaments in a polymer matrix, have made an upheaval in materials use that keeps on accelerate. Really, utilizing high-quality strands to strengthen a cheap matrix is not new.

A working meaning of a composite is a material that contains a physical mixture of at least two constituent phases that are synthetically unique and isolated by an unmistakable interface. Two-stage mixtures like eutectics (Pb-Sn) and eutectoids (Q:-Fe-Fe_3C) would qualify as composites in a broad sense. So would the characteristic composites surrounding us, wood being a prime illustration. It comprises adaptable cellulose fibres inserted in the stiffer undefined lignin polymer network. Yet, we are worried with misleadingly integrated composite structures, not those that frame normally.

Composites for the most part display a synergistic admixture of the attributes of the individual constituents with the goal that properties of the resultant blend are moved forward. In the chain of command of composites particulate composites are the most straightforward. Next are the sinewy composites, which speak to a more mind-boggling method for strengthening the network. In conclusion are basic composites, comprising sandwiches or covers of less complex composites. This segment is dedicated to the hypothetical reason for the reliance of the flexible and plastic quality of composites on molecule or fibre content.

6.6.1 Young's Modulus

Particle-Reinforced Composites: Consider a dispersion of isolated particles (*p*) in a matrix (*m*) as shown in Figure 6.15a. The volume fraction of particles is V_p and they have a Young's modulus of E_p; the corresponding values for the matrix are V_m and E_m. What is Young's modulus of the composite? This is part of a broader question of estimating net properties, not only mechanical but electrical, thermal, and so on, of a composite material containing a dispersed phase. There are two simple ways of averaging properties of mixtures. One is a linear law of mixtures expressed by the weighted average of individual properties or

$$E_c = V_p E_p + V_m E_m$$

where, E_c is the modulus of the composite. An alternate possibility is to assume that the inverses of the moduli are additive,

$$E_c = E_m E_p / (V_m E_p + V_p E_m)$$

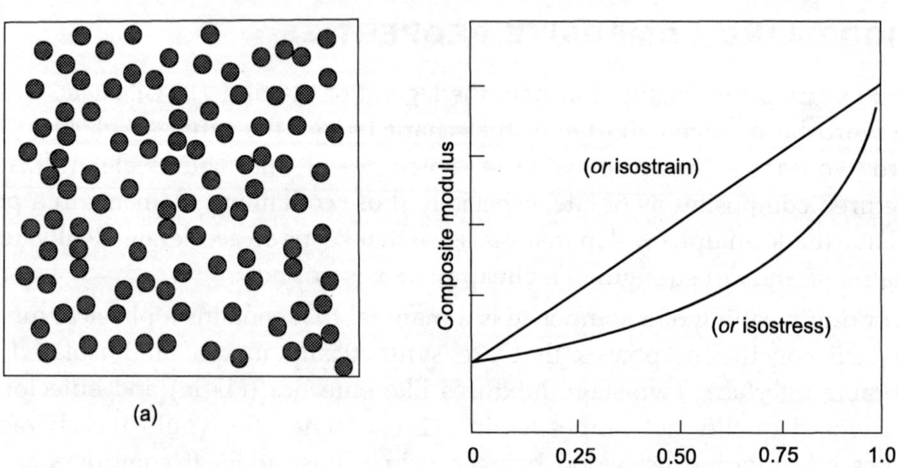

Figure 6.15 (a) Model of a composite containing a volume fraction V_p of dispersed particles. (b) Variation of E_c with V_p.

Fibrous Composites: These important composites consist of fibres, with a large length-to-diameter ratio, embedded in a matrix. For simplicity we consider long or continuous fibres that extend from one end to the other of a cube-shaped composite. With no loss in generality, the composite is equivalently visualized to consist of alternating parallel, rectangular layers of fibre and matrix as shown in Figure 6.16a. It is a simple matter to calculate Young's modulus for the case where the composite is loaded along the fibre direction with force F_c. Both the fibre and the matrix are assumed to behave elastically with respective moduli E_f and E_m. Static equilibrium implies that F_c is equal to the sum of the forces carried by each constituent, that is, F_f and F_m. These forces, in turn, are given the product of the corresponding stresses, σ_f and σ_m and the respective load bearing areas of fibre and matrix. But, for a composite of unit volume (a unit cube), these areas are equal to the respective volume fractions, V_f and V_m, in magnitude, where $V_f + V_m = 1$. Deformation compatibility requires that fibres and matrix extend by the same strain e. Therefore, with the help of Hooke's law ($\sigma = Ee$),

$$1 \times \sigma_c = V_f \sigma_f + V_m \sigma_m$$

$$\sigma_c = V_f E_f + V_m E_m e$$

But, Young's modulus for the composite is $E_c = \sigma_c / e$, with the result that

$$E_c = V_f E_f + V_m E_m$$

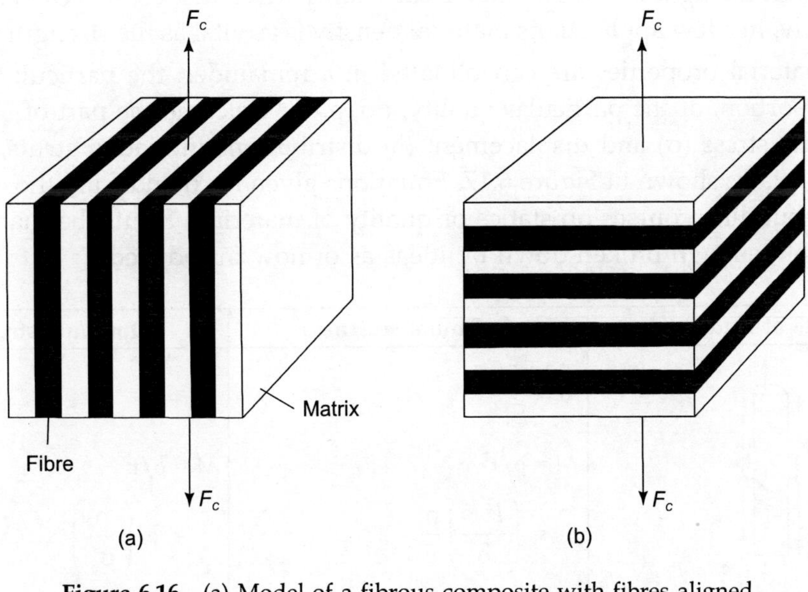

Figure 6.16 (a) Model of a fibrous composite with fibres aligned parallel to the loading direction. (b) Model of a fibrous composite with fibres aligned perpendicular to the loading direction.

Isostress deformation happens when the strands are situated opposite to the stacking hub (Figure 6.16b). Both constituents now bolster a similar load or stretch σ, yet extend or strain distinctive sums in light of the fact that their versatile moduli contrast. The aggregate composite strain e_c is given by the total of the individual strains (e_f and e_m) or

$$e_c = e_f + e_m = \sigma V_f / E_f + \sigma V_m / E_m$$

6.7 ENGINEERING COMPOSITES

6.7.1 Polymer Composites

Design: Through reinforcement with the strands recorded in Table 6.4, the demonstrated industrially created polymer composites result. That unreinforced polymers are extensively less firm (modulus) and solid (yield and crack quality) than other designing materials (metals and earthenware production) has been called attention to some time recently. Then again, the glass, carbon, and aramid (Kevlar) filaments generally utilized as a part of composites have firmness and quality properties similar to those of designing materials. The individual composites are generally known as GFRP, CFRP, and KFRP, where the last three letters allude to fibre-reinforcement polymer.

In designing structures for vehicles and planes, vital contemplations are to improve both firmness and quality for a given weight. The sympathy towards lessened weight is justifiable in the light of the fact that great vitality investment funds can come about. Subsequently, in a few applications material density is as vital as the strength properties.

Both material properties are consolidated in a remainder: the particular modulus, or E/ρ proportion, or the particular quality, σo/ρ. To welcome the part of E/ρ in plan considers the stress (σ) and displacement (δ) distributions in loaded struts, cantilever beams, and plates shown in Figure 6.17. Equations given for σ and δ are those regularly inferred in building courses on statics or quality of materials. Struts, be that as it may, are effectively enough broken down by ideas as of now introduced.

Mode of loading	Optimum sitiffness	Optimum strength
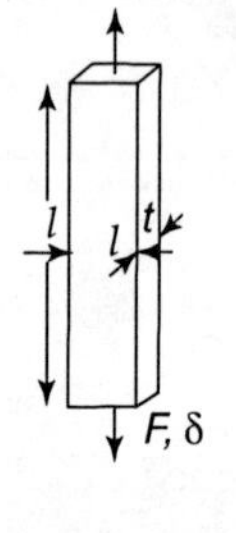	$\delta = \dfrac{Fl}{Et^2}$ $M = \rho/t^2$ $= \left(\dfrac{l^2 F}{\delta}\right)\dfrac{\rho}{E}$ Maximize $\dfrac{E}{\rho}$	$\sigma_o = \dfrac{F}{t^2}$ $M = \rho/t^2$ $= Fl\left(\dfrac{\rho}{\sigma_o}\right)$ Maximize $\dfrac{\sigma_o}{\rho}$

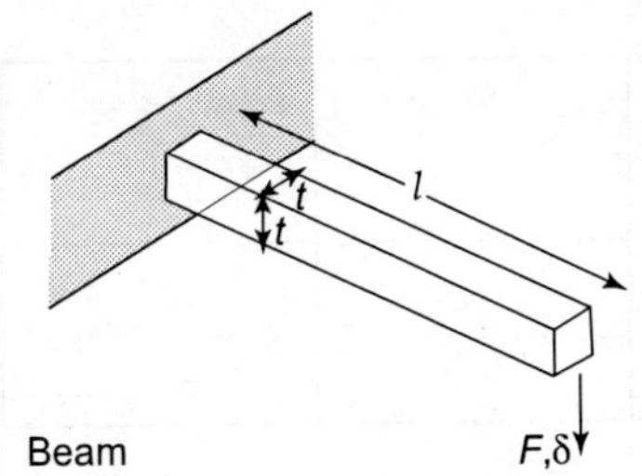 Beam	$\delta = \dfrac{4Fl^3}{Et^4}$ $M = \rho/t^2$ $= 2\left(\dfrac{Fl^5}{\delta}\right)^{1/2}\left(\dfrac{\rho}{E^{1/2}}\right)$ Maximize $\dfrac{E^{1/2}}{\rho}$	$\sigma_o = \dfrac{6Fl}{t^3}$ $M = \rho/t^2$ $= l(6Fl)^{2/3}\left(\dfrac{\rho}{\sigma_o^{2/3}}\right)$ Maximize $\dfrac{\sigma_o^{2/3}}{\rho}$
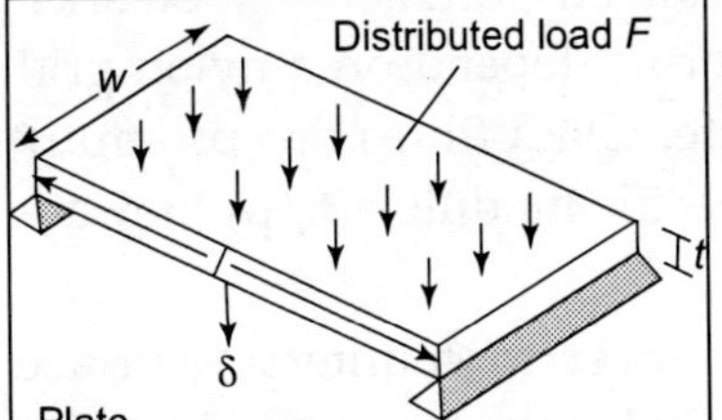 Plate	$\delta = \dfrac{5Fl^3}{32Ewt^3}$ $M = \rho/wt$ $= l^2\left(\dfrac{5Fw^2}{32\delta}\right)\left(\dfrac{\rho}{E^{1/3}}\right)$ Maximize $\dfrac{E^{1/3}}{\rho}$	$\sigma_o = \dfrac{3Fl}{4wt^2}$ $M = \rho/wt$ $= \left(\dfrac{3Fl^3 w}{4}\right)^{1/2}\left(\dfrac{\rho}{\sigma_o^{1/2}}\right)$ Maximize $\dfrac{\sigma_o^{1/2}}{\rho}$

Figure 6.17 Strut, beam, and plate geometries and the combination of properties that maximize the stiffness and strength to weight.

Table 6.4 Properties of reinforcing fibres, polymer matrices and composites

Material	Density (Mg/m^3)	E (GPa)	σ_{UTS} (GPa)	Fibre radius (μm)	Fracture Toughness, K_{IC} ($MPa\text{-}m^{1/2}$)
Fibre					
E-glass	2.56	76	1.4-2.5	10	
Carbon					
High-modulus	1.75	390	2.2	8.0	
High-strength	1.95	250	2.7	8.0	
Kevlar	1.45	125	3.2	12	
Matrix					
Epoxy	1.2-1.4	2.1-5.5	0.063		
Polyester	1.1-1.4	1.3-4.5	0.060		
Composites					
CFRP (58% C in epoxy)	1.5	189	1.05		32-45

Contd...

Contd...

GFRP (50% glass in polyster)	2	48	1.24		42-60
KFRP (60% Kevlar in epoxy)	1.4	76	1.24		

Applications: Fibreglass (GFRP): Produced in the biggest amounts of any polymer grid composite, fibreglass comprises glass filaments in arranged polymer frameworks, polyesters being the most well known. More up-to-date forms depend on a nylon grid and are by and large more grounded and more effect safe. Glass fibres are promptly drawn from the liquid state; notwithstanding their high specific modulus (£/p) they are chemically inert.

Yet, first they should be covered by a thin polymer layer to counteract surface imperfections and entrance by water. (Glass fibre for optical interchanges is likewise covered quickly in the wake of drawing for comparable reasons.) To deliver fibreglass composites it is then important to encompass the strands by the polymer lattice and afterward frame the two together into valuable shapes. Fibreglass applications incorporate car and marine bodies, and also dividers and floor materials in structures. Benefit temperatures don't develop much past 200°C when the matrix starts to stream.

Carbon and Other Fibre-Polymer Composites: Carbon (graphite) fibre has a higher particular modulus than glass; it is likewise more impervious to hoisted temperature and substance introduction, however is significantly more expensive. Along these lines, GFRP composites are held for all the more requesting applications that can legitimize the additional cost. In superior air ship they represent weight investment funds of – 25% in respect to metals. The broad utilization of graphite-reinforced polymers in wearing hardware is notable to everybody.

Tennis rackets, golf clubs, and sailboat poles are a portion of the applications. Less notable is their utilization in weaving gear parts. Their high firmness and low weight make conceivable the outline of fast weavers. Arranged polymers have likewise been impregnated with different strands, quite Kevlar (KFRP), boron, and silicon carbide. They have discovered their way into grouped applications including donning products and aviation, airplane, and marine parts.

Fabricating Polymer Composites: Numerous methods have been developed to fabricate fibre-reinforced polymer matrix composites as schematically indicated in Figure 6.18. The first step be subsequently consolidated into a laminated structure. To accomplish this, continuous fibre unwound from bobbins forms strands that are first coated with agents to promote good bonding to the matrix polymer (resin). These strands are then continuously immersed in resin baths that will become the matrix that surrounds the fibres (Figure 6.18a). Sheets can be compression molded or hand laid in molds to make laminates (Figure 6.18b). Tape prepregs can also be wound over large shapes and cured to make large lightweight containers (Figure 6.18c). By cofeeding and compacting cleaved fibre and resin (Figure 6.18d) sheet is created. Formed composite parts are normally injection shaped by constantly including chopped fibre (~ 1 cm long) alongside the polymer feedstock. Controlling the fibre introduction is a critical worry in this procedure.

6.7.2 Metal Matrix Composites

Particulate Composites: Albeit metal matrix composites (MMCs) have been of enthusiasm for quite a few years it is just as of late that techniques have been created to join extensive volume divisions (up to half) of oxide, carbide, boride, and different

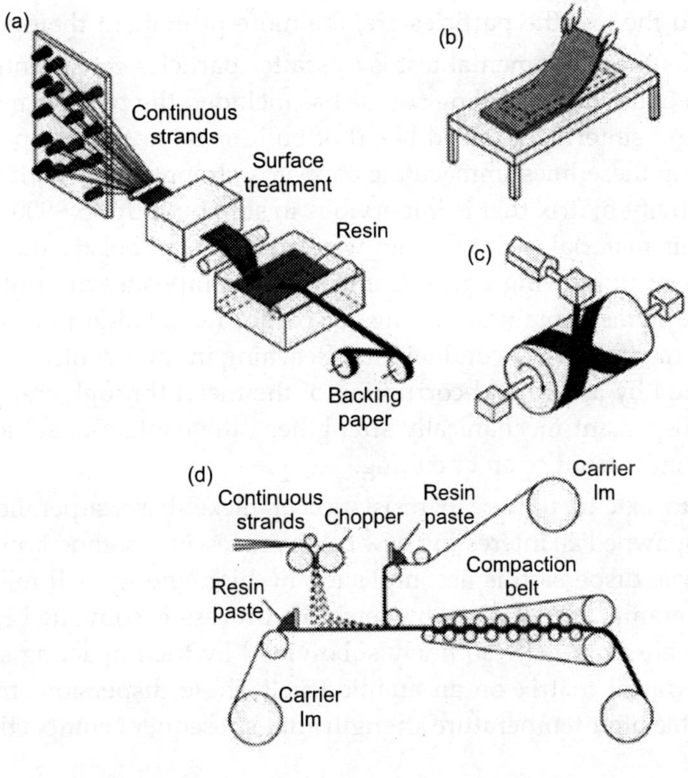

Figure 6.18 Methods for producing polymer matrix composites.
(a) Tape or sheet production using continuous fibres.
(b) Assembly into a laminated structure. (c) Filament winding.
(d) Sheet production using cut fibres.

particles (and filaments) into metal grids. Then again, dispersion strengthened combinations were the most punctual MMCs and they contained just a few volume percent particles, generally oxides.

Despite the fact that both sorts of MMCs have huge points of interest over the matrix metal, these extremes in molecule substance are promoted upon in various applications. By scattering oxide particles in a ductile lattice (e.g., ThO_2 in Ni, Al_2O_3 in Al) the oxide holds its high hardness at hoisted temperatures and limits the milder, bring down softening point metal from streaming. A critical change in imperviousness to thermal softening and creep can in this manner be accomplished in dispersion-strengthened alloys. The standards administering encourage size and dissemination for ideal strengthening of precipitation hardening combinations are additionally legitimate for dispersion-strengthened composites. Since incoherent interfaces create amongst dispersant and matrix, coherency stresses can't be depended upon. In their non-attendance, submicrometre (1 μm = 10^{-3} mm) measured particles are vital for blocking separation movement, and the less the particles are, the more prominent their adequacy.

In all MMCs the fundamental test is to scatter particles consistently all through the metal. The run of the mill handling course has included the powder metallurgy systems of squeezing and sintering a mixed blend of both metal and oxide powders, trailed by extrusion. Along these lines immaculate copper is strengthened with alumina to create a high-conductivity matrix that is impervious to softening up to ~900°C. Containing ~5 vol% Al_2O_3, this material is drawn into wire and used to bolster the exceptionally hot tungsten filament in glowing lights. Comparative composites are utilized for electrical contacts. These items contain scatterings of oxides (e.g., CdO) in a metal matrix (e.g., Ag) that has a high electrical conductivity. Reaching in high-control electrical circuits is frequently joined by arcing and corruption of the metal through cratering, pitting, and erosion. The dispersant mechanically strengthens the contact metal against such wear and shields circuits from open circuiting.

Attempts to extend the creep resistance of nickel-base superalloys to yet higher temperatures spawned an interesting new batch processing method known as **mechanical alloying.** Particle dispersion is accomplished by high energy ball milling of a mixture of metal and ceramic (e.g., Y_2O_3) powders. The process is so named because the brittle yttria particles are eventually so finely subdivided by the impacting steel balls that the alloy with the metal matrix on an atomic level! These dispersion-strengthened alloys show remarkable high-temperature strength after subsequent compaction, sintering, and extrusion.

Fibrous Composites: Strengthening metals by means of the fibrous composite course has additionally picked up momentum as of late. Carbon, boron, silicon carbide, and alumina, and in addition metal fibres, have been joined in measures of up to half by volume to harden Al, Mg, and Ti alloys for applications noted previously. Fibre or whisker mats that are penetrated with metals (press throwing) can be situated, giving quality and solidness focal points in respect to particulate MMCs; in any case, the fibre surfaces must be uniquely treated to improve metal wetting and interfacial grip, and in addition restrict undesirable responses that degrade bonding.

In respect to the base metal, half increments in rigidity and Young's modulus combined with a comparative reduction in thickness can be generally expected in both sorts of MMCs. Real property improvements are demonstrated in Table 6.5 and rely on the specific MMC and volume division of particles or fibres.

Table 6.5 Properties of some metal and ceramic matrix composites and their reinforcements.

Material	Density (Mg/m^3)	E (GPa)	σ_{UTS} (GPa)	Fibre radius (μm)
Fibre				
Silicon carbide				
Whisker	3.2	480	3	
Fibre	3.0	420	3.9	140
Alumina	3.2	100	1.0	3
Si_3N_4	3.2	380	2	
Boron	2.6	380	3.8	100-200
Composite				
2014-Al (20% Al_2O_3, particulate)		103	0.48	
2124-Al (30% SiC, particulate)		130	0.69	
6061-Al (continuous fibre-51% B)		231	1.42	–
601-Al (discontinuous fibre-20% SiC)		115	0.48	–
601-Al (no reinforcement)		69	0.31	–
Si_3N_4 (10% SiC whisker)			0.45	–
Al_2O_3 (10% SiC whisker)			0.45	–

6.8 STRENGTHENING AND TOUGHENING OF CERAMICS

6.8.1 Mechanical Properties of Ceramics

Table 6.6 is dedicated exclusively to the mechanical properties of pottery where their remarkable traits merit rehashing. Ceramics are seen to have high flexible moduli and low densities (and, in this manner, high E/ρ proportions), combined with high hardnesses and melting temperatures. Extra positive properties in numerous applications incorporate low warm conductivity and warm development coefficients.

Then again, ceramics are extremely powerless to warm stun. This imperative property reflects the propensity of materials to crack as an after-effect of rapid temperature (T) change, for the most part amid cooling.

6.8.2 Toughening Ceramics

In addition to thermal shock, the intrinsically brittle nature of ceramics is reflected in very low values of fracture toughness. The driving force to create fuel-efficient, high-temperature turbine engines has spawned a worldwide research effort to enhance the fracture toughness of ceramics. The main strategy seeks to engineer the microstructure so that cracks have a more tortuous path to execute as they advance through the matrix. Introducing compressive stress into the ceramic surface, thereby offsetting the influence of harmful applied tensile stresses, is another approach. How these toughening mechanisms have been practically implemented will be addressed in turn.

Composites: Ceramics can be toughened by the same principle applied successfully in polymers, namely, through the use of composites. Very short, small-diameter fibres known as whiskers are incorporated into ceramic bodies up to a volume fraction of 25%. Because energy is expended in pulling whiskers out of the matrix and in bending or breaking them, crack propagation is blunted and the fracture toughness is effectively raised. In one application, cordierite glass ceramic substrates, on which arrays of computer chips are mounted, have been toughened with 1 μm diameter, 10 μm long Si_3N_4 whiskers.

In another, SiC fibres have toughened Al_2O_3 cutting devices utilized as a part of machining superalloys. Various reinforcements have been economically accessible for quite a while including SiC, Al_2O_3, and boron strands; ordinary filaments are 10 to 20 μm in width. Some of their properties and those of the artistic frameworks they fortify are recorded in Table 6.5. Forecast of composite properties takes after the rules displayed before for fibre-reinforced polymers.

Achieving the right bond quality between the fibre and the matrix is the way to the accomplishment of harder ceramic composites. However, this is less demanding said than done. Lamentably, very much of the time change in mechanical properties of earthenware production comes to the detriment of simplicity in sintering. Porosity at the fibre matrix interface causes low bond quality and an absence of toughening. At the other outrageous, excessively great bond strength quality counteracts fibre pullout and prompts to fracture rather; again the outcome is low toughening.

Table 6.6 Properties of ceramics

Material	E (GPa)	ρ (Mg/m³)	T_M(K)	K(W/m-K)	α (10^{-60} C^{-1})	K_{IC} (MPa-$m^{1/2}$)	Modulus of rupture (MPa)
Diamond	1050	3.52		1100	1		
Tungsten carbide, WC	550	15.7		35	4	80	
Borides (Ti, Zr, Hf)	500	4.5		~30	7.8		
Silicon carbide, SiC	410	3.2	3110	84	5.3	3	200-500
Alumina, Al_2O_3	390	3.9	2323	~27	8.8	3.5	300-400
Beryllia, BeO	340	2.9	2570	~40	5	7.4	
Silicon nitride, Si_3N_4	310	3.2	2173	17	2.5	4	300-850
Magnesia, MgO	250	3.5	2620		10		
Zirconia, ZrO_2	200	5.5	2843	1.5	8	4-12	200-500
Mullite, $3Al_2O_3.2SiO_2$	150	3.1	1850			4.6	
Titanium carbide, TiC	460	4.9	3067	34	8.3	0.5	
Titanium nitride, TiN	590	5.4	2950	30	9.3		
Sialons (Si-Al-N)	300	3.2		23	3.2	5	500-830
Silica, SiO_2	94	2.2	1980	2	0.55	1	
Granite	70	2.6			8		23
Glass	70	2.4	~1000	1	~6	1	50
Concrete	40	2.4		2	12	0.2	7

Transformation Toughening: Unadulterated zirconia (ZrO_2) experiences an unconstrained tetragonal–monoclinic martensitic response at 1150°C; the related ≈9% volume extension makes creation forms troublesome in light of push incited breaking. Zirconia can be toughened, be that as it may, through a 3.6 wt% Y_2O_3 expansion. Diverse measures of CaO and MgO will likewise work. In such compounds at around 1100°C, tetragonal ZrO_2 is steady, yet this stage can be held at room temperature amid moderately rapid cooling.

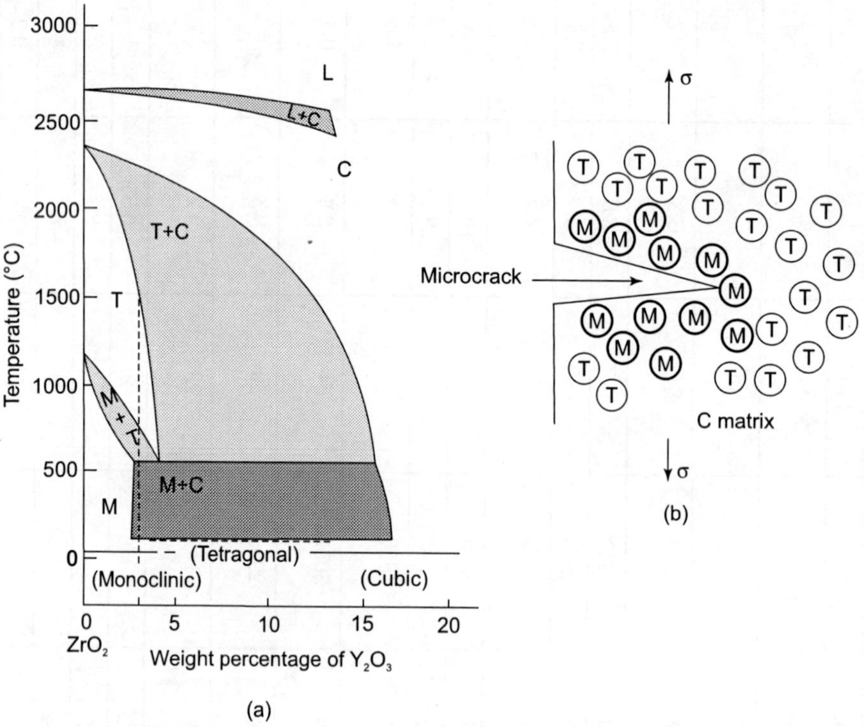

Figure 6.19 (a) Portion of the ZrO_2 –Y_2O_3 phase diagram.
(b) Mechanism of toughening in zirconia stabilized by yttria.
The stress-induced transformation of tetragonal (T) ZrO_2 grains to the monoclinic (M) structure expands the volume at the crack tip, squeezing it shut.

The stage outline of Fig. 6.19a demonstrates that the steady stage is a monoclinic ZrO_2-rich strong arrangement, which exists together with this now precarious tetragonal stage in metastable harmony. Change from the tetragonal to the monoclinic stage is drowsy however is speeded up by high temperature maturing or activated by splits exhibit in the grid. A ~ 3 % volume increment because of nuclear revamping goes with the change. All the while, a compressive push field creates around changed particles. The compressive hassles seal splits close and, in blunting their further progress, toughens the lattice as schematically shown in Figure 6.19b. Crack durability values drawing closer 10 MPa-m1/2, which are a few times higher than those for different earthenware production, have been measured in these incompletely settled zirconia (PZT) materials. This is the reason why they are called ceramic steels.

Surface Toughening: Presentation of desirable residual compressive stresses in ceramic surfaces can be achieved chemically. Indiffusion of metal ions bigger than those of the framework stuffs them into littler empty openings, compacting the diffused surface layers. Particle trade of this sort is widely drilled in toughening glasses. In such materials the typically damaging connected pliable burdens are adequately decreased to safe levels by the subtractive impact of the lingering residual compressive stresses. An allied method borrowed from semiconductor technology, known as *ion implantation,* has been employed on an experimental basis to toughen ceramic surfaces by the same principle. As the name implies very energetic ions (~ 50 keV) of the desired metals or nonmetals are propelled by large electric fields and implanted ~1 μm beneath the surface.

6.9 SUMMARY

It is the fate of every material to have a mix of desirable and undesirable properties. The main driving force to persistently enhance them is to capture, retain, and expand their business utilization. For mechanical properties this stances diverse difficulties relying upon the material included. Malleable materials like metals are as of now extreme furthermore tolerably firm and hard; the essential assignment is to harden, strengthen, and protect them from natural harm. At the other outrageous, pottery are hardened, hard, and solid; they should be toughened. Between these extremes are polymeric materials which are neither hardened or hard or strong and nor tough; be that as it may, they are light, cheap, and effortlessly shaped. High stiffness, strength, and toughness then turn into the non-specific objectives that we set for all materials. How nearly properties merge to these general goals relies on upon the specific material being referred to. When in general, wonderful advances have been and keep on being made in every material class.

On a microscopic level, plastic deformation corresponds to the motion of dislocations in response to an externally applied shear stress, a process termed "slip." Slip occurs on specific crystallographic planes and within these planes only in certain directions. A slip system represents a slip plane-slip direction combination, and operable slip systems depend on the crystal structure of the material.

The critical resolved shear stress is the minimum shear stress required to initiate dislocation motion; the yield strength of a single crystal depends on both the magnitude of the critical resolved shear stress and the orientation of slip components relative to the direction of the applied stress.

For polycrystalline materials, slip occurs within each grain along the slip systems that are most favourably oriented with the applied stress; furthermore, during deformation, grains change shape in such a manner that coherency at the grain boundaries is maintained.

Under some circumstances limited plastic deformation may occur in BCC and HCP metals by mechanical twinning. Normally, twinning is important to the degree that accompanying crystallographic reorientations make the slip process more favourable.

Metals

Metals begin off with may be the best starting supplement of mechanical properties, and that is the reason they have been the materials of decision in auxiliary and machine applications for so long. *Cold working, alloying, quenching, tempering* (for steels), and *precipitation hardening* treatments (for nonferrous alloys) are the customary approaches to fortify and solidify metals. In these cases there is some yield in flexibility and strength. Alloy preparation and solution then look to strike the ideal trade-off between yield strength and fracture toughness, two key properties that as a rule differ contrarily with each other.

Upsetting the lattice perfection and reducing the size of structural features in the matrix constitute the strategy employed to increase the strength of metals. The reason it works is that disengagement development is hindered. Be that as it may, confined disengagement movement likewise constrains the advancement of break tip plastic zones that are viable in blunting split engendering. The outcome decreases break durability. To meet the requests of cruel, high-temperature applications, unique stubborn metals (e.g., superalloys) containing attractive stages, encourages, and dispersants have been created. Not to be beaten by other material contenders, particulate and fibre-reinforced metal matrix composites of aluminium and magnesium are getting a charge out of expanding use.

Polymers

For a heap of non-load bearing applications (e.g., packaging, insulation, assorted moulded consumer goods, toys), unstrengthened polymers will do fine and dandy. In any case, for all the more requesting designing applications, especially in transportation where low weight is basic for speed or vitality protection, polymer composites have filled a key need. The reinforced polymer matrices have been at the same time solidified, reinforced, and toughened by scattering firm, solid fibres of glass, carbon, or polymer inside them. Coming about fibre/lattice admixtures have yielded improved composite flexible moduli and strength. The previous is expected specifically to the nearness of the filaments which convey most of the heap; the last happens in light of the fact that the trouble in hauling strands out captures break expansion. Ceaseless adjusted strands present most extreme advantages, however cut filaments are profitably utilized if they surpass a basic length. Without their fuse into composites both the filaments and the polymer matrices would have more restricted designing use.

Ceramics

Blessed with high stiffness, strength, and temperature capabilities, ceramics have two fatal attributes: low fracture toughness and high susceptibility to thermal shock. These shortcomings have been addressed by a number of strategies including the use of fibre composites, transformation toughening (in ZrO_2), and residual surface compressive stresses. Despite improvements, ceramics still largely lack sufficient toughness to be included in most fracture-safe structural designs. But, in the form of filaments and coatings, ceramics have transferred their desirable properties to *other engineering materials* in creating bulk and surface composites.

Since the ease with which a material is capable of plastic deformation is a function of dislocation mobility, restricting dislocation motion increases hardness and strength. On the basis of this principle, three different strengthening mechanisms were discussed. Grain boundaries serve as barriers to dislocation motion; thus refining the grain size of a polycrystalline material renders it harder and stronger. Solid-solution strengthening results from lattice strain interactions between impurity atoms and dislocations. Finally, as a material is plastically deformed, the dislocation density increases, as does also the extent of repulsive dislocation–dislocation strain field interactions; strain hardening is just the enhancement of strength with increased plastic deformation.

The microstructural and mechanical characteristics of a plastically deformed metal specimen may be restored to their predeformed states by an appropriate heat treatment, during which recovery, recrystallization, and grain growth processes are allowed to occur. During recovery there is a reduction in dislocation density and alterations in dislocation configurations. Recrystallization is the formation of a new set of grains that are strain free; in addition, the material becomes softer and more ductile. Grain growth is the increase in average grain size of polycrystalline materials, which proceeds by grain boundary motion.

EXERCISE

1. Name the heat treatments for 1080 steel that would affect the following conversions:
   (a) Pearlite to martensite
   (b) Martensite to coarse pearlite
   (c) Martensite to fine pearlite
   (d) Pearlite to bainite
   (e) Bainite to pearlite
   (f) Martensite to bainite
   (g) Spherodite to pearlite

Hint: Refer your treatments to the TTT curve.

2. Since time immemorial blacksmiths have been forging stone carving tools starting with essentially pure iron, containing perhaps 0.05 wt% carbon. Assuming you have such round bar blanks and a blacksmith's hearth, detail the steps you would follow to forge a flat-edged stone carving chisel. Note that chisels must be very hard and tough at the cutting edge, but soft where grasped by the hand to absorb mechanical impact and prevent a ringing sensation.
3. What would be the change in resulting hardness and distance profile of Jominy curve if a Jominy bar is austenitized end quenched by applying oil rather than water?
4. Vanadium, a common alloying element in tool steels, refines the grain size by reducing the prior austenitic grain size. Consider a steel with and without V. Jominy bars of both steels are austenitized at high temperatures and end quenched. Sketch the expected microstructures along the bar for each steel. (**Note:** Pearlite nucleates at austenite grain boundaries.)
5. The increase in carbon content increases the hardness of martensite, pearlite and spherodite. Why?
6. Values of fracture stress in ceramic materials are subjected to a wide standard deviation unlike the tensile strength of materials which shows little spread. Why?
7. Explain three methods of toughening ceramics. How can the toughened ceramics meet the toughness of metals?
8. Write in the increasing order of ability to withstand thermal shock BeO, SiC, ZrO_2, TiN, Si_3N_4.
9. A composite containing 25 vol% aligned continuous carbon fibres in an epoxy matrix fails at a tensile stress of 700 MPa. If the fibre tensile strength is 2.7 GPa what is the matrix yield strength?
10. Recommend the sequence of heat treatments for Al-Si phase diagram?
11. For each of edge, screw, and mixed dislocations cite the relationship between the direction of the applied shear stress and the direction of dislocation line motion.
12. Write a short note on strengthening of ceramics.
13. How can nonferrous metals be strengthened? Explain some of the methods in detail. Also provide relevant equations.
14. List the properties and applications of ferrous and nonferrous alloys.
15. What do you understand by the term mechanical working? What are the mechanisms involved in mechanical working? How does recrystallisation differ from these phenomena?

Multiple-Choice Questions

1. Time dependent permanent deformation is called ________.
   (a) Plastic deformation (b) Elastic deformation
   (c) Creep (d) Anelastic deformation
2. Figure out the odd point in the following:
   (a) Proportinal limit (b) Elastic limit
   (c) Yield point (d) Fracture point
3. If a material is subjected to two incremental true strains, namely, ε_1 and ε_2, then the total true strain is
   (a) $\varepsilon_1 * \varepsilon_2$ (b) $\varepsilon_1 - \varepsilon_2$
   (c) $\varepsilon_1 + \varepsilon_2$ (d) $\varepsilon_1 / \varepsilon_2$
4. Engineering stress-strain curve and true stress-strain curve are equal up to
   (a) Proportional limit (b) Elastic limit
   (c) Yield point (d) Tensile strength point
5. Value of Poisson's ratio for ionic solids in the range of
   (a) 0.1 (b) 0.2
   (c) 0.3 (d) 0.4
6. Hydrostatic stress results in the following:
   (a) Linear strain (b) Shear strain
   (c) Both linear and shear strains (d) None
7. High elastic modulus in materials arises from
   (a) High strength of bonds (b) Weak bonds
   (c) Combination of bonds (d) None
8. Change in elastic modulus for ordinary materials between 0K and melting point
   (a) 10-20% increase (b) 10-20% decrease
   (c) 80-90% decrease (d) 80-90% increase
9. Bauschinger effect
   (a) Hysteresis loss during loading and unloading
   (b) Anelastic deformation
   (c) Dependence of yield stress on path and direction
   (d) None

10. Sharp break in dislocation line that is in slip plane ________.
    (a) Jog
    (b) Kink
    (c) Either jog or kink
    (d) None
11. Minimum number of slip systems that must be operative during plastic deformation is
    (a) 3
    (b) 4
    (c) 5
    (d) 6
12. Which strengthening mechanism applies to multi-phase material?
    (a) Grain size reduction
    (b) Dispersion hardening
    (c) Solid solution strengthening
    (d) Strain hardening
13. If ASTM grain size number is 1, approximate grain diameter (in mm) is
    (a) 0.1
    (b) 0.2
    (c) 0.25
    (d) 10
14. If volume fraction of spherical shaped second phase particles is 50% with radius of 3 μm, interspacing of particles (in μm) is
    (a) 1
    (b) 2
    (c) 3
    (d) 4
15. Characteristic shape of martensite platelets is
    (a) Disc
    (b) Lenticular
    (c) Cylindrical
    (d) Spheroids
16. Recrystallization temperature of pure materials is ________ (in terms of homologous temperature)
    (a) 0.1
    (b) 0.2
    (c) 0.3
    (d) 0.4
17. Example for strengthening mechanism in single-phase material is
    (a) Strain hardening
    (b) Precipitation hardening
    (c) Fibre strengthening
    (d) Dispersion strengthening
18. Higher the degree of deformation, recrystallization temperature is
    (a) Higher
    (b) Lower
    (c) No effect
    (d) Either higher or lower
19. Electrical conductivity of insulators is in the range of ________.
    (a) $10\text{-}10(\Omega\text{-mm})^{-1}$
    (b) $10\text{-}10(\Omega\text{-cm})^{-1}$
    (c) $10\text{-}10(\Omega\text{-m})^{-1}$
    (d) $10\text{-}8(\Omega\text{-m})^{-1}$

20. Units for electric field strength are
    (a) A/cm^2 (b) mho/metre
    (c) $cm^2/V.s$ (d) V/cm
21. Energy band gap size for semiconductors is in the range of ________ eV.
    (a) 1-2 (b) 2-3
    (c) 3-4 (d) > 4
22. Energy band gap size for insulators is in the range of ________ eV.
    (a) 1-2 (b) 2-3
    (c) 3-4 (d) > 4
23. Flow of electrons is affected by the following:
    (a) Thermal vibrations (b) Impurity atoms
    (c) Crystal defects (d) All
24. Not a superconductive metallic element
    (a) Fe (b) Al
    (c) Ti (d) W
25. Fermi energy level for intrinsic semiconductors lies
    (a) At middle of the band gap (b) Close to conduction band
    (c) Close to valence band (d) None
26. Fermi energy level for *p*-type extrinsic semiconductors lies
    (a) At middle of the band gap (b) Close to conduction band
    (c) Close to valence band (d) None
27. Fermi energy level for *n*-type extrinsic semiconductors lies
    (a) At middle of the band gap (b) Close to conduction band
    (c) Close to valence band (d) None
28. ________ not an example for intrinsic semiconductor
    (a) Si (b) Al
    (c) Ge (d) Sn
29. Find the wrong statement: Specific heat of a material ________.
    (a) Constant for a material (b) Heat capacity per unit mass
    (c) Extrinsic property (d) Has units as J/kg-K.
30. Heat capacity has units as
    (a) J/kg.K (b) J/mol.K
    (c) $J.ohm/sec.K^2$ (d) W/m.K

31. Units for thermal conductivity
    (a) J/kg.K (b) J/mol.K
    (c) J.ohm/sec.K^2 (d) W/m.K
32. Lorentz constant has units as
    (a) J/kg.K (b) J/mol.K
    (c) J.ohm/sec.K^2 (d) W/m.K
33. Thermal expansion of a material has units as
    (a) J/kg.K (b) J/mol.K
    (c) J.ohm/sec.K^2 (d) 1/°C
34. Polymers have thermal conductivities in the range of
    (a) < 1 (b) 1-10
    (c) 10-100 (d) >100
35. Basic source of magnetism _________.
    (a) Charged particles alone (b) Movement of charged particles
    (c) Magnetic dipoles (d) Magnetic domains
36. Units for magnetic flux density
    (a) Wb/m^2 (b) Wb/A.m
    (c) A/m (d) Tesla/m
37. Magnetic permeability has units as
    (a) Wb/m^2 (b) Wb/A.m
    (c) A/m (d) Tesla/m
38. Magnetic permeability has units as
    (a) Tesla (b) Henry
    (c) Tesla/m (d) Henry/m
39. Magnetic field strength's units are
    (a) Wb/m^2 (b) Wb/A.m
    (c) A/m (d) Tesla/m
40. Example for diamagnetic materials
    (a) Superconductors (b) Alkali metals
    (c) Transition metals (d) Ferrites
41. Example for paramagnetic materials
    (a) Superconductors (b) Alkali metals
    (c) Transition metals (d) Ferrites

42. Example for ferromagnetic materials
    (a) Superconductors (b) Alkali metals
    (c) Transition metals (d) Ferrites
43. Example for anti-ferromagnetic materials
    (a) Salts of transition elements (b) Rare earth elements
    (c) Transition metals (d) Ferrites
44. Visible light's wavelength range ________.
    (a) 0.39 – 0.77 mm (b) 0.39 – 0.77 μm
    (c) 0.39 – 0.77 nm (d) 0.39 – 0.77 cm
45. Planck's constant
    (a) 6.62 × 10-34 J.sec (b) 6.62 × 10-34 J.min
    (c) 6.62 × 10-34 Cal.sec (d) 6.62 × 10-34 Cal.min
46. Sum of these is unity
    (a) Reflectivity
    (b) Reflectivity + Refractivity
    (c) Reflectivity + Refractivity + Transmitivity
    (d) Any
47. Metals can ________ the light beams.
    (a) Reflect (b) Refract
    (c) Transmit (d) Any
48. Metals are ________.
    (a) Transparent (b) Opaque
    (c) Translucent (d) None
49. Metals can transmit these ________.
    (a) Radiowaves (b) Visible light
    (c) Microwaves (d) X-rays
50. Reflectivity of metals is
    (a) 0.05 (b) 0.50
    (c) 0.95 (d) None
51. Shape of true stress-strain curve for a material depends on
    (a) Strain (b) Strain rate
    (c) Temperature (d) All

52. Toughness of a material is equal to the area under _________ part of the stress-strain curve.
   (a) Elastic (b) Plastic
   (c) Both (d) None

53. True stress-strain curve needs to be corrected after
   (a) Elastic limit (b) Yield limit
   (c) Tensile strength (d) No need to correct

54. Following condition represents onset of necking
   (a) $\varepsilon_u = n$ (b) $\varepsilon_u = 1 - n$
   (c) $\varepsilon_u = 1 + n$ (d) $\varepsilon_u = \ln (1 + n)$

55. As compared with conventional stress-strain curve, the true stress-strain curve is
   (a) Above and right (b) Below and right
   (c) Above and left (d) Below and left

56. According to distortion-energy criterion, yielding occurs when
   (a) Distortion energy reaches a critical value
   (b) Second invariant of the stress deviator exceeds some critical value
   (c) Octahedral shear stress reaches a critical value
   (d) All

57. von Mises and Tresca criteria give different yield stresses for
   (a) Uniaxial stress (b) Balanced biaxial stress
   (c) Pure shear stress (d) All

58. Plastic deformation results from the following:
   (a) Slip (b) Twinning
   (c) Both (d) None

59. Time dependent recoverable deformation under load is called _________ deformation.
   (a) Elastic (b) Anelastic
   (c) Elastic after-effect (d) Viscoelastic

60. Recrystallization rate varies in the following manner with temperature
   (a) Linearly increasing (b) Linearly decreasing
   (c) Exponential (d) Logarithmic

61. Methods to retard grain growth
    (a) Solute drag (b) Pinning action of particles
    (c) Both (d) None
62. Driving force for recrystallization process
    (a) Stored energy of cold work (b) Grain boundary energy
    (c) Both (d) Stacking-fault energy
63. Driving force for grain growth process
    (a) Stored energy of cold work (b) Grain boundary energy
    (c) Both (d) Stacking-fault energy
64. Fine grain size, usually, cannot be obtained during the following process:
    (a) Slow cooling (b) Increasing nucleation rate
    (c) Retarding grain growth (d) Fast cooling
65. Hardness during over-aging
    (a) Decreases (b) Increases
    (c) Constant (d) Decreases abruptly
66. Decrease in free energy during recrystallization is attributed to
    (a) Excess point defects (b) Excess dislocations
    (c) Grain boundaries (d) All
67. Decrease in free energy during recovery is attributed to
    (a) Excess point defects (b) Excess dislocations
    (c) Grain boundaries (d) All
68. In intrinsic semiconductors, number of electrons ________ number of holes.
    (a) Equal (b) Greater than
    (c) Less than (d) Cannot define
69. In n-type semiconductors, number of holes ________ number of electrons.
    (a) Equal (b) Greater than
    (c) Less than (d) Cannot define
70. In p-type semiconductors, number of holes ________ number of electrons.
    (a) Equal (b) Greater than
    (c) Less than (d) Twice

71. Mobility of holes is _________ mobility of electrons in intrinsic semiconductors.
    (a) Equal (b) Greater than
    (c) Less than (d) Cannot define
72. Fermi level for extrinsic semiconductor depends on
    (a) Donor element (b) Impurity concentration
    (c) Temperature (d) All
73. Value of dielectric constant for a material _________.
    (a) Equal to 1 (b) Greater than 1
    (c) Less than 1 (d) Zero
74. High dielectric constant material is must for _________.
    (a) Insulation of wires (b) Generators
    (c) Switch bases (d) None of the above
75. Dielectric constant for most polymers lies in the range of _________.
    (a) 1-3 (b) 2-5
    (c) 4-7 (d) 6-10.
76. Example for piezoelectric material
    (a) Rochelle salt (b) Lead zirconate
    (c) Potassium niobate (d) Barium titanium oxide
77. Example for ferroelectric material
    (a) Potassium niobate (b) Lead titanate
    (c) Lead zirconate (d) quartz
78. Polymers have thermal expansion coefficients in the range of _________ $\times 10^{-6}$.
    (a) 0.5-15 (b) 5-25
    (c) 25-50 (d) 50-400
79. Coefficient of thermal expansion for ceramics is in the range of _________ $\times 10^{-6}$.
    (a) 0.5-15 (b) 5-25
    (c) 25-50 (d) 50-400
80. Metals have thermal conductivities in the range of
    (a) < 1 (b) 1-5
    (c) 5-25 (d) 20-400
81. Heat capacity of most materials is approximately equal to _________.
    (a) R (b) 2R
    (c) 3R (d) R/2

82. With increase in temperature, thermal conductivity of a metal ________.

(a) Increases (b) Decreases

(c) Either (d) All, depending on metal

83. Thermal conductivity in polymers increases with ________.

(a) Increase in crystallinity (b) Decrease in crystallinity

(c) Either (d) None

84. Example for ferrimagnetic materials

(a) Salts of transition elements (b) Rare earth elements

(c) Transition metals (d) Ferrites

85. Magnetic susceptibility of paramagnetic materials is

(a) +10-5 (b) – 10-5

(c) 105 (d) 10-5 to 10-2

86. Magnetic susceptibility of diamagnetic materials is

(a) +10-5 (b) – 10-5

(c) 105 (d) 10-5 to 10-2

87. Magnetic susceptibility of ferromagnetic materials is

(a) +10-5 (b) – 10-5

(c) 105 (d) 10-5 to 10-2

88. Typical size of magnetic domains ________ (mm).

(a) 1-10 (b) 0.1-1

(c) 0.05 (d) 0.001

89. Typical thickness of Bloch walls is ________ (nm).

(a) 0.1-1 (b) 1-10

(c) 10-50 (d) 100

90. Example for soft magnet is

(a) 45 Permalloy (b) CrO_2

(c) Fe-Pd (d) Alnico

91. Example for hard magnet is

(a) 45 Permalloy (b) CrO_2

(c) Fe-Pd (d) Alnico

92. Example for magnetic material used in data storage devices is

(a) 45 Permalloy (b) CrO_2

(c) Cunife (d) Alnico

93. Refractive index of materials is approximately equal to square root of
    (a) electrical permittivity
    (b) magnetic permeability
    (c) electrical permittivity × magnetic permeability
    (d) None
94. Snell's law relates to ________.
    (a) Light relfection (b) Light refraction
    (c) Light transmission (d) Light absorption
95. Bouguer's law relates to ____________.
    (a) Light reflection (b) Light refraction
    (c) Light transmission (d) Light absorption
96. Sky looks blue because the sunlight is subjected to ________.
    (a) Rayleigh scattering (b) Compton scattering
    (c) Both (d) None
97. Luminescence is because of
    (a) Photons emitted while excited electrons drop down
    (b) Knocking out of electrons by photons
    (c) Photons stimulated by photons
    (d) All
98. Fluorescence occurs within ________.
    (a) 10-5 s. (b) 10-5 ms.
    (c) 10-5 μs. (d) 10-5 ns.
99. Electro-luminescence occurs in ________.
    (a) Electrical conductors (b) Electrical insulators
    (c) p-n junctions (d) All
100. Pyrometer works based on
    (a) Laser technology (b) Photo-conduction
    (c) Thermal emission (d) Tyndall effect

Answers

1. (c)	**2.** (d)	**3.** (c)	**4.** (c)	**5.** (b)
6. (d)	**7.** (a)	**8.** (b)	**9.** (c)	**10.** (b)
11. (c)	**12.** (b)	**13.** (c)	**14.** (d)	**15.** (b)
16. (c)	**17.** (a)	**18.** (b)	**19.** (a)	**20.** (c)
21. (b)	**22.** (c)	**23.** (d)	**24.** (b)	**25.** (c)
26. (b)	**27.** (b)	**28.** (a)	**29.** (c)	**30.** (b)
31. (d)	**32.** (c)	**33** (d)	**34.** (a)	**35.** (b)
36. (a)	**37.** (b)	**38.** (d)	**39.** (c)	**40.** (a)
41. (b)	**42.** (c)	**43.** (a)	**44.** (b)	**45.** (a)
46. (c)	**47.** (a)	**48.** (b)	**49.** (d)	**50.** (c)
51. (d)	**52.** (c)	**53.** (c)	**54.** (a)	**55.** (c)
56. (d)	**57.** (c)	**58.** (c)	**59.** (b)	**60.** (c)
61. (c)	**62.** (a)	**63.** (b)	**64.** (a)	**65.** (a)
66. (b)	**67.** (a)	**68.** (c)	**69.** (d)	**70.** (a)
71. (d)	**72.** (d)	**73.** (a)	**74.** (a)	**75.** (c)
76. (b)	**77.** (b)	**78.** (d)	**79.** (a)	**80.** (d)
81. (c)	**82.** (d)	**83.** (a)	**84.** (d)	**85.** (d)
86. (b)	**87.** (c)	**88.** (c)	**89.** (d)	**90.** (a)
91. (d)	**92** (b)	**93.** (a)	**94.** (b)	**95.** (d)
96. (a)	**97.** (a)	**98.** (b)	**99.** (c)	**100.** (c)

SECTION III

Degradation and Failure of Engineering Materials

7

Degradation of Engineering Materials

Learning Objectives

After going through this chapter the reader will be able to:

- Understand degradation of engineering materials.
- Know the types of wearing in engineering materials.
- Get an idea on fracture in various engineering materials.

7.1 INTRODUCTION

Let us begin by characterizing what we mean by corruption, degradation and failure. At the point when a part or framework stops to meet planned execution determinations we may comprehensively say that it has fizzled. A component does not really need to break into pieces for it to come up short, in spite of the fact that is absolutely the most prompt and unambiguous indication of failure.

Quality-control practices weed out components that fail inspection; however, we are not concerned with these, but rather with service failures. Wear in metal forming dies that results in parts that no longer maintain tolerance means that the tools have failed to meet their intended purpose. Similarly, turbine blades that stretch beyond the designed dimensions during operation have essentially failed. In these examples each component has technically "failed" but not broken, and may even be repairable so that it can be returned to service.

Here in this chapter, *failure* signifies mechanical fracture or breakage. All too frequently catastrophic fracture occurs suddenly and without warning. After postmortem analysis of the fractured components or structure it often happens that the mechanical design and stress analysis were not at fault. Rather, the wrong material was chosen, it was not processed or fabricated properly for the job, or the subtle progressive damage to it was undetected. Nothing will concentrate an engineer's mind more than involvement in a failure of some consequence and its analysis. This is, perhaps, the most compelling reason for engineers to understand materials.

Degradation denotes the progressive loss in the ability of a component, system, or structure to meet performance specifications. Corrosion and wear are two important mechanisms of degradation that are discussed in this chapter. Often one learns to live with such degradation if there is no danger of imminent failure. An example is corrosion damage to an automobile fender. Wear of varied items such as tools, machine parts, and auto components is also frequently tolerated up to a point. But all too often, degradation may ultimately lead to catastrophic failure (fracture) if the progressive damage is not recognized and halted in time.

7.2 CORROSION AND WEAR OF ENGINEERING MATERIALS

7.2.1 Corrosion

Corrosion may be broadly defined as an unintentional attack on a material through chemical reaction with the surrounding medium. According to this definition ceramics, concrete, and polymers also corrode but our discussion here is limited to metals. And of the metals, corrosion damage to iron and steel far exceeds that to other metals. As we shall see the typical corrosion scenario involves the equivalent of two somehow different metal electrodes, exposed to an aqueous ionic electrolyte and connected by a conductor of electrons; in the process metal is lost from one electrode, signifying that corrosion has occurred. This is an introduction to corrosion more about corrosion is stated and well explained in the next chapter.

7.2.2 Wear in Engineering Materials

The subject of wear is concerned with the interacting material surfaces that are in frictional contact and in relative motion. Wear is one component of a relatively new and interdisciplinary field of scientific and engineering study known as tribology. Derived from the Greek word *tribos* meaning "rubbing," tribology is concerned with adhesion, hardness, friction, wear, erosion, lubrication, and related phenomena. Both friction and wear effects are not only limited to engineering systems, they also arise in the movement of animal joints and lead to arthritis and degradation of bone-socket interfaces, necessitating replacement. But we also rely on controlling friction when we walk or participate in sports such as rock climbing, ice skating, and skiing.

Since the dawn of civilization humans have struggled to overcome friction when moving heavy loads. Consider the Egyptian bas-relief of Figure 7.1. It dates to 1880 BC and illustrates the use of rollers and sledges to facilitate transport of a large statue, estimated to exert a force of 6×10^5 N, along a wooden track. Closer examination reveals that a man, perhaps one of the first lubrication engineers, is pouring a liquid in the path of motion. Assuming that each of the 172 famished slaves pull with a force of 444 N (100 lb) the coefficient of friction (f), as defined in physics is $f = (172 \times 444)/ (6 \times 10^5)$

= 0.13. The four-millennia evolution of methods to support and move loads from sliding to rolling friction, from single-contact to multiple contact wheels, has been accompanied by a corresponding decrease in the coefficient of friction.

Wear degradation involves two different materials and mass loss from usually one of them. But, whereas mass transfer phenomena during corrosion are due to chemical driving forces, wear is produced by interfacial mechanical forces. Wear is of crucial importance in machine components such as ball bearings and computer disk heads where high relative velocities between the contacting surfaces (i.e., ball and race, head and disk) occur. Heads that fly over but very close to the surface of computer discs normally do not contact them except during starting and stopping of the drives. Occasionally, the contact is such that a "crash" occurs, with attendant loss of stored information. In this case as well as in bearings much damage can be produced by very little mass loss.

From an engineering outlook three separate classes of conduct can be recognized in the light of the relative extents of friction and wear.

1. *Low friction and wear.* Found in bearings, gears, cams, and slide-ways.
2. *High friction but low wear.* This combination is desired in devices that use friction to transmit power such as clutches, belt drives, and tyres.
3. *Low friction and wear of one body is high.* This is the situation that wins in material expulsion procedures, for example, machining, cutting, penetrating or drilling, and grounding. In these operations the apparatuses likewise endure wear and should be occasionally sharpened.

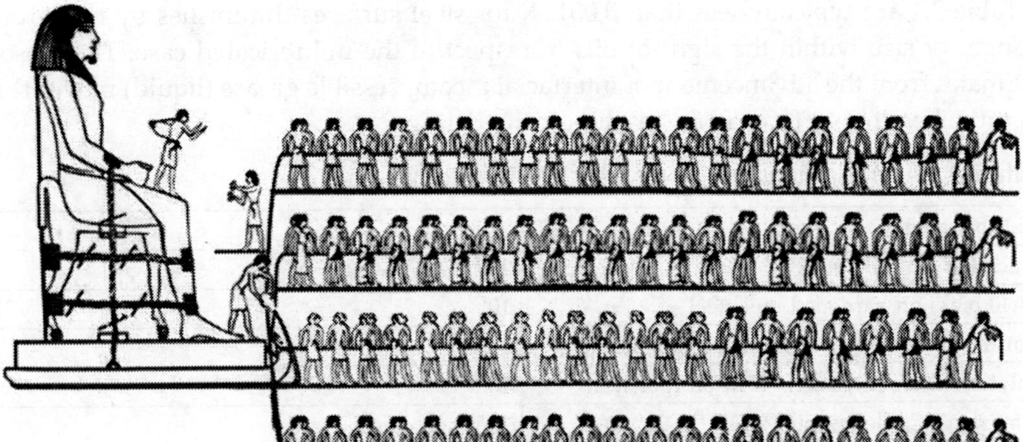

Figure 7.1 Egyptian bas-relief dating to 1880 BC, illustrating transport of an Egyptian colossus.

Adhesive Wear: Material surfaces that seem smooth on a plainly visible level really comprise a dissemination of minute ill tempers or pinnacles. At the point when bodies containing two such surfaces are united contact happens at moderately couple of severities. Under stacking the stretch develops at the severities, or more the yield push they twist and tend to weld together. Persistent distracting sliding causes the welds to shear. On the off chance that shear happens at the interface between ill tempers then there is no wear. All the more normally, be that as it may, shear happens far from the interface on the grounds that the interfacial bond quality surpasses the firm quality of one (or both) of the reaching bodies. For this situation, material is exchanged from one surface to the next, more often than not from the milder to the harder body. With further rubbing the exchanged material gets to be segregated to shape free wear particles. In extreme instances of cement wear, spreading, irking, and seizure of the surfaces may happen.

We can list three facts of adhesive wear as follows:

(i) The volume (*V*) of wear material is proportional to the distance (*L*) over which relative sliding occurs.

(ii) The volume of wear material is proportional to the load applied (*P*).

(iii) The volume of wear material is inversely proportional to the hardness (*H*) or yield stress of the softer material.

With these observations combined, the Archard expression for adhesive wear emerges as

$$V = KFL/3H$$

where, *K* is a dimensionless constant known as the wear coefficient. Values of *K* listed in Table 7.1 are typically less than 0.001. *K* for steel surfaces diminishes by numerous changes of size within the sight of oils in respect to the unlubricated case. The reason originates from the advancement of interfacial incompressible grease (liquid) moves that keep the bearing surfaces from reaching constricting.

Table 7.1 Approximate values of the wear coefficient (in air)

Unlubricated	
Mild steel on mild steel	10^{-2} - 10^{-3}
60-40 brass on hardened tool steel	10^{-3}
Hardened tool steel on hardened tool steel	10^{-4}
Polytetrafluoroethylene (PTFE) on tool steel	10^{-5}
Tungsten carbide on mild steel	10^{-6}
Lubricated	
52100 steel on 52100 steel	10^{-7} - 10^{-10}
Aluminium bronze on hardened steel	10^{-8}
Hardened steel on hardened steel	10^{-9}

Example 7.1 In a 2 cm diameter ball bearing composed of 52100 steel balls and races, how many revolutions would be required to produce a wear volume of 10^{-6} cm^3 under an applied load of 50 kg, if the hardness is 1000 kg/mm^2? Assume metal loss occurs by adhesive wear.

Solution: From Table 7.1 the value for K is chosen to be 10^{-9}. The relative distance L travelled along the race is

$$27\pi(10)\ N$$

where, N is the number of bearing revolutions. For a wear volume of $V = 10^{-6}$ cm^3 or 10^{-3} mm^3,

$$V = KFL/3H$$

$$10^{-3} = (10^{-9})\,(50)\ [2\pi\ (10)N]/3(1000)$$

$$\Rightarrow \quad N = 9.55 \times 10^5$$

Abrasive Wear: Abrasive wear covers two distinct circumstances. Both include the furrowing/plowing or gouging out of the milder material by a harder surface. In the primary case an unpleasantly rough, hard surface rubs against a milder material as exemplified by the activity of a document or an emery paper on a delicate metal. This sort of wear has been dispensed with in current hardware by minimizing surface unpleasantness. In the second kind of rough wear, airborne tidy, coarseness, and the results of glue or destructive wear figure out how to discover their way to the interface between the reaching materials. Fixing and filtration are the main down to earth safeguards against the entrance of airborne grating particles. The volume of material evacuated by rough wear is given by an equation like that of adhesive wear.

Fatigue Wear: Fatigue wear comes about when rehashed stacking and emptying of the reaching surfaces happen. Not at all like adhesive and abrasive wear in sliding contact, which are agents from the minute rubbing begins, surfaces that are greased up don't promptly endure such wear. This is the situation in very much composed ball and roller (containing round and hollow bearing surfaces) direction. Be that as it may, these headings are inclined to moving contact fatigue, which shows itself after long times if the adequacy of the turned around hassles is sufficiently huge. Wear starts at metallic considerations, pores, and small scale splits, and in the end brings about unstuck particles. The testing of substantial quantities of roller orientation has uncovered that their life is contrarily corresponding to F^3.

Corrosive Wear: Corrosive wear is a blend of both erosion and rubbing. In the destructive environment, either liquid or vaporous, the erosion items that frame at the reaching surfaces are inadequately disciple and may go about as rough particles to quicken wear. On account of oxidation wear, be that as it may, defensive oxides that are evacuated may promptly redevelop, constraining the degree of harm.

Fretting Wear: Fretting wear may be viewed as a type of fatigue wear that occurs under conditions of high-frequency vibratory motion of small amplitude (in the range 1-200 μm). Many sequential damage processes occur during fretting including breakup of protective films, adhesion and transfer of material, oxidation of metal wear particles, and nucleation of small surface cracks.

Delamination Wear: Delamination wear takes the form of regular detachments of plate like particles from wearing surfaces due to the influence of high tangential (friction) forces in the surface contact zone. During cyclic loading the cracks that develop propagate parallel to the surface at a depth governed by the stress, material properties, and coefficient of friction.

Erosion: At the point when a surge of abrasive solid particles is coordinated at a surface it dissolves at a rate that is firmly reliant on the impingement edge Φ as appeared in Figure 7.2. Moreover, the wear rate displays contradicting patterns relying on whether ductile (metal) or brittle (clay) surfaces are included. Low impingement edges extend bendable materials in strain, making them break, though typical effect peening packs the surface, reinforcing it. Then again, fragile materials endure more disintegration with expanding impingement edge.

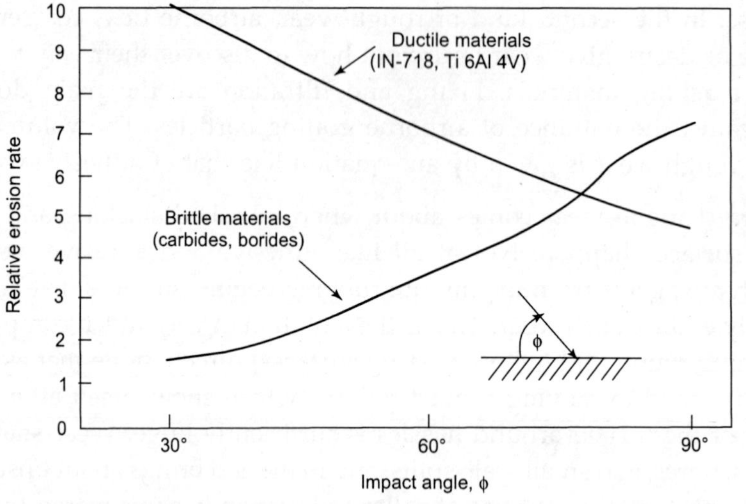

Figure 7.2 Relative erosion rates in ductile and brittle materials as a function of particle impact angle (ϕ).

Disintegration is normal in earth moving, mining, and farming hardware taking care of shake, particulates, and slurries. It is also a troublesome form of wear in gas turbine engines that ingest airborne particulates. Erosion also occurs during electrical arcing and leads to pitting degradation of electrical contacts. When arcing occurs between the copper commutator and brushes of motor, both metal and brush loss cause the contact to deteriorate. Spark erosion is not always deleterious, however. It is capitalized upon in spark cutting or electrical-discharge machining, a technique employed to cut hard metals through controlled incremental erosion.

Wear in Cutting Tools: Wear of cutting tools has a major impact on the costs sustained by the metal removal industries and figures prominently in the price of machined products. An instant in the high-speed cutting action of a tool against a workpiece is frozen in the depiction of Figure 7.3a. Severe local stresses are generated to continually shear and fracture the metal forming the chip. The friction generated between the chip and the tool can result in local temperatures approaching 1000°C or more depending on tool and workpiece materials, cutting speeds and feed rates, extent of lubrication, and so on (Figure 7.3b). Wear craters and material loss, accelerated by abrasive particles, thermal softening, and oxidation and chemical reactions, occur on the tool (rake) face, side or flank face, and nose (Figure 7.3c). The tool eventually fails and must be either sharpened or replaced because the work dimensions and surface finish no longer meet specifications. In 1907, F.W. Taylor formulated a widely used relationship between steel tool life (T_t) and machining velocity (V_t), namely, $V_t T_t^n = C$, where C and n are constants.

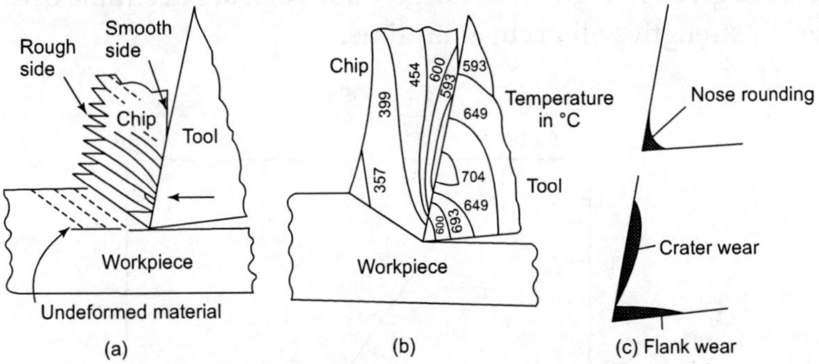

Figure 7.3 (a) Cutting action of a tool bit against a metal workpiece in a two-dimensional machining operation. (b) Typical temperature profiles within tool and chip. (c) Tool degradation caused by nose rounding (above), and flank and crater wear (below).

The life of a tool is directly proportional to the product of its fracture toughness (K_{1c}) and hardness. Although the two properties vary inversely with one another, the hundredfold tool life improvement in the past century has been realized largely by increasing hardness. Today, TiC, TiN, and Al_2O_3 coatings (5-10 μm thick), used singly or in combination, are the chief means of combating tool wear.

7.3 FRACTURE OF ENGINEERING MATERIALS

This section and succeeding ones on fatigue and creep are of considerable practical importance. The theoretical basis for fracture analysis is the Griffith crack theory. Here the discussion takes a practical turn to examine actual manifestations and morphologies of failure and, where possible, the operative mechanisms.

7.3.1 Fracture of Glass Fibres

It is fitting to start with fracture in glass fibres because they were originally studied by Griffith. More importantly, optical fibres composed of ~125 μm diameter, coaxial silica glass filaments that can be tens of kilometres long have revolutionized communications systems. Data on the strength dependence of these fibres as a function of crack size are presented in Figure 7.4. To withstand high service stresses it is essential that fibres are defect free or, if not, have very small surface flaws. In addition, glass generally exhibits the phenomenon of static fatigue and glass fibre is no exception. This means that if fibres are found to have a proof stress σ_p during a test time t_p, then they will safely support an applied stress σ ($\sigma < \sigma_p$) until failure at time t_f according to the relationship is

$$(\sigma/\sigma_p)^n = t_p/t_f$$

where, w is a constant known as the slow crack growth exponent. Static fatigue data for optical fibres are given in Figure 7.5. Large values of n are desirable because relatively small decreases in strength will occur with time.

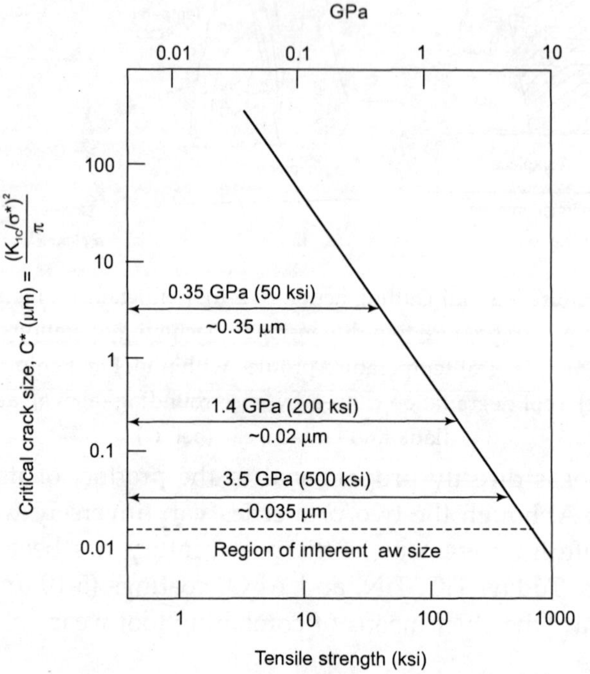

Figure 7.4 Size of critical crack considering strength of optical (glass) fibres.

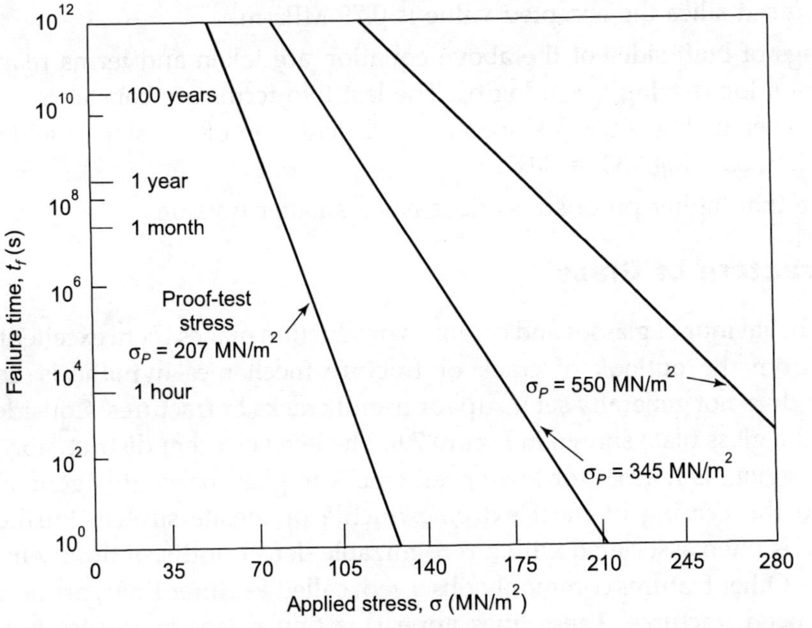

Figure 7.5 Static fatigue life of optical (glass) fibres having different proof stresses.

Example 7.2

(a) How well does Griffith crack theory describe the results displayed in Figure 7.4?

(b) What is the value of K_{1c} for silica employed in optical fibre?

(c) Using Figure 7.5 what is the value of the slow crack growth exponent for a fibre that was found to have a proof stress of 207 MN/m^2?

Solution:

(a) Griffith crack theory suggests that c^* varies as σ^{-2}. Therefore, a plot of log c^* versus log σ should have a slope of −2. Careful measurement with a ruler demonstrates that for every two-orders-of-magnitude drop in c^* there is very nearly a one-order-of-magnitude increase in σ. Therefore, the experimental data strongly support the Griffith dependence.

(b) As $K_{1c} = \sigma(\pi c^*)^{1/2}$ simple substitution and calculation using three pairs of indicated data yield the following:

σ (MPa)	c*	K_{1c}(MPa-m$^{1/2}$)
350	3.5	0.367
1400	2.0	1.11
3500	3.5	1.16

For fused silica the accepted value is 0.79 MPa-m$^{1/2}$.

(c) If logs of both sides of the above equation are taken and terms rearranged, log $t_f = -n \log \sigma + \log t_p + n \log \sigma_p$. The last two terms are constants, and a plot of log t_f versus log σ has a slope of $-n$. Calculation of the slope yields $n = -(12 - 0)/ (\log 53 - \log 130) = 30.8$.

Note that higher proof stress fibre has a smaller n value.

7.3.2 Fracture of Glass

The brittle behaviour of glasses and ceramics production makes them excellent illustrative materials from the outlook of crack or fracture mechanics hypothesis. In any case, hypothesis does not generally set us up for useful cracks or fractures. Consider the break surface of the glass plate shown in Figure 7.6. The level crescent district, normally called the reflect region, is normal for low-speed cracks in glass parts. It is generally situated opposite to the bearing of most extreme ductile or tensile stretch. Furthermore, the mirror area is often associated with a recognizable defect or discontinuity in the surface of the glass. Other features commonly observed, called **Wallner lines,** are also associated with low-speed fractures. These lines appear as tiny ridges or ripples in the fracture surface where the crack seems to have stopped, waited for the stress to realign, and then started again. Wallner lines are always concave towards the crack initiation point which can then be located by backtracking along the ripples. Lastly, are the **hackle marks,** a characteristic of high-speed fractures. When cracks travel at speeds higher than 5000 ft/s, these hackle marks are visible along the direction of propagation as tiny fractures in the newly formed surface. They are perpendicular to both mirror and Wallner lines (if present). A magnifying glass is frequently sufficient to reveal these telltale signs. Fractures in glass seldom travel at a constant velocity but speed up, slow down, and change direction before stopping.

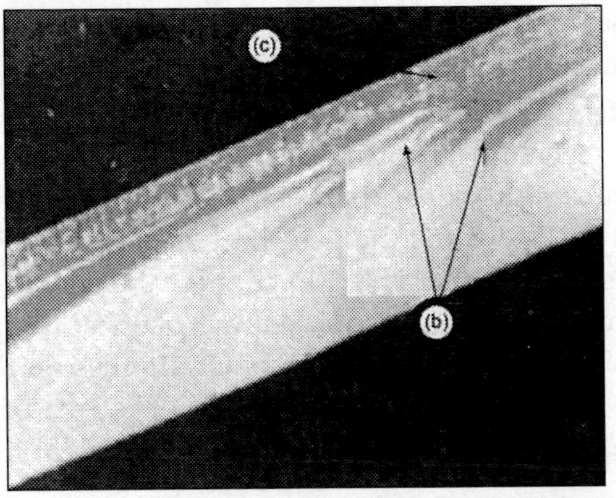

Figure 7.6 Fracture of glass plate showing mirror area a, Wallner lines b, and hackle marks c.

7.3.3 Ductile Fracture of Metals

Metals fracture in both ductile and brittle modes and sometimes in a combined ductile-brittle manner, as schematically indicated in Figure 7.7a. Although these are relative terms, ductile fracture is accompanied by large percentage elongation and extensive plastic deformation in the region of the crack tip. Tensile tests provide an unambiguous way to relate fracture appearance to the nature of loading. Ductile metals neck and show the regular container cone break morphology. The procedure of ductile fracture might

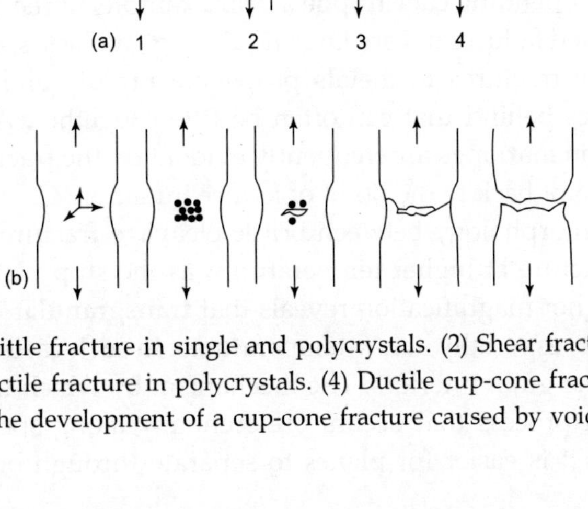

Figure 7.7 (a) (1) Brittle fracture in single and polycrystals. (2) Shear fracture in ductile single crystals. (3) Ideal ductile fracture in polycrystals. (4) Ductile cup-cone fracture in polycrystals. (b) Stages in the development of a cup-cone fracture caused by void coalescence.

be comprehensively seen regarding the consecutive small scale void nucleation and development forms, schematically demonstrated in Figure 7.7b.

Shrouded in mystery, the earliest stage of fracture spawns isolated microscopic cavities. These nucleate heterogeneously at inclusions, second-phase particles, and probably at grain boundary junctions. Micro-void coalescence then yields a flat elliptical crack that spreads outward towards the periphery of the neck. Finally, an overloaded outer ring of the material is all that is left to connect the specimen halves, and it fails by shear. Further examination reveals equiaxed or spherical **dimples** on the flat crater bottom loaded in tension and elongated ellipsoidal dimples on the shear lips oriented at 45° (Figure 7.8). In high-purity FCC and BCC metals free of inclusions, however, necking to ~ 100% area reduction (i.e., to a point) is possible.

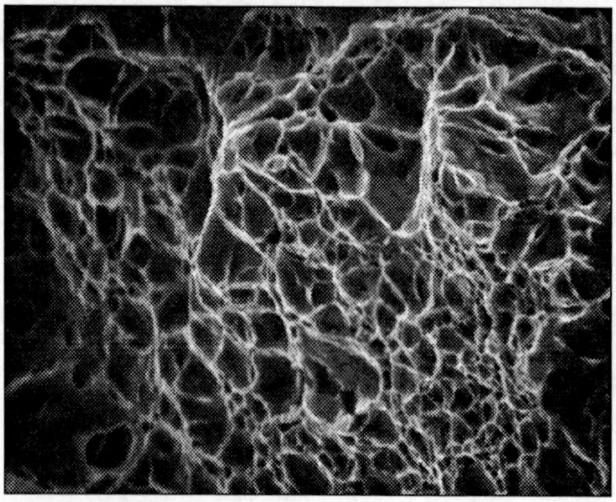

Figure 7.8 SEM image of dimples in the ductile tensile fracture of a stainless steel.

7.3.4 Brittle Fracture of Metals

Brittle fracture is the most feared one among all. It often occurs under static loading without warning of impending catastrophe as, for example, in the Liberty Ship fractures and numerous recorded failures of very large liquid (e.g., molasses, oil) storage containers made of steel. Brittle fractures in metals propagate rapidly and in their wake leave relatively flat surfaces behind that can often be fitted together. At low magnification, arrow-shaped chevron markings are frequently evident on the fracture surface and these often conveniently point back to the point of failure initiation. Chevron patterns actually represent transition morphology between brittle cleavage fracture at low temperatures and ductile shear fracture at higher temperatures, as the ship plate fractures in Figure 7.9 demonstrate. Higher magnification reveals that **transgranular** cracking often occurs during brittle fracture, especially at low temperatures. In such cases the crack propagates *across* and through the grain interiors. The mechanism by which transgranular fracture occurs is cleavage, a process that occurs preferentially along specific crystallographic planes. Under stress it is easier for planes to separate through bond breaking than to

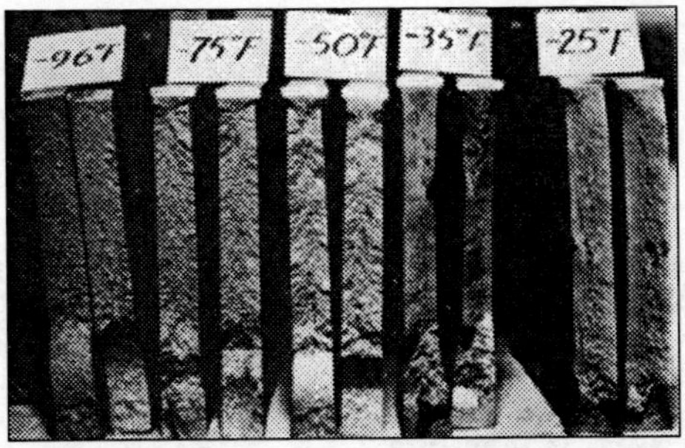

Figure 7.9 Chevron features in the fracture surface of ship plate steels tested at various temperatures.

deform by slip. Cleavage is observed in BCC as well as HCP metals and is quite common in ionic and covalent materials. Characteristics of such fractures include flat facets, like those seen in a 2.5% Si-Fe alloy that fractured at – 195°C (Figure 7.10a). River patterns are also common on the cleaved surfaces. Low temperatures and high strain rates are conducive to transgranular fracture. Fatigue cracks often propagate in a transgranular mode. When cracks propagate along grain boundaries then we speak of intergranular fracture. Such failures (Figure 7.10b) reveal clearly outlined grains that appear to be deeply etched or stand out in relief like rock candy. Grain boundary decohesion between contiguous grains, caused by the presence of harmful segregated impurities and second-phase particles or by penetration of aqueous corrodents, is a common feature. Intergranular embrittlement is also promoted by elevated temperature creep and stress corrosion. Such damage is generally made more visible by polishing flat metallographic sections that contain the intergranular cracks (Figure 7.11). Auger spectroscopy has often detected the presence of elements such as sulphur and phosphorus in postmortem failure analyses of embrittled steel grain boundaries.

(a) (b)

Figure 7.10 (a) SEM image of trans-granular fracture in Fe-2.5 wt% Si tested at – 195°C (b) SEM image of high-temperature intergranular fracture in a nickel-base superalloy.

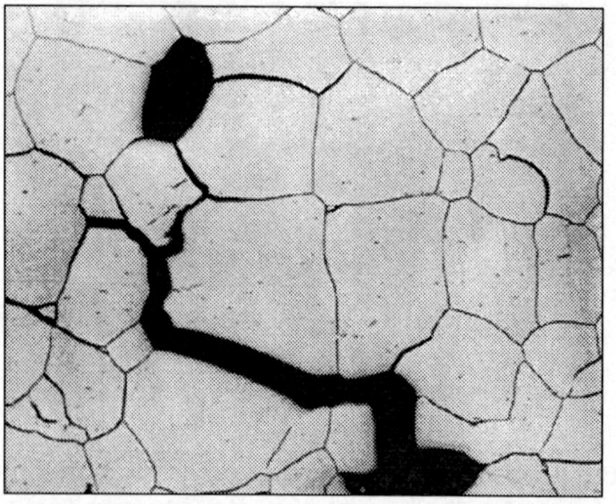

Figure 7.11 Intergranular stress cracking in a nickel-base alloy exposed to a sulphur-containing atmosphere at 1120°C.

Brittle fracture attributed to hydrogen has been accepted as one of the leading causes of failure in high-strength steels. Known as **hydrogen embrittlement,** the cracks that nucleate where excess (atomic) hydrogen accumulates propagate in an intergranular fashion. Although it may be incorporated during the melting of steel, hydrogen more often penetrates from aqueous and corrosive environments. Hydrogen is apparently drawn to incipient or pre-existent grain boundary cracks, stabilizing them against the tendency of the surrounding metal to blunt their advance. Softer, tougher steels are generally more resistant to hydrogen embrittlement.

7.3.5 Failure and Fracture of Polymers

Fracture of thermosetting polymers is much like that of other brittle solids. In the presence of surface flaws or sharp notches, critical levels of stress act to sever covalent bonds and cause fracture. But fracture of thermoplastic polymers is different and can be brittle, ductile, or some admixture of the two, in nature. It all depends on the service temperature, T, relative to the glass transition temperature, T_G. When $T < T_G$, brittle fracture occurs, whereas more ductile fracture modes are evident when $T > T_G$. Tensile (ductile) stretching is viscoelastic in nature, and during non-uniform plastic deformation, the interesting phenomenon of crazing occurs. Crazing is a kind of tortuous plastic flow that produces a large number of discontinuities, actually voids that are the seeds of the eventual failure. Crazes, like that shown in Figure 7.12, have a number of characteristics:

1. They appear to be surface cracks, but are not in reality.
2. The plane of the craze is normal to the tensile stress direction.
3. They reflect light in a manner similar to that of cracks.

As the localized tangled web of thin polymer fibrils in crazes entraps voids, crazes possess a kind of sponginess. Optimists would say that crazes are a source of toughness.

The fact that rubber toughening of commercial thermoplastics depends on the creation and stabilization of large craze fields supports this view. Pessimists, on the other hand, view crazes as the onset of brittle fracture. Surprisingly, the work per unit area required to cause fracture in amorphous thermoplastic polymers is many times larger than that in silica glasses and ceramics. Apparently, it takes more energy to align molecules within crazes than to break covalent bonds.

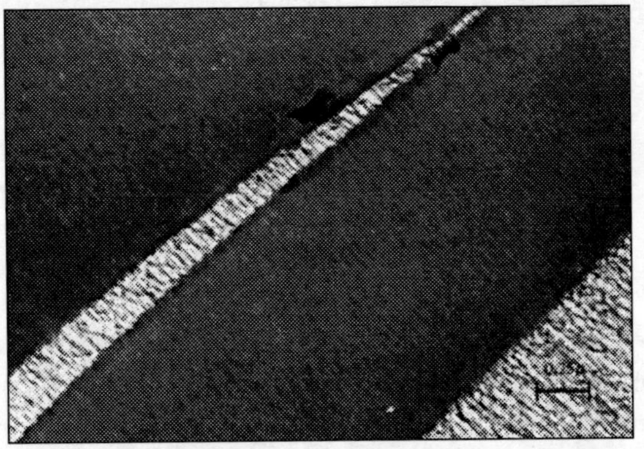

Figure 7.12 Transmission electron micrograph of a craze in poly (phenylene oxide).

7.4 PRINCIPLES OF FRACTURE MECHANICS

Extensive research endeavours over the past several decades have led to the evolution of the field of fracture mechanics. This subject allows quantification of the relationships between material properties, stress level, the presence of crack-producing flaws, and crack propagation mechanisms. Design engineers are now better equipped to anticipate, and thus prevent, structural failures. The present discussion centres on some of the fundamental principles of the mechanics of fracture.

The measured fracture strengths for most brittle materials are significantly lower than those predicted by theoretical calculations based on atomic bonding energies. This discrepancy is explained by the presence of very small, microscopic flaws or cracks that always exist under normal conditions at the surface and within the interior of a body of the material. These flaws are a detriment to the fracture strength because an applied stress may be amplified or concentrated at the tip, the magnitude of this amplification depending on crack orientation and geometry. This phenomenon is demonstrated in Figure 7.13, a stress profile across a cross section containing an internal crack. As indicated by this profile, the magnitude of this localized stress diminishes with distance away from the crack tip. At positions far removed, the stress is just the nominal stress σ_0 or the applied load divided by the specimen cross-sectional area (perpendicular to this load). Due to their ability to amplify an applied stress in their locale, these flaws are sometimes called **stress raisers.**

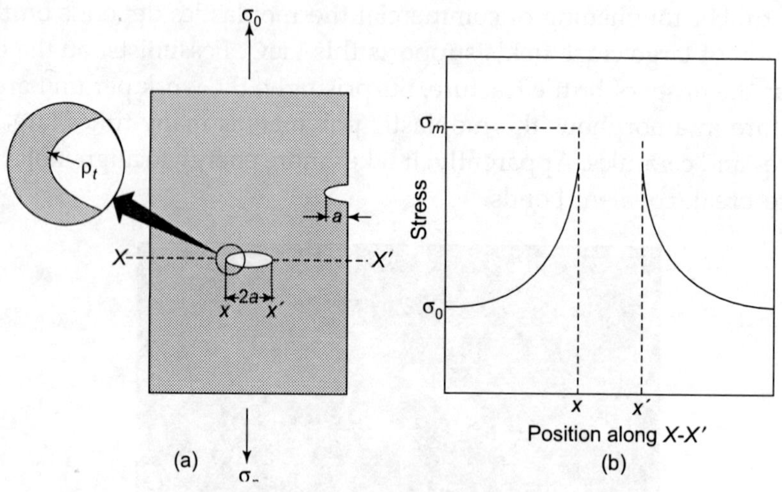

Figure 7.13 (a) The geometry of surface and internal cracks. (b) Schematic stress profile along the line X–X′ in (a), demonstrating stress amplification at crack tip positions.

If it is assumed that a crack is similar to an elliptical hole through a plate, and is oriented perpendicular to the applied stress, the maximum stress σ_m, occurs at the crack tip and may be approximated by

$$\sigma_m = 2\sigma_o \left(\frac{a}{\rho_t}\right)^{\frac{1}{2}}$$

where σ_o is the magnitude of the nominal applied tensile stress, ρ_t is the radius of curvature of the crack tip (Figure 7.13a), and a represents the length of a surface crack, or half of the length of an internal crack.

Using principles of fracture mechanics, it is possible to show that the critical stress required for crack propagation in a brittle material is described by the expression

$$\sigma_c = \left(\frac{2E\gamma_s}{\pi a}\right)^{\frac{1}{2}}$$

where,

E = modulus of elasticity

γ_s = specific surface energy

a = one half the length of an internal crack

All brittle materials contain a population of small cracks and flaws that have a variety of sizes, geometries, and orientations. When the magnitude of a tensile stress at the tip of one of these flaws exceeds the value of this critical stress, a crack occurs and then propagates, which results in fracture. Very small and virtually defect-free metallic and ceramic whiskers have been grown with fracture strengths that approach their theoretical values.

Example 7.3 A relatively large plate of glass is subjected to a tensile stress of 40 MPa. If the specific surface energy and modulus of elasticity for this glass are 0.3 J/m^2 and 69 GPa, respectively, determine the maximum length of the surface flaw that is possible without fracture.

Solution: Rearrangement of this expression such that a is the dependent variable, and realizing that

$$\sigma = 40 \text{ MPa}, \ \gamma_s = 0.3 \text{ J/m}^2, \quad \text{and} \quad E = 69 \text{ GPa leads to}$$

$$a = \frac{2E\gamma_s}{\pi\sigma^2}$$

$$= \frac{(2)(69 \times 10^9 \text{ N/m}^2)(0.3 \text{ N/m})}{\pi(40 \times 10^6 \text{ N/m}^2)^2}$$

$$= 8.2 \times 10^{-6} \text{ m} = 0.082 \text{ mm} = 8.2 \text{ µm}$$

7.4.1 Fracture Toughness

Furthermore, using fracture mechanical principles, an expression has been developed that relates this critical stress for crack propagation (σ_c) and crack length (a) as

$$K_c = Y\sigma_c\sqrt{\pi a}$$

In this expression K_c is the **fracture toughness,** a property that is a measure of a material's resistance to brittle fracture when a crack is present. Worth noting is that K_c has the unusual units of MPa $\sqrt{m}$. Furthermore, Y is a dimensionless parameter or function that depends on both crack and specimen sizes and geometries, as well as the manner of load application. Relative to this Y parameter, for planar specimens containing cracks that are much shorter than the specimen width, Y has a value of approximately unity.

For relatively thin specimens, the value K_c depends on specimen thickness. However, when specimen thickness is much greater than the crack dimensions, K_c becomes independent of thickness; under these conditions a condition of **plane strain** exists. By plane strain we mean that when a load operates on a crack in the manner represented in Figure 7.14a, there is no strain component perpendicular to the front and back faces. The K_c value for this thick-specimen situation is known as the **plane strain fracture toughness K_{Ic}** furthermore, it is also defined by

$$K_{Ic} = Y\sigma\sqrt{\pi a}$$

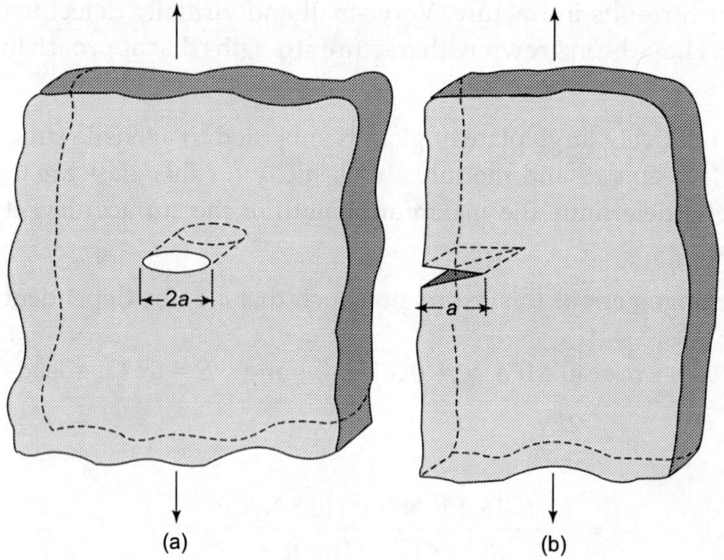

Figure 7.14 Schematic representations of (a) an interior crack in a plate of infinite width, and (b) an edge crack in a plate of semi-infinite width.

The plane strain fracture toughness K_{Ic} is a fundamental material property that depends on many factors, the most influential of which are temperature, strain rate, and microstructure. The magnitude of K_{Ic} diminishes with increasing strain rate and decreasing temperature. Furthermore, an enhancement in yield strength brought by solid solution or dispersion additions or by strain hardening generally produces a corresponding decrease in K_{Ic}. Furthermore, K_{Ic} normally increases with reduction in grain size as composition and other microstructural variables are maintained constant.

7.5 SUMMARY

Materials generally "fail" because of compositional or structural change brought about by chemical and mechanical driving forces. Other forces or stimuli can also serve as agents of degradation and failure, for example, radiation, light, electric fields, and some of these are discussed in connection with electronic materials.

The potential for aqueous metallic corrosion exists whenever metals acting as cathodes and anodes are surrounded by electrolyte and connected electronically so as to complete an electrical circuit.

The mechanical degradation and failure (fracture) mechanisms treated in the remainder of the chapter stem either from forces that act locally at the interface between materials in contact or from forces that are applied at a distance and stress the body throughout. Wear is an example of the first case; creep, fatigue, and fracture exemplify the second category. Although the details are different, at a fundamental level, fracture mechanics is applicable to both of these mechanically functional failures. While fracture in bulk solids is the result of tensile stresses, surfaces undergoing wear are nominally stressed in compression. But parasitic tensile stresses are also operative and they cause particle detachment. Because of the tendency of compressive stresses to seal cracks, wear usually leads not to catastrophic failure, but rather to surface degradation.

The central equation of fracture mechanics, $K_{1c} = \sigma(\pi c^*)^{1/2}$ very succinctly summarizes the roles that materials **processing, properties, and engineering design** play in influencing susceptibility to fracture. Processing or manufacturing is impossible without creating defects of size c in the product. Material properties are reflected in the fracture toughness K_{1c} and design stresses are given by σ. When the right-hand side of the equation just exceeds the left-hand side, an unsafe situation exists. Fracture in glass, ceramics, and metals can be profitably modelled by this equation, not only under static, but also cyclic loading conditions. In particular, a basis for predicting fatigue failure times is an outcome of fracture mechanics theory. Raising the fracture toughness, a subject addressed in other chapters, remains a primary materials goal.

Fracture, in response to tensile loading and at relatively low temperatures, may occur by ductile and brittle modes, both of which involve the formation and propagation of cracks. For ductile fracture, evidence will exist of gross plastic deformation at the fracture surface. In tension, highly ductile metals will neck down to essentially a point fracture; cup-and-cone mating fracture surfaces result for moderate ductility. Cracks in ductile materials are said to be stable (i.e., resist extension without an increase in applied stress); and inasmuch as fracture is noncatastrophic, this fracture mode is almost always preferred.

For brittle fracture, cracks are unstable, and the fracture surface is relatively flat and perpendicular to the direction of the applied tensile load. Chevron and ridge-like patterns are possible, which indicate the direction of crack propagation. Transgranular (through-grain) and intergranular (between-grain) fractures are found in brittle polycrystalline materials.

The significant discrepancy between actual and theoretical fracture strengths of brittle materials is explained by the existence of small flaws that are capable of amplifying an applied tensile stress in their vicinity, leading ultimately to crack formation. Fracture ensues when the theoretical cohesive strength exceeds at the tip of one of these flaws. The fracture toughness of a material is indicative of its resistance to brittle fracture when a crack is present. It depends on specimen thickness, and, for relatively thick specimens (i.e., conditions of plane strain), is termed the plane strain fracture toughness. This parameter is the one normally cited for design purposes; its value is relatively large for ductile materials (and small for brittle ones), and is a function of microstructure, strain rate, and temperature. With regard to designing against the possibility of fracture, consideration must be given to material (its fracture toughness), the stress level, and the flaw size detection limit.

Qualitatively, the fracture behaviour of materials may be determined using Charpy and Izod impact testing techniques. On the basis of the temperature dependence of measured impact energy (or appearance of the fracture surface), it is possible to ascertain whether or not a material experiences a ductile-to-brittle transition and the temperature range over which such a transition occurs. Low-strength steel alloys typify this behaviour, and, for structural applications, should be used at temperatures in excess of the transition range. Furthermore, low-strength FCC metals, most HCP metals, and high-strength materials do not experience this ductile-to-brittle transition.

Metal alloys that are especially resistant to creep have high elastic moduli and melting temperatures; these include the superalloys, the stainless steels, and the refractory metals. Various processing techniques are employed to improve the creep properties of these materials.

EXERCISE

1. The famous wrought iron pillar of Qutub Minar, Delhi, India, has not rusted in 1700 years of exposure to the ambient. In addition to iron the 7.2-m-high pillar contains 0.15 wt% carbon and 0.25 wt% phosphorus, and has a magnetic coating that is 50 to 600 μm thick. Provide possible reasons for the rust-free state of preservation.
2. Give a reason why static fatigue gets largely suppressed when glass is loaded in vacuum rather than in ambient conditions.
3. Surfaces of thermally toughened (tempered) windshield glass have a residual compressive stress while the interior is stressed in tension. When the thickness of the compressed layers is thin, surface damage due to particle impact or deep scratching leads to catastrophic fracture of the glass. Why?
4. The fracture toughness of a given steel is 60 MPa-$m^{1/2}$ and its yield strength is given by σ_o (MPa) = 1400 – 4T, where T is the temperature in degrees Kelvin. Surface cracks measuring 0.001 m are detected. Under these conditions determine the temperature at which there may be a ductile-brittle transition in this steel.
5. Write a short note on the following:
   (a) Fracture of glass
   (b) Fracture in polymers
   (c) Ductile and brittle fractures in metals
6. Consider a load of 50 kg applied to a tungsten carbide tool which slides over an unlubricated mild steel with hardness 1000 MPa. Determine the distance travelled between two contacting surfaces if the volume loss of about 0.0001 cm^3 (by adhesive wear)?

7. Consider a precision bearing with hardness of 1100 kg/mm^2. If it has to lose not more than 10^{-11} cm^3 when lubricated hardened steel surfaces contacts each other for a length of a mile, determine the maximum load that should be applied.
8. What is the magnitude of the maximum stress that exists at the tip of an internal crack having a radius of curvature of 1.9×10^{-4} mm and a crack length of 9.8×10^{-2} mm when a tensile stress of 140 MPa (20,000 psi) is applied?
9. Estimate the theoretical fracture strength of a brittle material if it is known that fracture occurs by the propagation of an elliptically shaped surface crack of length 0.5 mm (0.02 in.) and having a tip radius of curvature of 5×10^{-3} mm when a stress of 1035 MPa (150,000 psi) is applied.
10. If the specific surface energy for aluminium oxide is 0.90 J/m^2, compute the critical stress required for the propagation of an internal crack of length 0.40 mm.
11. An MgO component must not fail when a tensile stress of 13.5 MPa (1960 psi) is applied. Determine the maximum allowable surface crack length if the surface energy of MgO is 1.0 J/m^2.
12. Calculate the maximum internal crack length allowable for a Ti-6Al-4V titanium alloy component that is loaded to a stress one-half of its yield strength. Assume that the value of Y is 1.50.
13. After consultation of other references, write a brief report on one or two non-destructive test techniques that are used to detect and measure internal and/or surface flaws in metal alloys.
14. A structural component in the form of a wide plate is to be fabricated from a steel alloy that has a plane strain fracture toughness of 98.9 MPa $\sqrt{m}$ and a yield strength of 860 MPa (125,000 psi). The flaw size resolution limit of the flaw detection apparatus is 3.0 mm (0.12 in.). If the design stress is half of the yield strength and the value of Y is 1.0, determine whether or not a critical flaw for this plate is subject to detection.
15. Some aircraft component is fabricated from an aluminium alloy that has a plane strain fracture toughness of 40 MPa $\sqrt{m}$. It has been determined that fracture results at a stress of 300 MPa (43,500 psi) when the maximum (or critical) internal crack length is 4.0 mm (0.16 in). For this same component and alloy, will fracture occur at a stress level of 260 MPa (38,000 psi) when the maximum internal crack length is 6.0 mm (0.24 in.)? Why or why not?

8

Corrosion in Engineering Materials

Learning Objectives

After going through this chapter the reader will be able to:

- Understand corrosion of engineering materials.
- Explain electrochemical corrosion of metals.
- Define the corrosion rates.
- Describe the process of oxidation.
- Understand cathodic protection for corrosion control.

8.1 INTRODUCTION

To some degree, most materials experience some kind of association with an expansive number of various situations. Frequently, such connections impede a material's helpfulness as a consequence of the crumbling of its mechanical properties (e.g., ductility and strength), other physical properties, or appearance. Once in a while, to the mortification of an outline architect, the debasement conduct of a material for some application is disregarded, with unfriendly outcomes.

Deteriorative components are diverse for the three material sorts. In metals, there is genuine material misfortune either by dissolution (**corrosion**) or by the development of non-metallic scale or film (oxidation). Ceramic materials are generally impervious to disintegration, which as a rule happens at lifted temperatures or in rather extraordinary situations; the procedure is as often as possible additionally called corrosion. For polymers, systems and results contrast from those for metals and pottery, and the term corruption is most of the time utilized. Polymers may disintegrate when presented to a fluid dissolvable, or they may ingest the dissolvable and swell; likewise, electromagnetic radiation (essentially bright) and warmth may bring about modifications in their atomic structures.

The deterioration of each of these material sorts is examined in this part, with respect to component, imperviousness to assault by different situations, and measures to anticipate or diminish corruption or degradation.

8.2 ELECTROCHEMICAL CORROSION OF METALS

Corrosion is characterized as the destructive and unintentional assault of a metal. It is electrochemical and usually starts at the surface. The issue of metallic corrosion is one of huge extent in financial terms, it has been assessed that around 5% of an industrialized country's wages are spent on consumption counteractive action and the upkeep or substitution of items lost or debased as an after-effect of erosion responses. The outcomes of consumption are basic. Natural illustrations incorporate the rusting of car body boards and radiator and fumes segments. Consumption procedures are sometimes used to advantage. For instance, carving strategies make utilization of the specific synthetic reactivity of grain limits or different miniaturized scale auxiliary constituents.

For metallic materials, the erosion procedure is regularly electrochemical, that is, a substance response in which there is exchange of electrons starting with one compound then onto the next. Metal atoms naturally lose or surrender electrons in what is called an oxidation response. For instance, the theoretical metal M that has a valence of *n* (or *n* valence electrons) may encounter oxidation as per the response.

$$M \rightarrow M^{n+} + ne^-$$

where, M becomes an $n+$ positively charged ion and in the process loses its *n* valence electrons; e^- is used to symbolize an electron, for example, metals which oxidize are:

$$Fe \rightarrow Fe^{2+}\ 2e^-$$

$$Al \rightarrow Al^{3+} + 3e^-$$

The site at which oxidation happens is known as the anode; oxidation is once in a while called an anodic reaction. The electrons produced from every metal particle that is oxidized must be exchanged to and turn into a part of another concoction species in what is named a reduction response. For instance, a few metals experience consumption in corrosive arrangements, which have a high grouping of hydrogen (H^+) particles; the H^+ particles are decreased as takes after:

$$2H^+ + 2e^- \rightarrow H_2$$

Here hydrogen gas (H_2) is evolved.

Other reduction reactions are possible, depending on the nature of the solution to which the metal is exposed. For an acid solution having dissolved oxygen, reduction according to

$$O_2 + 4H^+ + 4e^- \rightarrow 2H_2O$$

will probably occur. Or, for a neutral or basic aqueous solution in which oxygen is also dissolved,

$$O_2 + 2H_2O + 4e^- \rightarrow 4(OH^-)$$

Any metal ions present in the solution may also be reduced; for ions that can exist in more than one valence state (multivalent ions), reduction may occur by

$$M^{n+} + e^- \rightarrow M^{(n-1)+}$$

where, the metal ion decreases its valence state by accepting an electron. Or a metal may be totally reduced from an ionic to a neutral metallic state according to

$$M^{n+} + ne^- \rightarrow M$$

The location at which lessening happens is known as the cathode. Besides, it is feasible for at least two of the lessening responses above to happen at the same time. A general electrochemical response must comprise no less than one oxidation and one reduction response, and will be equal; regularly the individual oxidation and reduction responses are named half-reaction. There can be no net electrical charge aggregation from the electrons and particles; that is, the aggregate rate of oxidation must be equivalent to the aggregate rate of reduction, or all electrons created through oxidation must be devoured by reduction.

For instance, consider zinc metal inundated in a corrosive arrangement containing H^+ particles. At a few areas on the metal surface, zinc will encounter oxidation or erosion as outlined in Figure 8.1, and as indicated by the reaction

$$Zn \rightarrow Zn^{2+} + 2e^-$$

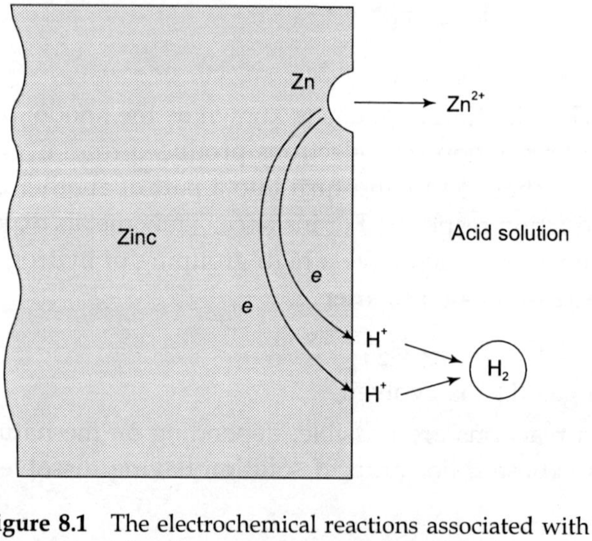

Figure 8.1 The electrochemical reactions associated with the corrosion of zinc in an acid solution.

Since zinc is a metal, and therefore a good electrical conductor, these electrons may be transferred to an adjacent region at which the H^+ ions are reduced according to

$$2H + 2e^- \rightarrow H_2 \text{ (gas)}$$

Combining the above two equations, we get the total electrochemical reaction as follows,

$$Zn + 2H^+ \rightarrow Zn^{2+} + H_2 \text{ (gas)}$$

Another illustration is the oxidation or rusting of iron in water, which contains broken up oxygen. This procedure happens in two stages; in principle, Fe is oxidized to Fe^{2+} [as $Fe(OH)_2$]

$$Fe + \frac{1}{2}O_2 + H_2O \rightarrow Fe^{2+}\ 2OH^- \rightarrow Fe\,(OH)_2$$

and, in the second stage, to Fe^{3-} [as $Fe(OH)_3$] according to

$$2Fe\,(OH)_2 + \frac{1}{2}O_2 + H_2O \rightarrow 2Fe\,(OH)_3$$

The compound $Fe(OH)_3$ is the all too familiar rust. As a consequence of oxidation, the metal ions may either go into the corroding solution as ions or they may form an insoluble compound with non-metallic elements.

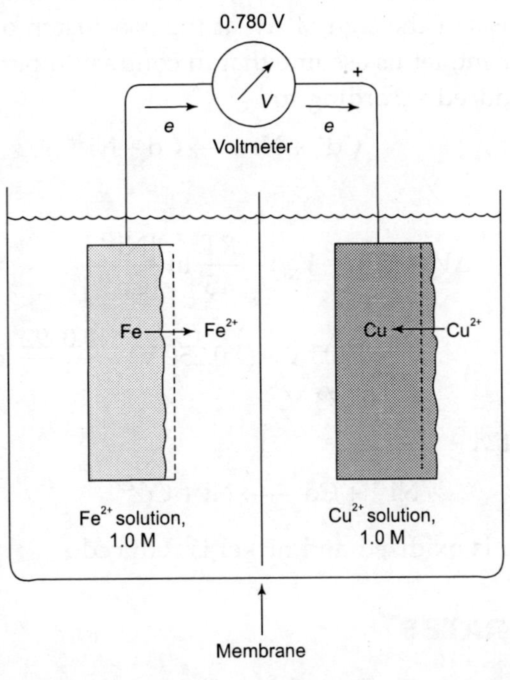

Figure 8.2 An electrochemical cell consisting of iron and copper electrodes, each of which is immersed in a 1 M solution of its ion. Iron corrodes while copper electrodeposits.

Example 8.1 One-half of an electrochemical cell consists of a pure nickel electrode in a solution of Ni^{2+} ions; the other half is a cadmium electrode immersed in a Cd^{2+}solution.

(a) If the cell is a standard one, write the spontaneous overall reaction and calculate the voltage that is generated.

(b) Compute the cell potential if the $Cd^{2+}$and Ni^{2+}concentrations are 0.5 and 10^{-3}M, respectively. Is the spontaneous reaction direction still the same as for the standard cell?

Solution:

(a) The cadmium electrode will be oxidized and nickel reduced because cadmium is lower in the emf series; thus, the spontaneous reactions will be

$$Cd \longrightarrow Cd^{2+}\ 2e^-$$

$$\frac{Ni^{2+} + 2e^- \longrightarrow Ni}{Ni^{2+} + Cd \longrightarrow Ni + Cd^{2+}}$$

The half-cell potentials for cadmium and nickel are, respectively, – 0.403 and – 0.250 V.

(b) For this portion of the problem, since the half-cell solution concentrations are no longer 1 M. At this point it is necessary to make a calculated guess as to which metal species is oxidized (or reduced). This choice will either be affirmed or refuted on the basis of the sign of ΔV at the conclusion of the computation. For the sake of argument, let us assume that in contrast to part (a), nickel is oxidized and cadmium reduced according to

$$Cd^{+} + Ni \longrightarrow Cd + Ni^{2+}$$

Thus,

$$\Delta V = (V^0_{Cd} - V^0_{Ni}) - \frac{RT}{n\Im} \ln \frac{[Ni^{2+}]}{[Cd^{2+}]}$$

$$= -0.403\ V - (-0.250\ V) - \frac{0.0592}{2} \log\left(\frac{10^{-3}}{0.50}\right)$$

$$= -0.073\ V$$

Since ΔV is negative,

$$Ni^{2+} + Cd \longrightarrow Ni + Cd^{2+}$$

That is, cadmium is oxidized and nickel is reduced.

8.3 CORROSION RATES

The half-cell potentials recorded in Table 8.1 are thermodynamic parameters that identify with systems at equilibrium. For instance, for examining electrochemical responses, it was implicitly expected that there was no movement through the outer circuit. Genuine consuming frameworks are not at harmony; there will be a stream of electrons from anode to cathode which implies that the half-cell potential parameters (Table 8.1) can't be connected.

Table 8.1 The standard emf series

	Electrode Reaction	Standard Electrode Potential, V^0(V)
↑	$Au^{3+} + 3e^- \rightarrow Au$	+1.420
	$O_2 + 4H^+ 4e^- \rightarrow 2H_2O$	+1.229
	$Pt^{2+} + 2e^- \rightarrow Pt$	~+1.2
	$Ag^+ + e^- \rightarrow Ag$	+0.800
Increasingly (cathodic)	$Fe^{3+} + e^- \rightarrow Fe^{2+}$	+0.771
	$O_2 + 2H_2O + 4e^- \rightarrow 4(OH^-)$	+0.401
	$Cu^{2+} + 2e^- \rightarrow Cu$	+0.340
	$2H^+ + 2e^- \rightarrow H_2$	0.000
	$Pb^{2+} + 2e^- \rightarrow H_2$	−0.126
	$Sn^{2+} + 2e^- \rightarrow Sn$	−0.136
	$Ni^{2+} + 2e^- \rightarrow Ni$	−0.250
	$Co^{2+} + 2e^- \rightarrow Co$	−0.277
	$Cd^{2+} + 2e^- \rightarrow Cd$	−0.403
	$Fe^{2+} + 2e^- \rightarrow Fe$	−0.440
Increasingly active (anodic)	$Cr^{3+} + 3e^- \rightarrow Cr$	−0.744
	$Zn^{2+} + 2e^- \rightarrow Zn$	−0.763
	$Al^{3+} + 3e^- \rightarrow Al$	−1.662
	$Mg^{2+} + 2e^- \rightarrow Mg$	−2.363
↓	$Na^+ + e^- \rightarrow Na$	−2.714
	$K^+ + e^- \rightarrow K$	−2.924

Moreover, these half-cell potential speak to the magnitude of a main force, or the propensity for the event of the specific half-cell response. In any case, it ought to be noticed that in spite of the fact that these possibilities might be utilized to decide unconstrained response bearings, they give no data as to erosion rates. That is, despite the fact that a ΔV potential processed for a particular erosion circumstance is a generally extensive positive number, the response may happen at just an inconsequential moderate rate. From the designing point of view, we are keen on anticipating the rates at which frameworks consume; this requires the use of different parameters, as discussed below.

The corrosion rate, or the rate of material removal as a consequence of the chemical action, is an important corrosion parameter. This may be expressed as the **corrosion penetration rate (CPR),** or the thickness loss of material per unit of time. The formula for this calculation is

$$\text{CPR} = \frac{KW}{\rho A t}$$

where, W is the weight loss after exposure time t; ρ and A represent the density and exposed specimen area, respectively, and K is a constant, its magnitude depends on the system of units used. The CPR is conveniently expressed in terms of either mils per year (mpy) or millimetres per year (mm/yr). In the first case, K = 534 to give CPR in mpy (where 1 mil = 0.001 in), and W, ρ, A, and t are specified in units of milligrams, grams per cubic centimetre, square inches, and hours, respectively. In the second case, K = 86.7 for mm/yr, and units for the other parameters are the same as for mils per year, except that A is given in square centimetres. For most applications a corrosion penetration rate less than about 20 mpy (0.50 mm/yr) is acceptable.

While there is an electric current connected with electrochemical corrosion, we can likewise express corrosion rate in terms of this current, or, all the more particularly, current thickness, that is, the current per unit surface range of material eroding which is assigned i. The rate r, in units of mol/m^2-s, is resolved utilizing the expression

$$r = \frac{i}{n\Im}$$

where, again, n is the number of electrons associated with the ionization of each metal atom, and $\Im$ is 96,500 C/mol.

Some normally active metals and alloys, under particular environmental conditions, lose their chemical reactivity and become extremely inert. This phenomenon, termed **passivity,** is displayed by chromium, iron, nickel, titanium, and many of their alloys. It is felt that this passive behaviour results from the formation of a highly adherent and very thin oxide film on the metal surface, which serves as a protective barrier to further corrosion. Stainless steels are highly resistant to corrosion in a rather wide variety of atmospheres as a result of passivation. They contain at least 11% chromium that, as a solid-solution alloying element in iron, minimizes the formation of rust; instead, a protective surface film forms in oxidizing atmospheres. (Stainless steels are susceptible to corrosion in some environments, and therefore are not always "stainless.") Aluminium is highly corrosion resistant in many environments because it also passivates. If damaged, the protective film normally reforms rapidly. However, a change in the character of the environment (e.g., alteration in the concentration of the active corrosive species) may cause a passivated material to revert to an active state. Subsequent damage to a pre-existing passive film could result in a substantial increase in corrosion rate, by as much as 100,000 times.

This passivation phenomenon may be explained in terms of polarization potential–log current density curves discussed in the preceding section. The polarization curve for a metal that passivates will have the general shape. At relatively low potential values, within the "active" region the behaviour is linear as it is for normal metals. With increasing potential, the current density suddenly decreases to a very low value that remains independent of potential; this is termed the "passive" region. Finally, at even higher potential values, the current density again increases with potential in the "transpassive" region.

8.4 OXIDATION

The discussion in Section 8.2 treated the erosion of metallic materials regarding electrochemical responses that happen in watery arrangements. Moreover, oxidation of metal compounds is additionally conceivable in vapourous air, regular air, wherein an oxide layer or scale shapes on the surface of the metal. This marvel is much of the time named scaling, tarnishing, or dry corrosion. In this area we will examine conceivable instruments for this sort of erosion, the sorts of oxide layers that can shape, and the energy of oxide arrangement.

Similarly, as with aqueous corrosion, the procedure of oxide layer development is an electrochemical one, which might be communicated, for divalent metal M, by the accompanying reaction

$$M + \frac{1}{2}O_2 \rightarrow MO$$

Furthermore, the above reaction consists of oxidation and reduction half-reactions. The former, with the formation of metal ions,

$$M \rightarrow M^{2+} + 2e^-$$

occurs at the metal-scale interface. The reduction half-reaction produces oxygen ions as follows:

$$\frac{1}{2}O_2 + 2e^- \rightarrow O^{2-}$$

and takes place at the scale gas interface. A schematic representation of this metal scale gas system is shown in Figure 8.3. For the oxide layer to increment in thickness, it is vital that electrons be directed to the scale–gas interface, and soon thereafter the diminishment response happens; furthermore, Mg^{2+} particles must diffuse far from the metal-scale interface, or potentially O^2 particles must diffuse towards this same interface (Figure 8.3). In this way, the oxide scale serves both as an electrolyte through which particles diffuse and as an electrical circuit for the entry of electrons. Moreover, the scale may shield the metal from quick oxidation when it goes about as a boundary to ionic dispersion and additionally electrical conduction; most metal oxides are exceedingly electrically insulative.

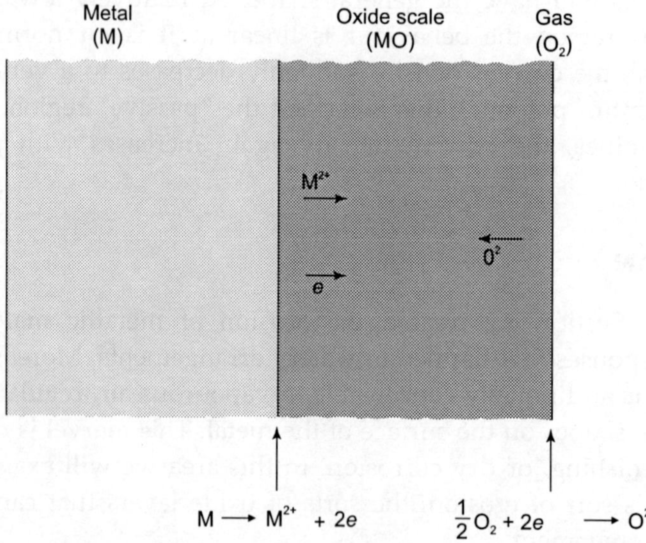

Figure 8.3 Schematic representation of processes that are involved in gaseous oxidation at a metal surface.

8.4.1 Scales Types

Rate of oxidation (i.e., the rate of film thickness increase) and the tendency of the film to protect the metal from further oxidation are related to the relative volumes of the oxide and metal. The ratio of these volumes, termed the **Pilling-Bedworth ratio,** may be determined from the following expression:

$$\text{P-B ratio} = \frac{A_o \, \rho_M}{A_M \, \rho_o}$$

where A_o is the molecular (or formula) weight of the oxide, A_M is the atomic weight of the metal, and ρ_o and ρ_M are the oxide and metal densities, respectively. For metals having P-B ratios less than unity, the oxide film tends to be porous and unprotective because it is insufficient to fully cover the metal surface. In the event that the proportion is more prominent than solidarity, compressive stresses result in the film as it structures. For a proportion more prominent than 2-3 the oxide covering may crack and drop off, constantly uncovering a new and unprotected metal surface. The perfect P-B proportion for the arrangement of a defensive oxide film is solidarity. Table 8.2 presents P-B proportions for metals that shape protective coatings and for those that don't. Note that protective coatings ordinarily frame for metals having P-B proportions somewhere around 1 and 2, though non-defensive ones generally result when this proportion is under 1 or more prominent than around 2. Notwithstanding the P-B proportion, different variables likewise impact the oxidation resistance bestowed by the film; these incorporate a high level of adherence amongst film and metal, practically identical coefficients of warm extension for metal and oxide, and, for the oxide, a moderately high dissolving point and great high temperature versatility or plasticity.

A few systems are accessible for enhancing the oxidation resistance of a metal. One includes utilization of a defensive surface covering of another material that holds or adheres fast to the metal furthermore is itself impervious to oxidation. In a few cases, the expansion of alloying components will frame a more adherent and defensive oxide scale by uprightness of delivering a more ideal Pilling-Bedworth proportion and additionally enhancing other scale qualities.

8.4.2 Kinetics

One of the essential concerns in respect to metal oxidation is the rate at which the response advances. In the light of the fact that the oxide scale response item regularly stays at first glance, the rate of response might be controlled by measuring the weight gain per unit zone as a component of time. At the point when the oxide that structures is nonporous and sticks to the metal surface, the rate of layer development is controlled by ionic diffusion. A *parabolic* relationship exists between the weight gain per unit area W and the time t as follows:

$$W^2 = K_1 t + K_2$$

where, K_1 and K_2 are time-independent constants at a given temperature. This weight gain-time conduct is plotted schematically in Figure 8.4. The oxidation of iron, copper, and cobalt takes after this rate expression. In the oxidation of metals for which the scale is permeable or drops off (i.e., for P-B proportions not exactly around 1 or more prominent than around 2), the oxidation rate expression is straight; that is,

$$W = K_3 t$$

where, K_3 is a constant. Under these conditions oxygen is constantly accessible for response with an unprotected metal surface in the light of the fact that the oxide does not go about as a response hindrance. Sodium, potassium, and tantalum oxidize as per this rate expression and, unexpectedly, have P-B proportions essentially not the same as solidarity (Table 8.2). Linear growth rate kinetics is additionally spoken to in Figure 8.4. Still a third response rate law has been watched for thin oxide layers (by and large under 100 nm) that frame at generally low temperatures. The reliance of weight gain on time is logarithmic and takes the frame

$$W = K_4 \log (K_5 t + K_6)$$

Again, the *K*'s are constants. This oxidation behaviour, also shown in Figure 8.4, has been observed for aluminium, iron, and copper at near-ambient temperatures.

Table 8.2 Pilling-Bedworth ratios for a number of metals

Protective		Nonprotective	
Ce	1.16	K	0.45
Al	1.28	Li	0.57
Pb	1.40	Na	0.57
Ni	1.52	Cd	1.21
Be	1.59	Ag	1.59
Pd	1.60	Ti	1.95
Cu	1.68	Ta	2.33
Fe	1.77	Sb	2.35
Mn	1.79	Nb	2.61
Co	1.99	U	3.05
Cr	1.99	Mo	3.40
Si	2.27	W	3.40

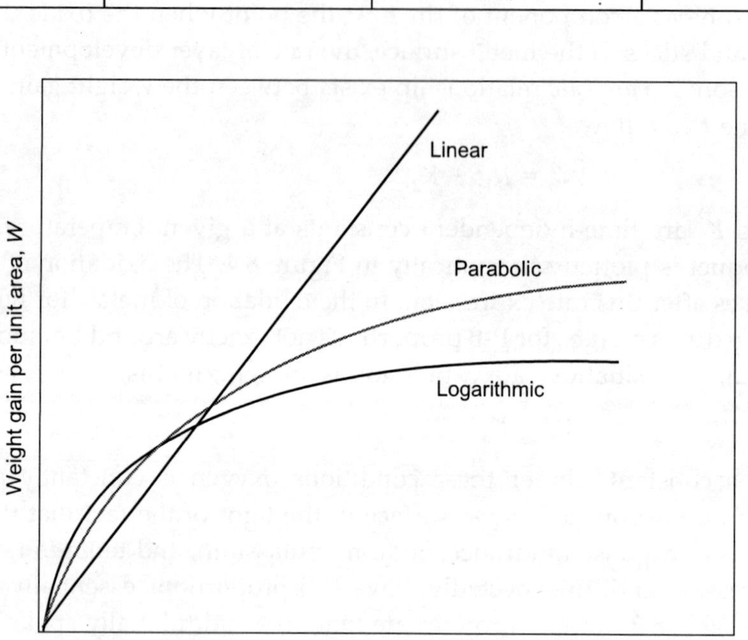

Figure 8.4 Oxidation film growth curves for linear, parabolic, and logarithmic rate laws.

8.4.3 Corrosion of Ceramic Materials

Ceramic materials, being compounds amongst metallic and nonmetallic components, might be considered having of now been corroded. Accordingly, they are exceedingly invulnerable to erosion by all situations, particularly at room temperature. Corrosion of ceramic materials for the most part includes straightforward compound disintegration, rather than the electrochemical procedures found in metals, as portrayed previously.

Ceramic materials are every now and again used in light of their imperviousness to consumption. Glass is regularly used to contain fluids therefore. Refractory ceramics production must not just withstand high temperatures and give warm protection in any case, on numerous occasions, should likewise oppose high-temperature assault by liquid metals, salts, slags, and glasses. A portion of the new innovation plans for changing over vitality starting with one frame then onto the next that is more valuable require generally high temperatures, destructive airs, and weights over the surroundings. Ceramic materials are greatly improved suited to withstand the vast majority of these situations.

8.4.4 Corrosion of Polymers

Polymeric materials likewise encounter crumbling by the method for ecological communications. Be that as it may, an undesirable communication is indicated as corruption instead of erosion in light of the fact that the procedures are essentially unique. While most metallic erosion responses are electrochemical, by difference, polymeric corruption is physiochemical; that is, it includes physical and in addition synthetic or chemical phenomena.

Moreover, a wide assortment of responses and antagonistic results are feasible for polymer debasement or degradation. Polymers may break down by swelling and dissolution. Covalent bond break, as an aftereffect of heat energy, chemical reactions, and radiation is additionally conceivable, conventionally with a chaperon decrease in mechanical integrity. It ought to likewise be specified that in light of the substance many-sided quality of polymers, their degradation mechanisms are not surely knew.

To quickly refer to a few cases of polymer degradation, polyethylene, if presented to high temperatures in an oxygen atmosphere, endures an impedance of its mechanical properties by getting to be brittle. Likewise, the utility of poly (vinyl chloride) might be constrained on the grounds that this material may get to be shaded when presented to high temperatures, albeit such situations don't influence its mechanical qualities.

8.5 FORMS OF CORROSION

It is convenient to classify corrosion according to the manner in which it is manifest. Metallic corrosion is sometimes classified into eight forms: uniform, galvanic, crevice, pitting, intergranular, selective leaching, erosion-corrosion, and stress corrosion. The causes and means of prevention of each of these forms are discussed briefly. In addition, we will discuss hydrogen embrittlement in this section. Hydrogen embrittlement is, in a strict sense, a type of failure rather than a form of corrosion; however, it is often produced by hydrogen that is generated from corrosion reactions.

8.5.1 Uniform Attack

Uniform attack is a form of electrochemical corrosion that occurs with equivalent intensity over the entire exposed surface and often leaves behind a scale or deposit. In a microscopic sense, the oxidation and reduction reactions occur randomly over the surface. Some familiar examples include general rusting of steel and iron and tarnishing of silverware. This is probably the most common form of corrosion. It is also the least objectionable because it can be predicted and designed for with relative ease.

8.5.2 Galvanic Corrosion

Galvanic corrosion occurs when two metals or alloys having different compositions are electrically coupled while exposed to an electrolyte. This is a type of corrosion or dissolution. The less noble or more reactive metal in the particular environment will experience corrosion; the more inert metal, the cathode, will be protected from corrosion. For example, steel screws corrode when in contact with brass in a marine environment; or if copper and steel tubing are joined in a domestic water heater, the steel will corrode in the vicinity of the junction.

Again, the galvanic series indicates the relative reactivities, in seawater, of a number of metals and alloys. When two alloys are coupled in seawater, the one lower in the series will experience corrosion. Some of the alloys in the table are grouped in brackets. Generally, the base metal is the same for these bracketed alloys, and there is little danger of corrosion if alloys within a single bracket are coupled. It is also worth noting from this series that some alloys are listed twice (e.g., nickel and the stainless steels), in both active and passive states.

The rate of galvanic attack depends on the relative anode-to-cathode surface areas that are exposed to the electrolyte, and the rate is related directly to the cathode-anode area ratio; that is, for a given cathode area, a smaller anode will corrode more rapidly than a larger one. The reason for this is that corrosion rate depends on current density the current per unit area of corroding surface, and not simply the current. Thus, a high current density results for the anode when its area is small relative to that of the cathode.

A number of measures may be taken to significantly reduce the effects of galvanic corrosion. These include the following:

1. If coupling of dissimilar metals is necessary, choose the two that are close together in the galvanic series.

2. Avoid an unfavourable anode-to-cathode surface area ratio; use an anode area as large as possible.
3. Electrically insulate dissimilar metals from each other.
4. Electrically connect a third anodic metal to the other two; this is a form of **cathodic protection**.

8.5.3 Crevice Corrosion

Crevice corrosion may also occur as a consequence of concentration differences of ions or dissolved gases in the electrolyte solution, and between two regions of the same metal piece. For such a *concentration cell,* corrosion occurs in the locale that has the lower concentration. A good example of this type of corrosion occurs in crevices and recesses or under deposits of dirt or corrosion products where the solution becomes stagnant and there is localized depletion of dissolved oxygen. Corrosion preferentially occurring at these positions is called **crevice corrosion**. The crevice must be wide enough for the solution to penetrate, yet narrow enough for stagnancy; usually the width is several thousandths of an inch. Crevice corrosion may be prevented by using welded instead of riveted or bolted joints, using nonabsorbing gaskets when possible, removing accumulated deposits frequently, and designing containment vessels to avoid stagnant areas and ensure complete drainage.

8.5.4 Pitting

Pitting is another form of localized corrosion attack in which small pits or holes form. They ordinarily penetrate from the top of a horizontal surface downward in a nearly vertical direction. It is an extremely insidious type of corrosion, often going undetected and with very little material loss until failure occurs. The mechanism for pitting is probably the same as for crevice corrosion in that oxidation occurs within the pit itself, with complementary reduction at the surface. It is supposed that gravity causes the pits to grow downward, the solution at the pit tip becoming more concentrated and dense as pit growth progresses. A pit may be initiated by a localized surface defect such as a scratch or a slight variation in composition. In fact, it has been observed that specimens having polished surfaces display a greater resistance to pitting corrosion. Stainless steels are somewhat susceptible to this form of corrosion; however, alloying with about 2% molybdenum enhances their resistance significantly.

8.5.5 Intergranular Corrosion

As the name suggests, intergranular corrosion occurs preferentially along grain boundaries for some alloys and in specific environments. The net result is that a macroscopic specimen disintegrates along its grain boundaries. This type of corrosion is especially prevalent in some stainless steels. When heated to temperatures between

500 and 800°C for sufficiently long periods, these alloys become sensitized to intergranular attack. It is believed that this heat treatment permits the formation of small precipitate particles of chromium carbide by reaction between the chromium and carbon in the stainless steel. These particles form along the grain boundaries. Both the chromium and the carbon must diffuse to the grain boundaries to form the precipitates, which leaves a chromium-depleted zone adjacent to the grain boundary. Consequently, this grain boundary region is now highly susceptible to corrosion.

Intergranular corrosion is a severe problem in the welding of stainless steels, when it is often termed **weld decay.** Stainless steels may be protected from intergranular corrosion by the following measures:

1. Subjecting the sensitized material to a high-temperature heat treatment in which all the chromium carbide particles are redissolved.
2. Lowering the carbon content below 0.03 wt% C so that carbide formation is minimal.
3. Alloying the stainless steel with another metal such as niobium or titanium, which has a greater tendency to form carbides than does chromium so that the Cr remains in solid solution.

8.5.6 Selective Leaching

Selective leaching is found in solid solution alloys and occurs when one element or constituent is preferentially removed as a consequence of corrosion processes. The most common example is the dezincification of brass, in which zinc is selectively leached from a copper–zinc brass alloy. The mechanical properties of the alloy are significantly impaired, since only a porous mass of copper remains in the region that has been dezincified. In addition, the material changes from yellow to a red or copper colour. Selective leaching may also occur with other alloy systems in which aluminium, iron, cobalt, chromium, and other elements are vulnerable to preferential removal.

8.5.7 Erosion-corrosion

Erosion-corrosion arises from the combined action of chemical attack and mechanical abrasion or wear as a consequence of fluid motion. Virtually, all metal alloys, to one degree or another, are susceptible to erosion-corrosion. It is especially harmful to alloys that passivate by forming a protective surface film. The abrasive action may erode away the film, leaving exposed a bare metal surface. If the coating is not capable of continuously and rapidly reforming as a protective barrier, corrosion may be severe. Relatively soft metals such as copper and lead are also sensitive to this form of attack. Usually, it can be identified by surface grooves and waves having contours that are characteristic of the flow of the fluid. The nature of the fluid can have a dramatic influence on the corrosion behaviour. Increasing fluid velocity normally enhances the rate of corrosion.

Also, a solution is more erosive when bubbles and suspended particulate solids are present. Erosion-corrosion is commonly found in piping, especially at bends, elbows, and abrupt changes in pipe diameter—positions where the fluid changes direction or flow suddenly becomes turbulent. Propellers, turbine blades, valves, and pumps are also susceptible to this form of corrosion. One of the best ways to reduce erosion–corrosion is to change the design to eliminate fluid turbulence and impingement effects. Other materials may also be utilized that inherently resist erosion. Furthermore, removal of particulates and bubbles from the solution lessens its ability to erode.

8.5.8 Stress Corrosion

Stress corrosion, sometimes termed stress corrosion cracking, results from the combined action of an applied tensile stress and a corrosive environment; both influences are necessary. In fact, some materials that are virtually inert in a particular corrosive medium become susceptible to this form of corrosion when a stress is applied. Small cracks form and then propagate in a direction perpendicular to the stress (see the chapter-opening photograph for this chapter), with the result that failure may eventually occur. Failure behaviour is characteristic of that for a brittle material, even though the metal alloy is intrinsically ductile. Furthermore, cracks may form at relatively low stress levels, significantly below the tensile strength. Most alloys are susceptible to stress corrosion in specific environments, especially at moderate stress levels. For example, most stainless steels stress corrode in solutions containing chloride ions, whereas brasses are especially vulnerable when exposed to ammonia.

The stress that produces stress corrosion cracking need not be externally applied; it may be a residual one that results from rapid temperature changes and uneven contraction, or for two-phase alloys in which each phase has a different coefficient of expansion. Also, gaseous and solid corrosion products that are entrapped internally can give rise to internal stresses.

Probably the best measure to take in reducing or totally eliminating stress corrosion is to lower the magnitude of the stress. This may be accomplished by reducing the external load or increasing the cross-sectional area perpendicular to the applied stress. Furthermore, an appropriate heat treatment may be used to anneal out any residual thermal stresses.

8.5.9 Hydrogen Embrittlement

Various metal alloys, specifically some steels, experience a significant reduction in ductility and tensile strength when atomic hydrogen (H) penetrates into the material. This phenomenon is aptly referred to as **hydrogen embrittlement;** the terms *hydrogen-induced cracking* and *hydrogen stress cracking* are sometimes also used. Strictly speaking, hydrogen embrittlement is a type of failure; in response to applied or residual tensile stresses, brittle fracture occurs catastrophically as cracks grow and rapidly propagate.

Hydrogen in its atomic form (H as opposed to the molecular form, H_2) diffuses interstitially through the crystal lattice, and concentrations as low as several parts per million can lead to cracking. Furthermore, hydrogen induced cracks are most often transgranular, although intergranular fracture is observed for some alloy systems. A number of mechanisms have been proposed to explain hydrogen embrittlement; most of them are based on the interference of dislocation motion by the dissolved hydrogen.

Hydrogen embrittlement is similar to stress corrosion in that a normally ductile metal experiences brittle fracture when exposed to both a tensile stress and a corrosive atmosphere. However, these two phenomena may be distinguished on the basis of their interactions with applied electric currents. Whereas cathodic protection reduces or causes setting in of stress corrosion, it may, on the other hand, lead to initiation or enhancement of hydrogen embrittlement.

For hydrogen embrittlement to occur, some source of hydrogen must be present, and, in addition, the possibility for the formation of its atomic species. Situations wherein these conditions are met include the following: pickling of steels in sulphuric acid; electroplating; and the presence of hydrogen-bearing atmospheres (including water vapour) at elevated temperatures such as during welding and heat treatments. Also, the presence of what are termed "poisons" such as sulphur (i.e., H_2S) and arsenic compounds accelerate hydrogen embrittlement. These substances retard the formation of molecular hydrogen and thereby increase the residence time of atomic hydrogen on the metal surface. Hydrogen sulphide, probably the most aggressive poison, is found in petroleum fluids, natural gas, oil-well brines, and geothermal fluids.

High-strength steels are susceptible to hydrogen embrittlement, and increasing strength tends to enhance the material's susceptibility. Martensitic steels are especially vulnerable to this type of failure; bainitic, ferritic, and spheroiditic steels are more resilient. Furthermore, FCC alloys (austenitic stainless steels, and alloys of copper, aluminum, and nickel) are relatively resistant to hydrogen embrittlement, mainly because of their inherently high ductilities. However, strain hardening these alloys will enhance their susceptibility to embrittlement.

Some of the techniques commonly used to reduce the likelihood of hydrogen embrittlement include reducing the tensile strength of the alloy via heat treatment, removal of the source of hydrogen, "baking" the alloy at an elevated temperature to drive out any dissolved hydrogen, and substitution of a more embrittlement resistant alloy.

8.6 CORROSION CONTROL

Some corrosion anticipation strategies were dealt with in respect to the eight types of consumption; be that as it may, just the measures particular to each of the different corrosion sorts were talked about. Presently, some more broad systems are displayed; these incorporate material determination, natural change, outline, coatings, and cathodic assurance. Maybe the most widely recognized and least demanding method for counteracting consumption is through the sensible determination of materials once the erosion environment has been portrayed. Standard corrosion references are useful in this regard. Here, cost might be a critical variable. It is not generally monetarily doable to utilize the material that gives the ideal consumption resistance; at times, either another alloy and additionally some other measure must be utilized.

Changing the character of the earth, if conceivable, may likewise fundamentally impact corrosion. Bringing down the liquid temperature and additionally velocity more often than not creates a decrease in the rate at which consumption happens. Ordinarily, expanding or diminishing the convergence of a few physical groups in the arrangement will have a beneficial outcome; for instance, the metal may encounter passivation.

Inhibitors are substances that, when included moderately lower fixations to the environment, diminish its destructiveness or corrosiveness. Obviously, the particular inhibitor depends both on the amalgam and on the destructive environment. There are a few systems that may represent the viability of inhibitors. Some respond with and practically take out a synthetically dynamic species categories in the arrangement, (for example, dissolved oxygen). Other inhibitor molecules append themselves to the consuming surface and meddle with either the oxidation or the lessening response, or shape a thin defensive covering. Inhibitors are typically utilized as a part of shut frameworks, for example, vehicle radiators and steam boilers.

A few parts of outline thought have as of now been talked about, particularly as to galvanic and fissure consumption and erosion-corrosion. What's more, the outline design ought to take into consideration finish waste on account of a shutdown, and simple washing. Since broken up oxygen may improve the destructiveness of numerous arrangements, the plan ought to, if conceivable, incorporate arrangement for the avoidance of air.

Physical hindrances to corrosion are connected on surfaces as movies and coatings. Expansive differing qualities of metallic and non-metallic covering materials are accessible. The covering keeps up a high level of surface attachment, which without doubt requires some pre-application surface treatment. By and large, the covering must be for all intents and purposes nonreactive in the destructive environment and impervious to mechanical harm that uncovered the exposed metal to the destructive environment. Every one of the three material sorts metals, ceramics, and polymers are utilized as coatings for metals.

8.6.1 Cathodic Protection

A standout amongst the best method for corrosion counteractive action is cathodic protection. It can be utilized for every one of the eight distinct types of consumption as examined above, and may, in a few circumstances, totally stop erosion. Once more, oxidation or corrosion of a metal M happens by the summed up response

$$M \rightarrow M^{n+} + ne^-$$

Cathodic protection essentially includes providing, from an outer source, electrons to the metal to be ensured, making it a cathode; the response above is along these lines constrained in the invert (or decrease) course.

One cathodic protection system utilizes a galvanic couple: the metal to be ensured is electrically associated with another metal that is more receptive in the specific environment. The last encounters oxidation, and, after surrendering electrons, shields the main metal from consumption. The oxidized metal is regularly called a conciliatory anode, and magnesium and zinc are generally utilized. All things considered in the light of the fact that they lie at the anodic end of the galvanic arrangement. This type of galvanic protection, for structures covered in the ground, is delineated in Figure 8.5a.

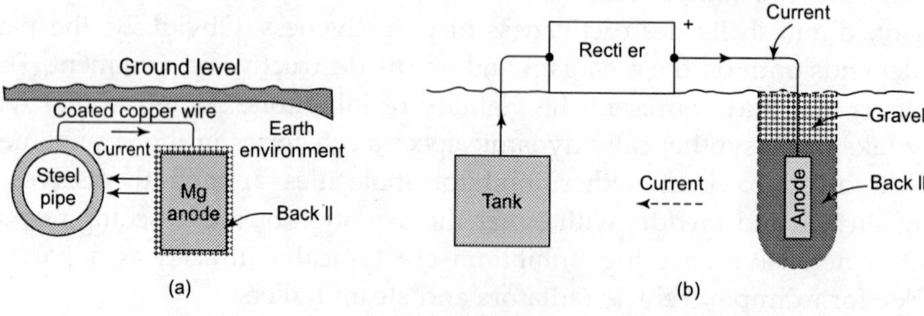

Figure 8.5 Cathodic protection of (a) an underground pipeline using a magnesium sacrificial anode, and (b) an underground tank using an impressed current.

The process of *galvanizing* is simply one in which a layer of zinc is connected to the surface of steel by hot dipping. In the atmosphere and most aqueous situations, zinc is anodic to and will accordingly cathodically ensure the steel if there is any surface harm (Figure 8.6). Any corrosion of the zinc covering will continue to a great degree moderate rate in light of the fact that the proportion of the anode-to-cathode surface range is vast.

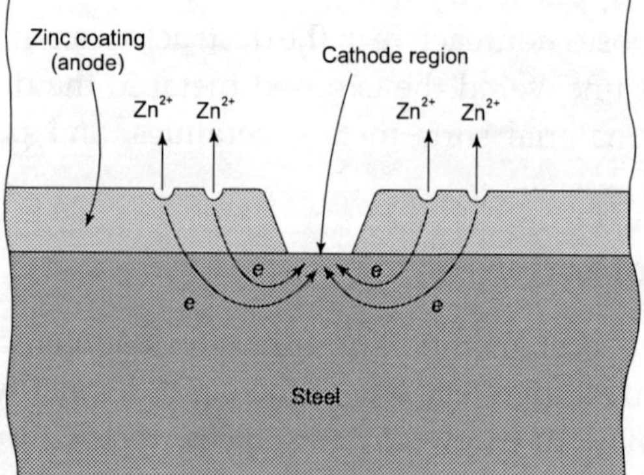

Figure 8.6 Galvanic protection of steel as provided by a coating of zinc.

For another strategy for cathodic protection, the source of electrons is an impressed current from an outside dc control source, as spoke to in Figure 8.5b for an underground tank. The negative terminal of the power source is associated with the structure to be secured. The other terminal is joined to a dormant anode (regularly graphite), which is, for this situation, covered in the dirt; high-conductivity inlay material gives great electrical contact between the anode and encompassing soil. Another way is between the cathode and anode through the interceding soil, finishing the electrical circuit. Cathodic protection is particularly valuable in averting erosion of water radiators, underground tanks and pipes, and marine hardware equipment.

8.7 SUMMARY

Electrochemical Considerations

Metallic corrosion is normally electrochemical, including both oxidation and lessening responses. Oxidation is the loss of the metal atom's valence electrons; the subsequent metal particles may either go into the consuming arrangement or shape an insoluble compound. Amid decrease, these electrons are exchanged to no less than one other substance physical groups. The character of the erosion environment directs which of a few conceivable diminishment responses will happen.

Not all metals oxidize with a similar level of simplicity, which is shown with a galvanic couple; when in an electrolyte, one metal (the anode) will consume, while a diminishment response will happen at the other metal (the cathode). The size of the electric potential that is built up amongst anode and cathode is demonstrative of the main thrust for the corrosion reaction.

The standard emf and galvanic arrangement are essentially rankings of metallic materials on the premise of their inclination to consume when coupled to different metals. For the standard emf arrangement, positioning depends on the size of the voltage created when the standard cell of a metal is coupled to the standard hydrogen terminal at the galvanic arrangement comprises the relative reactivities of metals and alloys in seawater.

The half-cell possibilities in the standard emf arrangement are thermodynamic parameters that are substantial just at harmony; corroding frameworks are not in equilibrium. Moreover, the extents of these possibilities give no sign with regards to the rates at which corrosion reactions happen.

Passivity

A number of metals and alloys passivate, or lose their chemical reactivity, under some environmental circumstances. This phenomenon is thought to involve the formation of a thin protective oxide film. Stainless steels and aluminium alloys exhibit this type of behaviour. The active-to-passive behaviour may be explained by the alloy's S-shaped electrochemical potential versus log current density curve. Intersections with reduction polarization curves in active and passive regions correspond, respectively, to high and low corrosion rates.

Forms of Corrosion

Metallic corrosion is sometimes classified into eight different forms: uniform attack, galvanic corrosion, crevice corrosion, pitting, intergranular corrosion, selective leaching, erosion-corrosion, and stress corrosion. Hydrogen embrittlement, a type of failure sometimes observed in corrosion environments, was also discussed.

Corrosion Rates

The rate of corrosion might be communicated as corrosion penetration rate, that is, the thickness loss of material per unit of time. Mils every year and millimetres every year are the basic units for this parameter. On the other hand, rate is relative to the present thickness connected with the electrochemical response.

Corrosion Control

The measures that might be taken to forestall, or possibly decrease, corrosion incorporate material choice, natural modification, utilization of inhibitors, design changes, use of coatings, and cathodic protection.

Oxidation

Oxidation of metallic materials by electrochemical activity is likewise conceivable in dry, vaporous airs. An oxide film frames at first glance which may go about as a boundary to further oxidation if the volumes of metal and oxide film are comparative, that is, if the Pilling-Bedworth proportion is close solidarity. The energy of film development may take after illustrative, straight, or logarithmic rate laws.

Corrosion of Ceramic Materials

Ceramic materials, being inherently corrosion resistant, are frequently utilized at elevated temperatures and/or in extremely corrosive environments. Polymeric materials deteriorate by noncorrosive processes. Upon exposure to liquids, they may experience degradation by swelling or dissolution. With swelling, solute molecules actually fit into the molecular structure. Scission, or the severance of molecular chain bonds, may be induced by radiation, chemical reactions, or heat. This results in reduction of molecular weight and a deterioration of the physical and chemical properties of the polymer.

EXERCISE

1. How do oxidation and reduction electrochemical reactions differ from each other?
2. Mention possible oxidation and reduction half-reactions that take place when magnesium is immersed in each of the following solutions:
   (a) HCl,
   (b) HCl solution with dissolved oxygen,
   (c) HCl solution with dissolved oxygen and, in addition, Fe^{2+} ions.
3. An electrochemical cell is constructed such that on one side a pure Zn electrode is in contact with a solution containing Zn^{2+} ions at a concentration of 10^{-2}M. The other cell half consists of a pure Pb^{2+} electrode immersed in a solution of ions that has a concentration of 10^{-4}M. At what temperature will the potential between the two electrodes be +0.568 V?
4. For the following pairs of alloys that are coupled in seawater, predict the possibility of corrosion; if corrosion is probable, note which metal/alloy will corrode.
   (a) Aluminium and cast iron
   (b) Inconel and nickel
   (c) Cadmium and zinc
   (d) Brass and titanium
   (e) Low-carbon steel and copper
5. A thick sheet is exposed to air near the ocean with an area of 100 in^2. After a gap of about one year 485 g of it was lost due to corrosion. Determine the rate of corrosion in both mpy and mm/yr.
6. (a) Explain the role of inhibitors?
   (b) List the possible mechanisms for the effectiveness of inhibitors?
7. Describe in short any two techniques for galvanic protection.
8. In an ocean vessel a corroded metal was found to be submerged. Considering the fact that original area of the plate being 800 cm^2 and about 7.6 kg had corroded because of the submersion, determine the time of submersion in years. Take density of alloy as 4.5 g/cm^3 and corrosion penetration rate of 4 mm/yr for this alloy in seawater.
9. An electrochemical cell consists of pure lead immersed in 5×10^{-2} M solution of Pb^{2+} ions and pure tin in a 0.25 M solution of Sn^{2+} ions. Find the voltage at 25°C and also mention the spontaneous electrochemical reaction.
10. Cite the major differences between activation and concentration polarizations.
11. Under what conditions is activation polarization rate is controlled?
12. Under what conditions is concentration polarization rate is controlled?

13. Briefly describe the phenomenon of passivity. Name two common types of alloy that passivate.

14. Why does chromium in stainless steels make them more corrosion resistant in many environments than plain carbon steels?

15. Briefly explain why, for a small anode to cathode area ratio, the corrosion rate is higher than for a large ratio.

9

Failure of Engineering Materials: Fatigue and Creep

Learning Objectives

After going through this chapter the reader will be able to:

- Gather an idea of failure of engineering materials.
- Understand the phenomenon of creep.
- Describe the elevated temperature creep degradation and failure.
- Know the role of fatigue in engineering materials.

9.1 INTRODUCTION

The failure of designing or engineering materials is quite often an undesirable occasion for a few reasons; these incorporate human lives that are placed in peril, financial misfortunes, and the obstruction with the accessibility of items and administrations. Despite the fact that the reasons for disappointment and the conduct of materials might be known, aversion of disappointments is hard to ensure. The standard causes are inappropriate materials choice and handling and insufficient plan of the part or its abuse. It is the obligation of the specialist to suspect and plan for conceivable disappointment and, if disappointment occurs, to evaluate its cause and after that take fitting preventive measures against future occurrences. The following topics are addressed in this chapter: Phenomenon of fatigue and creep, fundamentals of fracture mechanics, elevated temperature creep degradation and failure and fatigue in engineering materials.

9.2 PHENOMENON OF CREEP

Materials are frequently set in administration at lifted temperatures and presented to static mechanical burdens (e.g., turbine rotors in fly motors and steam generators that experience radial stresses, and high-weight steam lines). Twisting under such conditions is named crawl. Characterized as the time-needy and changeless disfigurement of materials when subjected to a steady load or stress, creep is ordinarily an undesirable wonder and is frequently the restricting element in the lifetime of a section or part. It is seen in all materials sorts; for metals it gets to be imperative just for temperatures more prominent than around 0.4 T_m (T_m outright softening temperature). Indistinct polymers, which incorporate plastics and rubbers, are particularly touchy to the creep deformation.

Creep is the inelastic response of materials loaded at high temperatures. Temperature has important effects on deformation phenomena. Microstructural defect rearrangement processes are often accelerated at high temperatures. Since these processes tend to soften the material they counteract the strain hardening produced by plastic deformation. Elasticity and plasticity are mechanical responses to loading which are independent of time. As soon as the load is applied, the corresponding level of strain sets in. In contrast, during creep the mechanical response is time dependent. This is somewhat analogous to viscoelastic behaviour except that during creep an often significant portion of the strain is permanent and remains after unloading.

With appropriate modifications, the standard tension test can be applied to investigate the mechanical response of materials deforming at high temperatures. Easiest to perform are tension tests at constant load and temperature. The resulting strain is measured as a function of time. This is often called the creep test and the resulting strain-time curves are called creep curves. If the creep test extends over a sufficiently long period, a stage may be reached at which the sample weakens rapidly and eventually breaks. Tests aimed at investigating this final stage of the creep process are called creep rupture tests.

Cyclic loading is often associated with material fatigue processes which eventually may cause the failure of the loaded component. At high temperatures, creep phenomena appear and they have an important influence on the fatigue process. Under cyclic loading, the resulting creep deformation and relaxation phases affect the fatigue response of the material depending on their relative duration according to the specific form of the cycling pattern. Typical strain-time curves obtained from creep tests at constant load over sufficiently long periods often exhibit three characteristic stages.

Primary or Transient Creep. Following the setting in of the instantaneous elastic strain 0, the material deforms rapidly but at a decreasing rate. The duration of this stage is typically relatively short in relation to the total creep curve. In secondary or steady-state creep the creep strain rate reaches a minimum value and remains approximately constant over a relatively long period of time. Tertiary creep is the stage where the creep strain rate accelerates rapidly. **Rupture** is the material unable to withstand the load anymore and breaks.

Atoms in solids vibrate at high frequencies ($\approx 10^{13}s^{-1}$) about their mean positions. Ocasionally ($\approx 1/10^6$ times), the amplitude of the vibration is large enough for the atom to move out of its current location and into a neighbouring site. This process is called solid-state diffusion. Diffusion is a thermally activated process. Thermal energy provides the necessary activation energy required to overcome the potential energy barrier preventing atomic displacements. Therefore, solid-state diffusion processes are enhanced significantly at high temperatures. Since atomic movements are directly related to microstructural reorganization processes it is then natural to expect that creep phenomena will be directly related to solid-state diffusion phenomena, particularly, at the highest temperatures. If the temperature is relatively low, diffusion is somewhat less important and the creep deformation process at the atomic scale is most influenced by slip and glide phenomena characteristic of low temperature plastic deformation.

9.2.1 Generalized Creep Behaviour

An average creep test comprises subjecting an example to a consistent load or stretch while keeping up the temperature steady; disfigurement or strain is measured and plotted as a component of slipped by time. Most tests are the steady load sort, which yield data of a engineering nature; consistent push tests are utilized to give a superior comprehension of the systems of creep.

Figure 9.1 is a schematic representation of the ordinary steady load creep conduct of metals. Endless supply of the heap there is an immediate distortion, as demonstrated in the figure, which is for the most part elastic. The subsequent creep curve comprises three districts, each of which has its own unmistakable strain-time include. Essential or transient creep happens initially, epitomized by a constantly diminishing at slow rate; that is, the incline of the bend reduces with time. This proposes the material is encountering an expansion in crawl resistance or strain solidifying twisting turns out to be more troublesome as the material is strained. For auxiliary crawl, now and again named relentless state crawl, the rate is consistent; that is, the plot gets straight. This is frequently the phase of crawl that is of the longest term.

The consistency of creep rate is clarified on the premise of a harmony between the contending procedures of strain solidifying and recuperation. Recuperation being the procedure whereby a material gets to be milder and holds its capacity to experience deformation. At last, for tertiary creep, there is quickening of the rate and extreme failure. This failure is often named rupture and results from smaller scale basic as well as metallurgical changes; for instance, grain limit division, and the development of inner splits, depressions, and voids. Likewise, for ductile burdens, a neck may shape sooner or later inside the disfigurement area. These all prompt diminishing in the compelling cross-sectional region and an expansion in strain rate.

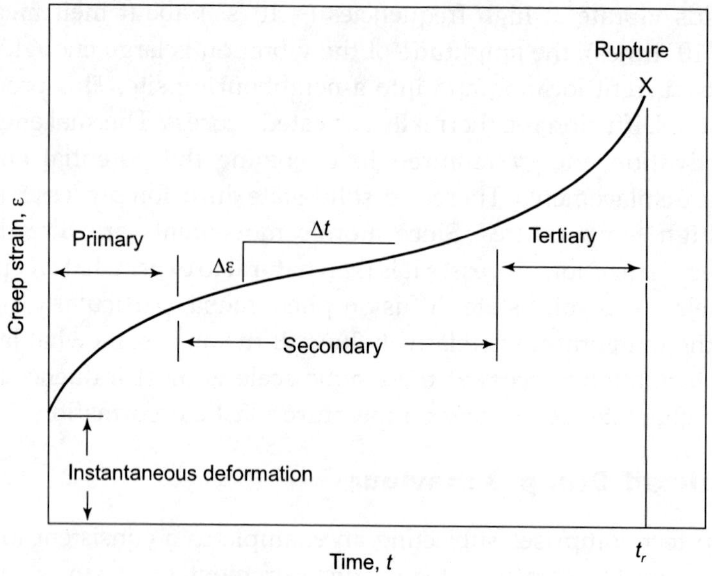

Figure 9.1 Typical creep curve of strain versus time at constant stress and constant elevated temperature. The minimum creep rate $\Delta\varepsilon/\Delta t$ is the slope of the linear segment in the secondary region. Rupture lifetime t_r is the total time to rupture.

For metallic materials most creep tests are directed in uniaxial strain utilizing an example having an indistinguishable geometry from for pliable tests (Figure 6.2). Then again, uniaxial pressure tests are more fitting for weak materials. These give a superior measure of the characteristic crawl properties provided there is no stress enhancement and crack propagation, as with tensile loads. Compressive test examples are generally right cylinders or parallelepipeds having length-to-diameter proportions extending from around 2 to 4. For most materials creep properties are for all intents and purposes free of stacking course or loading directions.

Conceivably the most critical parameter from a creep test is the slant of the auxiliary segment of the creep curve or bend ($\Delta\varepsilon/\Delta t$ in Figure 9.1). This is frequently called the base or unfaltering steady-state creep rate. It is the building outline parameter that is considered for long-life applications, for example, an atomic power plant part that is planned to work for very long, and when disappointment or a lot of strain are not alternatives. Then again, for some generally short-life creep circumstances (e.g., turbine sharp edges in military flying machine and rocket engine spouts), time to burst, or the crack lifetime, is the predominant plan thought; it is additionally shown in Figure 9.1. Obviously, for its assurance, creep tests must be led to the point of failure; these are termed *creep rupture* tests. Thus, a knowledge of these creep characteristics of a material allows the design engineer to ascertain its suitability for a specific application.

9.3 ELEVATED TEMPERATURE CREEP DEGRADATION AND FAILURE

Several defect mechanisms for thermal softening and elevated temperature degradation under stress, that is, creep, are depicted in Figure 9.2. Irrespective of material class, the plastic strain rate in each of the mechanisms varies with stress and temperature according to the equation below, which is reproduced here

$$\dot{\varepsilon} = A\sigma^m \exp(-E_c/RT)$$

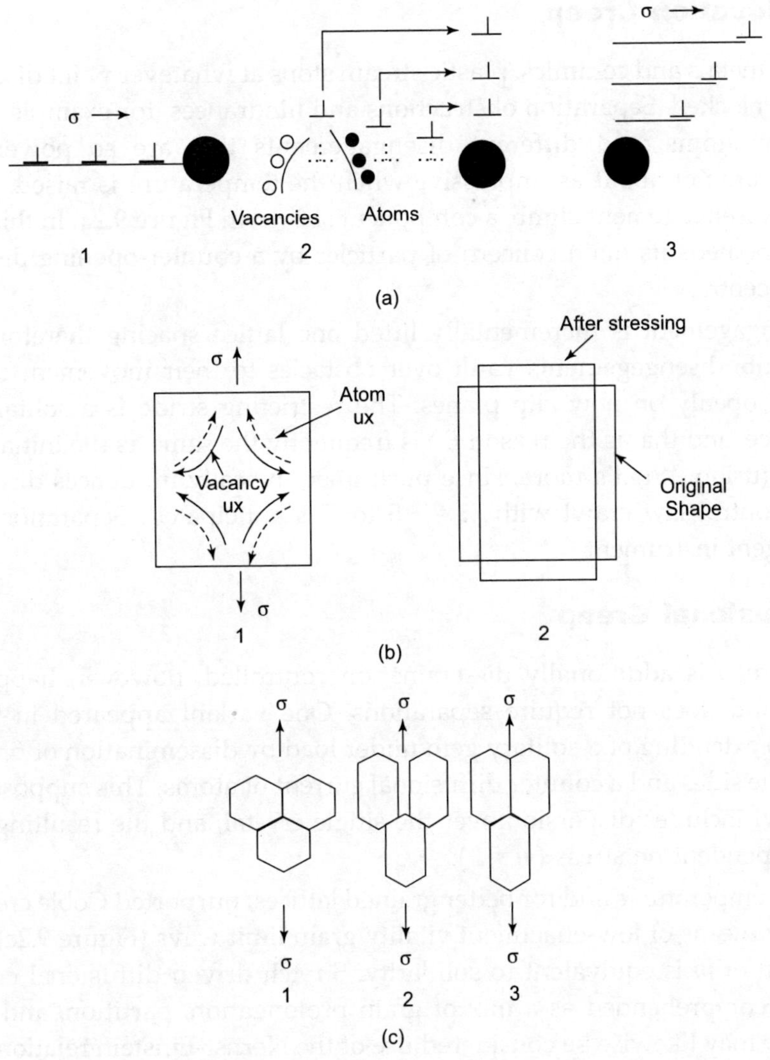

Figure 9.2 Microscopic mechanisms of creep: (a) Dislocation climb. (b) Herring-Nabarro creep. (c) Coble creep.

This equation applies strictly to steady-state creep. But, with little loss in generality, an equation of the sort

$$1/t_R = B\sigma^{m^*} \exp(-E_{CR}/RT)$$

is expected to also predict creep rupture failure times (t_R). A new set of constants are B, m^* and E_{CR} the effective activation energy for creep rupture hold. These can, in principle, be determined from accelerated creep rupture testing.

9.3.1 Dislocation Creep

In crystalline metals and ceramics, plastic stream stops at whatever point disengagement movement is blocked. Separation obstructions and hindrances, for example, accelerates, broken down atoms, and different disengagements that are so powerful at low temperatures are not about as impressive when the temperature is raised. One reason is because of disengagement climb, a component shown in Figure 9.2a. In this procedure the separation sheds its main concern of particles by a counter-opening dissemination flux into the centre.

The disengagement is incrementally lifted one lattice spacing therefore. Through rehashed climb, disengagements vault over obstacles to their movement and, by and by, can skim openly on new slip planes. The restricting stride is a solitary diffusive nuclear bounce, and that is the reason $E^\wedge$ is frequently the same as the initiation vitality for lattice diffusion. What's more, since push unequivocally influences disengagement movement, control law crawl with m = ~5 to 7 is anticipated. Separation crawl was clearly the agent instrument.

9.3.2 Diffusional Creep

Diffusional creep is additionally dissemination controlled, however, happens at low push levels and does not require separations. One variant appeared in Figure 9.2b imagines hub extending of a solitary gem under load by dissemination of opening from the pivot to the sides and a counter diffusional current of atoms. This supposed Herring-Nabarro crawl includes dispersion over the single crystal, and the resulting strain rate is linearly dependent on stress (m = 1).

At lower temperatures and for better grained lattices, purported Coble creep is agent. It happens by means of low-enactment vitality grain limit ways (Figure 9.2c), and again the estimation of m is equivalent to solidarity. Stretch driven diffusional creep can by and large be comprehended as a mix of grain prolongation, partition, and grain limit sliding. Creep may likewise be considered use of the Nernst-Einstein relationship where velocity is supplanted by the strain rate, and forces by stress.

9.3.3 Creep Failure

Perceptions of creep failure frequently uncover the predominant part of grain restriction engaged in the phenomena. Between granular fracture is a main indication of creep harm. Disappointment is the outgrowth of various interrelated procedures, for instance, cavitation or void nucleation, their development and linkage at grain limits, and gooey sliding of grains in respect to nearby ones. The last component has been exhibited by taking note of balances in lines scribed crosswise over limits between grains that have creeped. Employing single-crystal turbine blades to minimize these deleterious effects of grain boundaries during high-temperature and stress service is a popular strategy; superalloy, metal matrix, ceramic matrix, and carbon-carbon composites are alternative options in addressing general creep problems.

9.4 FATIGUE IN ENGINEERING MATERIALS

Fatigue is a type of failure that happens in structures subjected to dynamic and fluctuating stresses (e.g., extensions, airplane, and machine segments). Under these conditions it is workable for inability to happen at a stretch level impressively lower than the tractable or yield strength for a static load. The expression "fatigue" is utilized in the light of the fact that this kind of failure ordinarily happens after a long time of rehashed stress or strain cycling. Fatigue is vital in light of the fact that it is the single biggest reason for failure in metals, assessed to involve around 90% of every metallic failures; polymers and ceramics production (with the exception of glasses) are likewise vulnerable to this sort of failure. Besides, fatigue is catastrophic and insidious, happening abruptly. Fatigue failure is brittle like in nature even in regularly ductile metals, in that there is practically nothing, assuming any, gross plastic distortion connected with failure. The procedure happens by the start and propagation of cracks, and normally the fracture surface is opposite to the course of a connected tensile stress.

9.4.1 Morphology and Mechanism of Fatigue

The fracture surface of shafts that have suffered fatigue is distinctive and not hard to recognize. Such a fracture is shown in Figure 9.3 and consists of three main regions. The first is associated with the fracture initiation site, where local stress concentration (point C) nucleates a tiny circular crack. Next, most of the sectional area is covered with **beachmarks** that spread in concentric circular or semicircular ridges away from the crack nucleus. Beachmarks denote the instantaneous positions of the advancing crack and give the fracture surface a clamshell appearance. Finally, only a small area of the cross section at the shaft keyway remains to support the load, and it ruptures by a ductile overload fracture.

Figure 9.3 Fracture surface.

Machined grooves, threads, and deep fillets also serve as crack nuclei. Even dislocation models of crack nucleation that occur without benefit of a surface flaw, have been proposed. One such mechanism involves the alternate inward and outward glide of dislocations with stress until slip band *intrusions* and *extrusions* develop on the surface. Once nucleated at an intrusion, the incipient crack grows slowly (perhaps only 10^{-12} cm per cycle on average) across several grains via slip planes. This is *stage I* fatigue propagation, and results in a small, flat, featureless crack that acts as the springboard for further damage.

In *stage II,* a well-defined crack now begins to propagate at a much faster rate (10^{-4} cm per cycle) perpendicular to the direction of maximum tensile stress. Advance of the crack proceeds by repetitive sharpening and blunting of the crack tip as tension changes to compression. Each stress cycle leaves a striation behind that is easily visible in the scanning electron microscope (Figure 9.4). In forensic investigations these are often analyzed to determine the crack origin and propagation rate. There may be hundreds or even thousands of microscopic striations in a single macroscopic beachmark.

9.4.2 Fracture Mechanics and Fatigue

Observation of discrete striations has supported the fracture mechanics approach to fatigue. We now amplify on the predictive capability inherent in this approach. It should be recalled that through experiment the instantaneous crack length c is measured as a function of N, the number of stress cycles. Therefore, dc/dN can be extracted as a result. Fatigue-crack growth behaviour for a number of metals and non-metals is reproduced in Figure 9.5. Despite the limited data it is evident that crack growth rates in polymers

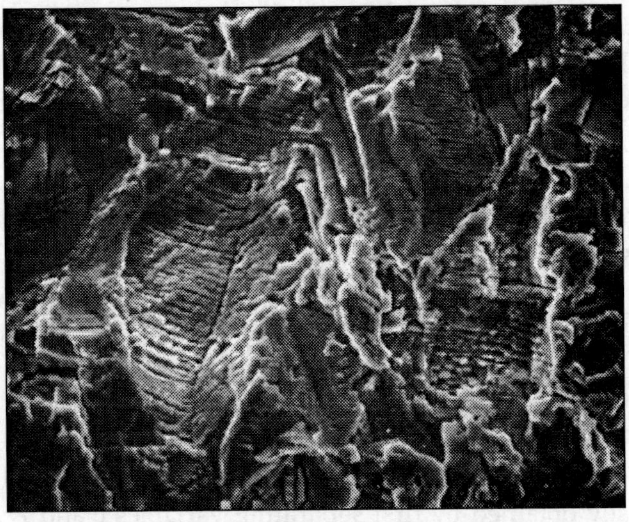

Figure 9.4 Fatigue striations in a nickel-base superalloy.

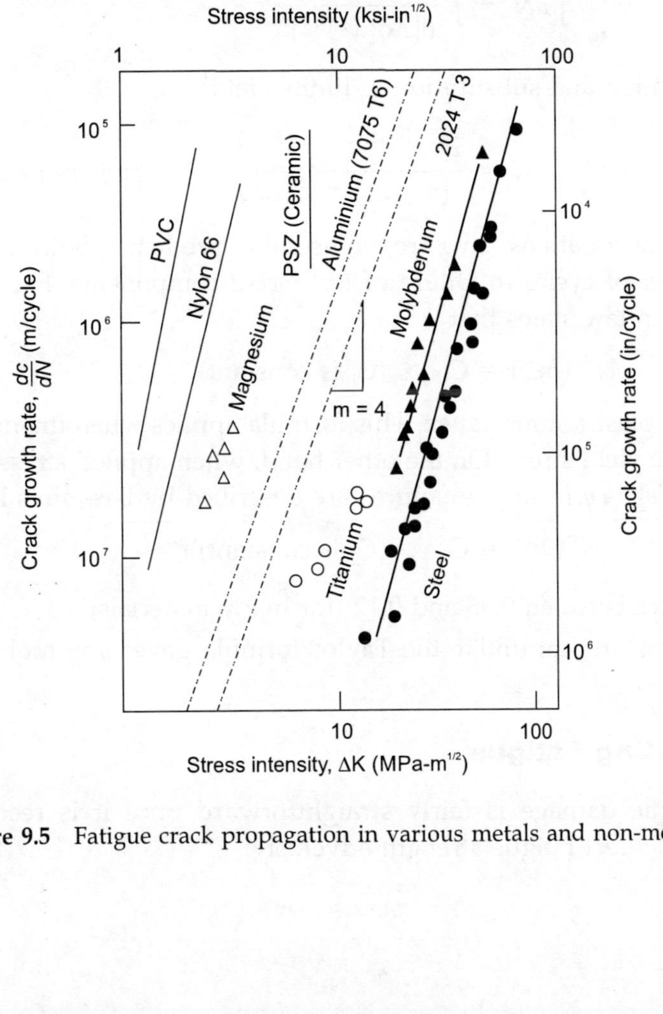

Figure 9.5 Fatigue crack propagation in various metals and non-metals.

and ceramics typically exceed those in metals by several orders of magnitude for the same value of ΔK. The data are consistent with a crack extension that depends on some power of the applied stress. Hence, a frequently employed expression is

$$dc/dN = B(\Delta K)^m = B[\Delta\sigma(\pi c)^{1/2}]^m$$

where, B and m are constants derived directly from experiment. Physically, the crack growth rate is necessarily proportional to the amplitude of the applied peak-to-peak stress, $\Delta\sigma$. $\Delta\sigma$ is equal to the difference between the maximum and minimum applied stress and is proportional to ΔK. Corrections for cracks of different geometry alter the constants slightly. This differential equation then directly links the crack size to applied stress and material constants. Below the indicated stress levels, cracks do not grow; above them crack growth no longer obeys the above equation.

If the crack size is initially c_o, the number of cycles, N_f, required to extend it to a final length c_f is easily obtained by first separating variables c and N:

$$\int_0^{N_f} dN = \int_{c_o}^{c_f} \frac{dc}{B[\Delta\sigma(\pi c)^{1/2}]^{m^*}}$$

Direct integration and substitution of limits yield

$$N_f = \frac{2[c_o^{-(m-2)/2} - c_f^{-(m-2)/2}]}{(m-2)\,B\pi^{m/2}\,(\Delta\sigma)^m} \qquad (m \text{ is not equal to } 2)$$

Two additional equations have proven useful in predicting both low and high cycle fatigue or number of cycles to failure of *un-cracked* components. For *low cycle* fatigue, the Coffin-Manson law states that

$$N_f^{0.5}(\Delta\varepsilon_p) = C_1 \qquad (C_1 \text{ is constant})$$

where, $\Delta\varepsilon_p$ is the plastic strain range. This formula applies when the maximum applied stress exceeds the yield stress. On the other hand, when applied stresses do not exceed the yield stress, *high-cycle* fatigue failures are described by Basquin's law:

$$N_f^{\alpha}(\Delta\sigma) = C_2 \qquad (C_2 \text{ is constant})$$

Constant α lies between 0.06 and 0.12 (for many materials)

Both equations are not unlike the Taylor formula governing tool life given earlier (i.e., $V_t T_t n = C$).

9.4.3 Combating Fatigue

Combating fatigue damage is fairly straightforward once it is recognized that the following factors affect fatigue strength adversely:

Stress Concentrators: Keyways on shafts, holes, abrupt changes in cross sections, sharp corners, and other stress raisers are design features that should be avoided in components subjected to repetitive loading. Stress concentrators mean operation at higher stress levels on the *S-N* curve. This, in turn, reduces the number of stress cycles to failure.

Rough Surfaces: Crack nucleation is facilitated at micro-crevices and grooves on rough surfaces. Grinding, honing, and polishing of surfaces remove these sources of potential cracks.

Corrosion and Environmental Attack: Localized chemical attack will always reduce fatigue life. Metal removed from exposed grains and grain boundaries leaves a generally rough surface, pits, and corrosion products behind that serve as incipient cracks.

Residual Tensile Stresses: To produce equivalent fatigue damage, less additional load is required when the surface is stressed in residual tension relative to the unstressed state. Therefore, by introducing desirable residual compressive stresses in surface layers by peening, higher levels of applied tension can be tolerated. As a result, fatigue strength and life can be increased.

Low-Strength Surfaces: Strengthening surfaces can enhance fatigue resistance; carburization and deposition of hard coatings are ways to achieve this.

9.5 *S-N* CURVE

As with other mechanical characteristics, the fatigue properties of materials can be determined from laboratory simulation tests. A test apparatus should be designed to duplicate as nearly as possible the service stress conditions (stress level, time frequency, stress pattern, etc.). A schematic diagram of a rotating-bending test apparatus, commonly used for fatigue testing, is shown in Figure 9.18. The compression and tensile stresses are imposed on the specimen as it is simultaneously bent and rotated. Tests are also frequently conducted using an alternating uniaxial tension-compression stress cycle.

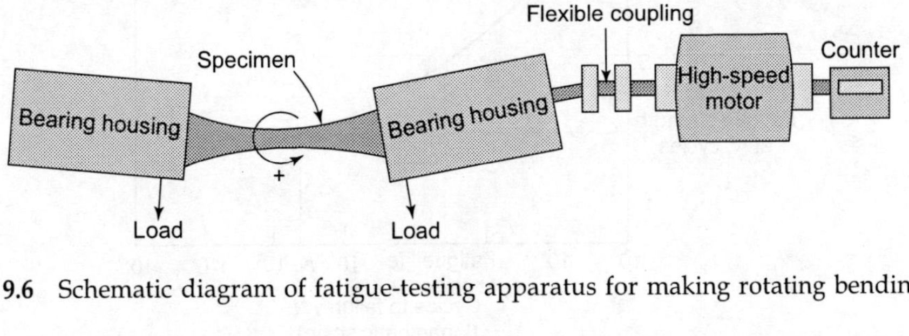

Figure 9.6 Schematic diagram of fatigue-testing apparatus for making rotating bending tests.

A series of tests are commenced by subjecting a specimen to the stress cycling at a relatively large maximum stress amplitude (σ_{max}), usually on the order of two-thirds of the static tensile strength; the number of cycles to failure is counted. This procedure is repeated on other specimens at progressively decreasing maximum stress amplitudes. Data are plotted as stress *S* versus the logarithm of the number *N* of cycles to failure for each of the specimens. The values of *S* are normally taken as stress amplitudes on occasion σ_{max} or σ_{min} values may be used.

Two distinct types of *S-N* behaviour are observed, which are represented schematically in Figure 9.7. As these plots indicate, the higher the magnitude of the stress, the smaller the number of cycles the material is capable of sustaining before failure. For some ferrous (iron base) and titanium alloys, the *S-N* curve (Figure 9.7a) becomes horizontal at higher *N* values; or there is a limiting stress level, called the **fatigue limit** (also sometimes the *endurance limit*), below which fatigue failure will not occur. This fatigue limit represents the largest value of fluctuating stress that will *not* cause failure for essentially an infinite number of cycles. For many steels, fatigue limits range between 35% and 60% of the tensile strength.

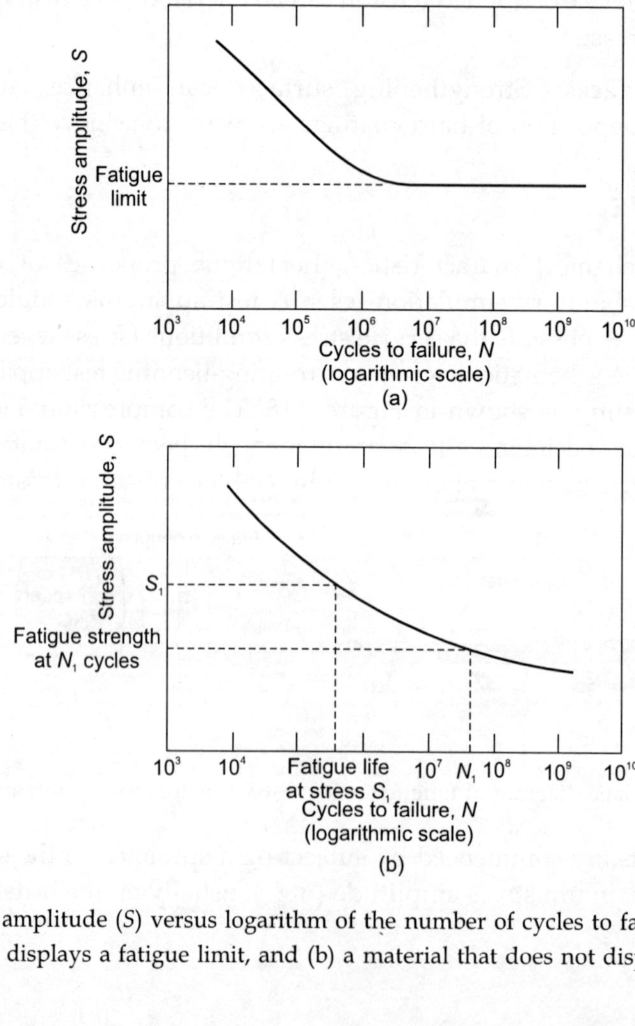

Figure 9.7 Stress amplitude (*S*) versus logarithm of the number of cycles to fatigue failure (*N*) for (a) a material that displays a fatigue limit, and (b) a material that does not display a fatigue limit.

Most nonferrous alloys (e.g., aluminum, copper, magnesium) do not have a fatigue limit, in that the *S-N* curve continues its downward trend at increasingly greater *N* values (Figure 9.7b).Thus, fatigue will ultimately occur regardless of the magnitude of the stress. For these materials, the fatigue response is specified as **fatigue strength,** which is defined as the stress level at which failure will occur for some specified number of cycles (e.g., cycles). The determination of fatigue strength is also demonstrated in Figure 9.7b.

Another important parameter that characterizes a material's fatigue behaviour is **fatigue life N_f** It is the number of cycles to cause failure at a specified stress level, as taken from the *S-N* plot.

Unfortunately, there always exists considerable scatter in fatigue data, that is, a variation in the measured *N* value for a number of specimens tested at the same stress level. This variation may lead to significant design uncertainties when fatigue life and/ or fatigue limit (or strength) are being considered. The scatter in results is a consequence of the fatigue sensitivity to a number of tests and material parameters that are impossible to control precisely. These parameters include specimen fabrication and surface preparation, metallurgical variables, specimen alignment in the apparatus, mean stress, and test frequency.

Several statistical techniques have been developed to specify fatigue life and fatigue limit in terms of probabilities. One convenient way of representing data treated in this manner is with a series of constant probability curves, several of which are plotted in Figure 8.20. The *P* value associated with each curve represents the probability of failure. For example, at a stress of 200 MPa (30,000 psi), we would expect 1% of the specimens to fail at about 10^6 cycles and 50% to fail at about 2×10^7 cycles, and so on. Remember that *S-N* curves represented in the literature are normally average values, unless noted otherwise.

The fatigue behaviours represented in Figures 9.7a and 9.7b may be classified into two domains. One is associated with relatively high loads that produce not only elastic strain but also some plastic strain during each cycle. Consequently, fatigue lives are relatively short; this domain is termed *low-cycle fatigue* and occurs at less than about 10^4 to 10^5 cycles. For lower stress levels wherein deformations are totally elastic, longer lives result. This is called *high-cycle fatigue* inasmuch as relatively large number of cycles are required to produce fatigue failure. High-cycle fatigue is associated with fatigue lives greater than about 10^4 to 10^5 cycles.

9.6 CRACK INITIATION AND PROPAGATION

The process of fatigue failure is characterized by three distinct steps: (1) crack initiation, wherein a small crack forms at some point of high stress concentration; (2) crack propagation, during which this crack advances incrementally with each stress cycle; and

(3) final failure, which occurs rapidly once the advancing crack has reached a critical size. Cracks associated with fatigue failure almost always initiate (or nucleate) on the surface of a component at some point of stress concentration. Crack nucleation sites include surface scratches, sharp fillets, keyways, threads, dents, and the like. In addition, cyclic loading can produce microscopic surface discontinuities resulting from dislocation slip steps that may also act as stress raisers, and therefore as crack initiation sites.

The region of a fracture surface that formed during the crack propagation step may be characterized by two types of markings termed *beachmarks* and *striations*. Both these features indicate the position of the crack tip at some point in time and appear as concentric ridges that expand away from the crack initiation site(s), frequently in a circular or semicircular pattern. Beachmarks (sometimes also called "clamshell marks") are of macroscopic dimensions and may be observed with the unaided eye. These markings are found for components that experienced interruptions during the crack propagation stage, for example, a machine that operated only during normal work-shift hours. Each beachmark band represents a period of time over which crack growth occurred.

On the other hand, fatigue striations are microscopic in size and subject to observation with the electron microscope (either TEM or SEM). Each striation is thought to represent the advance distance of a crack front during a single load cycle. Striation width depends on, and increases with, increasing stress range.

At this point it should be emphasized that although both beachmarks and striations are fatigue fracture surface features having similar appearances, they are nevertheless different, both in origin and size. There may be literally thousands of striations within a single beachmark.

Often the cause of failure may be deduced after examination of the failure surfaces. The presence of beachmarks and/or striations on a fracture surface confirms that the cause of failure was fatigue. Nevertheless, the absence of either or both does not exclude fatigue as the cause of failure. One final comment regarding fatigue failure surfaces. Beachmarks and striations will not appear on that region over which the rapid failure occurs. Rather, the rapid failure may be either ductile or brittle; evidence of plastic deformation will be present for ductile, and absent for brittle, failure.

9.7 FACTORS AFFECTING FATIGUE LIFE

The fatigue behaviour of engineering materials is highly sensitive to a number of variables. Some of these factors include mean stress level, geometrical design, surface effects, and metallurgical variables, as well as the environment. This section is devoted to a discussion of these factors and, in addition, to measures that may be taken to improve the fatigue resistance of structural components.

9.7.1 Mean Stress

The dependence of fatigue life on stress amplitude is represented on the *S-N* plot. Such data are taken for a constant mean stress σ_m, often for the reversed cycle situation Mean stress, however, will also affect fatigue life; this influence may be represented by a series of *S-N* curves, each measured at a different σ_m as shown in Figure 9.8. As may be noted, increasing the mean stress level leads to a decrease in fatigue life.

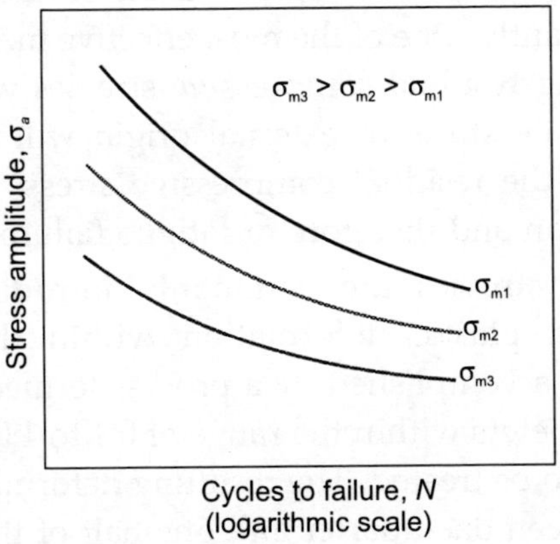

Figure 9.8 Demonstration of the influence of mean stress σ_m on *S-N* fatigue behaviour.

9.7.2 Surface Effects

For many common loading situations, the maximum stress within a component or structure occurs at its surface. Consequently, most cracks leading to fatigue failure originate at surface positions, specifically at stress amplification sites. Therefore, it has been observed that fatigue life is especially sensitive to the condition and configuration of the component surface. Numerous factors influence fatigue resistance, proper management of which will lead to an improvement in fatigue life. These include design criteria as well as various surface treatments.

9.7.3 Design Factors

The design of a component can have a significant influence on its fatigue characteristics. Any notch or geometrical discontinuity can act as a stress raiser and fatigue crack initiation site; these design features include grooves, holes, keyways, threads, and so on. The sharper the discontinuity that is more severe stress concentration. The probability of fatigue failure may be reduced by avoiding (when possible) these structural irregularities, or by making design modifications whereby sudden contour changes leading to sharp corners are eliminated, for example, calling for rounded fillets with large radii of curvature at the point where there is a change in diameter for a rotating shaft.

9.7.4 Surface Treatments

During machining operations, small scratches and grooves are invariably introduced into the workpiece surface by cutting tool action. These surface markings can limit the fatigue life. It has been observed that improving the surface finish by polishing will enhance fatigue life significantly. One of the most effective methods of increasing fatigue performance is by imposing residual compressive stresses within a thin outer surface layer. Thus, a surface tensile stress of external origin will be partially nullified and reduced in magnitude by the residual compressive stress. The net effect is that the likelihood of crack formation and therefore of fatigue failure is reduced.

Residual compressive stresses are commonly introduced into ductile metals mechanically by localized plastic deformation within the outer surface region. Commercially, this is often accomplished by a process termed *shot peening*. Small, hard particles (shot) having diameters within the range of 0.1 to 1.0 mm are projected at high velocities onto the surface to be treated. The resulting deformation induces compressive stresses to a depth of between one-quarter and one-half of the shot diameter.

Case hardening is a technique by which both surface hardness and fatigue life are enhanced for steel alloys. This is accomplished by a carburizing or nitriding process whereby a component is exposed to a carbonaceous or nitrogenous atmosphere at an elevated temperature. A carbon- or nitrogen-rich outer surface layer (or "case") is introduced by atomic diffusion from the gaseous phase. The case is normally on the order of 1 mm deep and is harder than the inner core of material. The improvement of fatigue properties results from increased hardness within the case, as well as the desired residual compressive stresses the formation of which attends the carburizing or nitriding process.

9.8 SUMMARY

Fracture, in response to tensile loading and at relatively low temperatures, may occur by ductile and brittle modes, both of which involve the formation and propagation of cracks. For ductile fracture, evidence will exist of gross plastic deformation at the fracture surface. In tension, highly ductile metals will neck down to essentially a point fracture; cup-and-cone mating fracture surfaces result for moderate ductility. Cracks in ductile materials are said to be stable (i.e., resist extension without an increase in applied stress); and inasmuch as fracture is non-catastrophic, this fracture mode is almost always preferred.

For brittle fracture, cracks are unstable, and the fracture surface is relatively flat and perpendicular to the direction of the applied tensile load. Chevron and ridge-like patterns are possible, which indicate the direction of crack propagation. Transgranular (through-grain) and intergranular (between-grain) fractures are found in brittle polycrystalline materials.

The huge inconsistency amongst real and hypothetical fracture strengths of brittle materials is clarified by the presence of little blemishes that are fit for increasing a connected tensile stress their region, driving at last to crack formation. Fracture results when the hypothetical firm quality is surpassed at the tip of one of these imperfections. The fracture toughness of a material is demonstrative of its imperviousness to brittle fracture when a split is available. It relies on example thickness, and, for moderately thick examples (i.e., states of plane strain), is named the plane strain fracture toughness. This parameter is the one ordinarily referred to for design purposes; its esteem is generally substantial for ductile materials (and little for brittle ones), and is an element of microstructure, strain rate, and temperature. With respect to outlining against the likelihood of fracture, thought must be given to material (its break strength), the stress level, and the defect measure discovery restrain.

Fatigue is a common type of catastrophic failure wherein the applied stress level fluctuates with time.Test data are plotted as stress versus the logarithm of the number of cycles to failure. For many metals and alloys, stress diminishes continuously with increasing number of cycles at failure. Fatigue strength and fatigue life are the parameters used to characterize the fatigue behaviour of these materials. On the other hand, for other metals/alloys, at some point, stress ceases to decrease with, and becomes independent of the number of cycles. The fatigue behaviour of these materials is expressed in terms of fatigue limit.

The processes of fatigue crack initiation and propagation were discussed. Cracks normally nucleate on the surface of a component at some point of stress concentration. Two characteristic fatigue surface features are beachmarks and striations. Beachmarks form on components that experience applied stress interruptions; they normally may be observed with the naked eye. Fatigue striations are of microscopic dimensions, and each is thought to represent the crack tip advance distance over a single load cycle.

Measures that may be taken to extend fatigue life include: (1) reducing the mean stress level; (2) eliminating sharp surface discontinuities; (3) improving the surface finish by polishing; (4) imposing surface residual compressive stresses by shot peening; and (5) case hardening by using a carburizing or nitriding process.

Metal alloys that are especially resistant to creep have high elastic moduli and melting temperatures; these include the superalloys, the stainless steels, and the refractory metals. Various processing techniques are employed to improve the creep properties of these materials.

EXERCISE

1. Considering the terms size and origin, differentiate between fatigue striations and beachmarks.
2. For metals listed below: tin, molybdenum, iron, gold, zinc, and chromium, provide the temperatures with an approximation at which creep becomes an important consideration.
3. Mention any four validations that may considered for increasing the resistance of fatigue of metal alloy.
4. Write any five factors that may lead to scatter in fatigue life data.
5. If a component fabricated from a low carbon-nickel alloy is to be exposed to a tensile stress of 31 MPa (4500 psi) at 649°C, calculate its rupture lifetime.
6. Write a short note on one or two non-destructive test techniques that are used for detection and measurement of internal and/or surface flaws in metal alloys.
7. A cylindrical component constructed from a low carbon-nickel alloy has a diameter of 19.1 mm (0.75 in). Determine the maximum load that may be applied for it to survive 10,000 h at 538°C.
8. Write a short note on combating fatigues.
9. Explain in brief the phenomenon of fatigue and creep.
10. What is the concept behind elevated temperature creep degradation and failure? Explain with relevant diagram.
11. Write short notes on the following:
    (a) Dislocation creep
    (b) Diffusion creep
    (c) Creep failure
12. A fatigue test was conducted in which the mean stress was 70 MPa (10,000 psi), and the stress amplitude was 210 MPa (30,000 psi).
    (a) Compute the maximum and minimum stress levels.
    (b) Compute the stress ratio.
    (c) Compute the magnitude of the stress range.
13. Briefly explain the difference between fatigue striations and beachmarks both in terms of **(a)** size and **(b)** origin.
14. List four measures that may be taken to increase the resistance to fatigue of a metal alloy.
15. Cite three metallurgical/processing techniques that are employed to enhance the creep resistance of metal alloys.

Multiple-Choice Questions

1. Failure due to excessive deformation is controlled by _________.
   (a) Material properties (b) Design & dimensions
   (c) Both (d) None
2. Failure due to excessive deformation is controlled by _________.
   (a) Yield strength (b) Tensile strength
   (c) Young's modulus (d) All
3. Time dependent yield is known as
   (a) Fracture (b) Fatigue
   (c) Buckling (d) Creep
4. Cleavage fracture appears
   (a) Bright (b) Dull
   (c) Difficult to identify (d) None
5. Usually, materials with following crystal structure fail in ductile mode
   (a) FCC (b) BCC
   (c) HCP (d) None
6. Brittle fracture is more dangerous than ductile fracture because _________.
   (a) No warning sign
   (b) Crack propagates at very high speeds
   (c) No need for extra stress during crack propagation
   (d) All
7. Fracture voids usually form at
   (a) Inclusions (b) Second phase particles
   (c) Grain boundary triple points (d) All
8. Fracture stress (σ_f) is proportional to
   (a) crack length (b) 1/crack length
   (c) (crack length)$^{1/2}$ (d) (crack length)$^{-1/2}$
9. Fracture toughness is measured in terms of
   (a) Strain energy release rate (b) Stress concentration factor
   (c) Both (d) None

10. In fracture mode-II, fracture surfaces
    (a) Shear parallel to edge of crack
    (b) Shear perpendicular to edge of crack
    (c) Displace normal to each other
    (d) None
11. Fracture toughness, K_{IC}, decreases with
    (a) Increasing temperature (b) Increasing strain rate
    (c) Increase in yield strength (d) Increase in grain size
12. DBTT for ceramics is in the range of _________ × T_m.
    (a) 0.1-0.2 (b) 0.2-0.3
    (c) 0.3-0.5 (d) 0.5-0.7
13. Following impurity decreases DBTT for steels
    (a) Mn (b) P
    (c) Si (d) Mo
14. Fatigue strength for non-ferrous materials is defined at _________ stress cycles.
    (a) 10^3 (b) 10^5
    (c) 10^7 (d) 10^9
15 The following equation defines *S-N* curve
    (a) Paris equation (b) Basquin equation
    (c) Andrede equation (d) Garofalo equation
16. Creep rate in ternary stage _________.
    (a) Decreases (b) Constant
    (c) Increases (d) None
17. Ternary stage creep is associated with _________.
    (a) Strain hardening (b) Recovery
    (c) Necking (d) None
18. Total strain range in a creep test
    (a) <1% (b) Around 10%
    (c) Around 50% (d) > >50%

19. Creep mechanism that is operational at stresses $10^{-2} > \sigma/G > 10^{-4}$ is ________.
    (a) Dislocation creep (b) Dislocation glide
    (c) Diffusion creep (d) GB sliding
20. Most often machine components fail by
    (a) Buckling (b) Creep
    (c) Fatigue (d) All
21. If the surface crack causing fracture in a brittle material is made twice as deep, the fracture strength will
    (a) Decrease by a factor of $\sqrt{2}$ (b) Decrease by a factor of 2
    (c) Decrease by a factor of 22 (d) No change
22. Corrosion of metals involves
    (a) Physical reactions (b) Chemical reactions
    (c) Both (d) None
23. The following factors play a vital role in corrosion process
    (a) Temperature (b) Solute concentration
    (c) Both (d) None
24. Following equation is related to corrosion rate
    (a) Nernst equation (b) Faraday's equation
    (c) Either (d) Neither
25. Passivity is due to
    (a) Higher EMF (b) Lower EMF
    (c) Oxide film (d) All
26. Passivity is not reason for inertness of the following:
    (a) Au (b) Al
    (c) Ti (d) Ni
27. Difficult to monitor and very dangerous form of corrosion
    (a) Galvanic (b) Pitting
    (c) Crevice (d) Stress
28. This form of corrosion occurs due to concentration difference in a component
    (a) Uniform (b) Galvanic
    (c) Intergranular (d) Stress

29. Main form of ceramic degradation
    (a) Corrosion (b) Weathering
    (c) Dissolution (d) Swelling
30. The following influences deterioration of polymers
    (a) Weather (b) Radiation
    (c) Temparature (d) All
31. Following is not the main form of polymer deterioration
    (a) Corrosion (b) Swelling and dissolution
    (c) Weathering (d) Scission
32. When Pt and Co are electrically connected, which one gets corroded
    (a) Pt (b) Co
    (c) None (d) Can't decide
33. Which of the following can be used for cathodic protection:
    (a) Al (b) Cd
    (c) Cu (d) Either
34. Slow plastic deformation of metals under a constant stress is known as
    (a) Fatigue (b) Proof deformation
    (c) Gradual deformation (d) Creep
35. In which of the following cases creep is an important consideration
    (a) Cast iron water pipe (b) Cycle chain
    (c) Gas turbine blade (d) All the above
36. The rollers of a cycle chain are subjected to
    (a) Compressive stresses (b) Tensile stresses
    (c) Shear stresses (d) Fatigue stresses
37. Corrosion can be prevented by
    (a) Alloying (b) Tinning
    (c) Galvanizing (d) All of the above
38. As compared to iron, aluminium has
    (a) higher tendency to oxidize (b) less tendency to oxidize
    (c) equal tendency to oxidize (d) none of the above

39. Chemical action during corrosion converts metal into metallic component as _________

(a) Hydroxide (b) Oxide

(c) Sulphate (d) All of the above

40. Which type of chemical reaction is observed at cathode in electrochemical corrosion?

(a) Oxidation reaction (b) Peretectic reaction

(c) Reduction rection (d) None of above

41. Which of the following corrosions are caused due to velocity of fluid flow in pipes?

(a) Bimetal corrosion (b) Cavitation corrosion

(c) Galvanic corrosion (d) Intergranular corrosion

42. Corrosion fatigue is a combined effect of _________

(a) Corrosive environment and mechanical stresses

(b) Cyclic loading and corrosion

(c) Velocity and mechanical stresses

(d) None of the above.

43. The intergranular corrosion can be prevented using _________

(a) Stabilized grade of stainless steel containing titanium and niobium as an alloying element

(b) Low carbon grade of stainless steel

(c) Both a and b

(d) None of the above

44. Knife line attack is a type of _________

(a) Bimetal corrosion (b) Cavitation corrosion

(c) Galvanic corrosion (d) Intergranular corrosion

45. The erosion corrosion can be controlled by _________

(a) Increasing the flow velocity through pipes

(b) Increasing temperature of fluid

(c) Minimizing turbulence

(d) All of the above

46. Which of the following statement(s) is/are true for erosion corrosion?
   (a) The dissolution rate of protective film increases due to turbulent flow
   (b) Erosion corrosion occurs due to high turbulent fluid flow
   (c) The corrosion rate can be reduced by reducing the velocity of fluid
   (d) All of the above
47. Which of the following methods is not used to coat metallic elements to prevent corrosion?
   (a) Electroplating (b) Metal cladding
   (c) Metal spraying (d) None of the above
48. Volatile oxidation corrosion product of a metal is,
   (a) Fe_2O_3 (b) MoO_3
   (c) Fe_3O_4 (d) FeO
49. Electrochemical corrosion takes place on
   (a) Anodic area (b) Cathodic area
   (c) Near cathode (d) Near anode
50. Which of the following metals could provide cathodic protection to Fe?
   (a) Al & Cu (b) Al & Zn
   (c) Zn & Cu (d) Al & Ni

Answers

1. (c)	**2.** (c)	**3.** (d)	**4.** (a)	**5.** (a)
6. (d)	**7.** (d)	**8.** (d)	**9.** (c)	**10.** (b)
11. (b)	**12.** (a)	**13.** (a)	**14.** (c)	**15.** (b)
16. (c)	**17.** (c)	**18.** (a)	**19.** (a)	**20.** (c)
21. (a)	**22.** (b)	**23.** (c)	**24.** (b)	**25.** (c)
26. (a)	**27.** (b)	**28.** (c)	**29.** (c)	**30.** (d)
31. (a)	**32.** (b)	**33.** (a)	**34.** (d)	**35.** (c)
36. (d)	**37.** (d)	**38.** (a)	**39.** (d)	**40.** (c)
41. (b)	**42.** (b)	**43.** (c)	**44.** (d)	**45.** (c)
46. (d)	**47.** (d)	**48.** (b)	**49.** (a)	**50.** (b)

SECTION IV

Properties and Processing of Engineering Materials

10

Polymers, Glasses and Ceramics

Learning Objectives

After going through this chapter the reader will be able to:

- Introduce polymers, glasses and ceramics.
- Understand polymer morphology.
- Have an idea on inorganic glasses.
- Cite some light on ceramics.

10.1 INTRODUCTION

This chapter introduces two of the four important classes of engineering materials: polymers and ceramics. These include some of the oldest as well as some of the newest materials used by humans. Fired clays, glasses, and ceramic materials that contain mixtures of glassy and crystalline constituents were produced and used even before historical records were kept; synthetic polymers, on the other hand, date only to the last century. The former materials are *inorganic* in nature and generally produced at high temperatures; in contrast, polymers have *organic* compositions and are formed at low temperatures. Nevertheless, both display some common structural features. For example, ceramics are usually crystalline but can be disordered or amorphous; polymers are usually amorphous but can exhibit crystallinity. The closest parallels exist between amorphous inorganic glasses and polymers. In addition, both classes of materials are poor conductors of electricity and best suited as insulators and dielectrics. Their mechanical properties range from hard and brittle ceramics to very soft and extendable rubbery polymer materials. These electrical and mechanical properties, as well as the methods used to process polymers and ceramics into useful shapes, are dealt with in subsequent chapters. Although ordered crystalline states of matter have been the focus of the book until now, this chapter deals with non-crystalline materials and mixtures of both.

Polymeric materials have made and continue to make great inroads into applications traditionally met by metal usage. As Figure 10.1a indicates, polymer production and usage have grown at a far more rapid rate than production and use of metals in the recent past. Due largely to their ease of processing, high strength-to-weight ratios, low density (typically - 1.5 mg/m^3 compared with - 7.5 mg/m^3 for metals), and competitive costs, polymers and polymer-based composite materials continue to find expanded usage, particularly in automobiles and aircraft. Polymers have also replaced glass and ceramic materials in food containers, dinnerware, and assorted kitchen and toilet accessories.

Ceramic materials have a number of advantages relative to other materials. These include high melting points, great hardness, low densities, and chemical and environmental stability. However, ceramics are severely handicapped by lack of toughness. Their intrinsic inability to absorb shock loading causes them to fracture in a brittle manner rather than deform in a ductile fashion the way metals do. Although they do not normally compete with metals, ceramics stand poised to potentially replace metals in selected applications, for example, high-temperature gas turbine engines and cutting tools. Ceramic materials still retain their traditional usage in pottery, bricks tiles, whiteware, and kitchen as well as toilet fixtures. But, in addition, there are exciting new high-tech uses for so-called new or advanced ceramics in assorted mechanical, electronic, magnetic, and optical applications. As Figure 10.1b shows, markets for these high added value

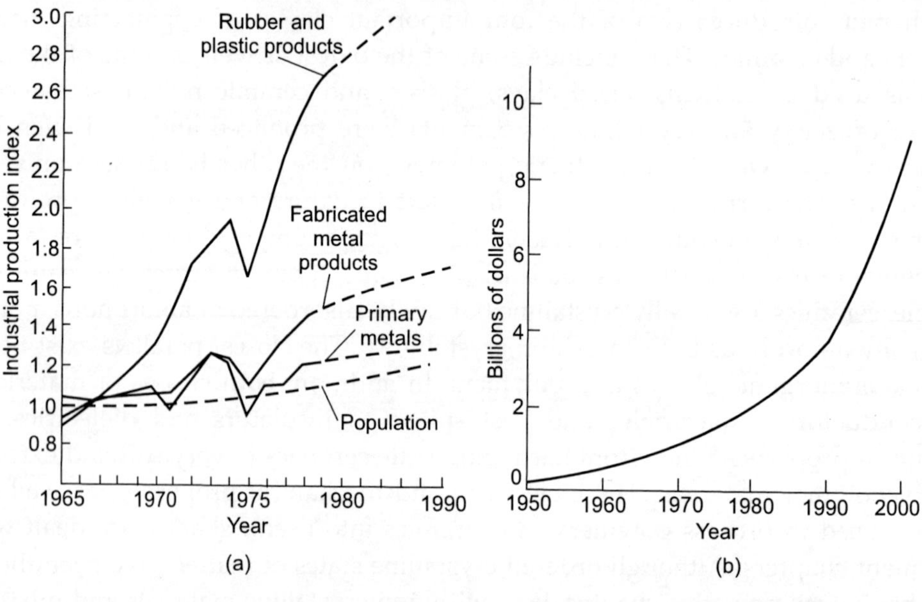

Figure 10.1 (a) Industrial production indices for metals and polymeric materials compared with growth in population. (b) Approximate annual value of advanced ceramics in billions of dollars.

materials are growing rapidly; in contrast, traditional ceramics industries are expanding more slowly. Both ceramics and polymers are clearly distinct from metals and from one another. Treating these materials together in this as well as several subsequent chapters sharply emphasizes their different as well shared properties. Nevertheless, organization into separate sections also ensures that the individual attributes of each class of material are preserved and emerge undiminished.

10.2 POLYMERS

The origin of synthetic polymers can be traced to the American Wild West, of all places. In the mid-19th century, towns in Colorado attracted a motley crowd of adventurers, gamblers, saloon ladies, and other characters intent on acquiring wealth easily and quickly. In an effort to provide saloon patrons with enough billiard balls, John Hyatt sought a substitute for the scarce and expensive ivory spheres that had been used until that time. His early efforts resulted in balls consisting of a core of ivory dust bonded with shellac and an exterior coating of the somewhat unstable collodion. The latter frequently exploded when the balls collided, prompting every gunman in the saloon to instinctively draw his six-shooter. In 1868, Hyatt successfully adapted a material developed some 14 years earlier by Alexander Parkes in England. It consisted of a mixture of cellulose nitrate and camphor. This first synthetic plastic known as celluloid was widely used until recently in making toys, particularly dolls. Its dangerous flammability spawned substitutes, and when Baekeland patented the process for making Bakelite in 1907 the age of plastics formally arrived.

A polymer is a large molecule containing hundreds to many thousands of small molecular units or mers linked together into either chainlike or network structures. Although our concern here is with *synthetic* organic polymers, the definition is broad enough to include inorganic polymers based on silicon. The word *synthetic* is important because polymers are contained in animals (wool, leather), insects (silk), and vegetables (wood, cotton) products. Cellulose and lignin are natural polymers found in vast quantities in wood. Some common polymers are so very much composed by nature that they have not yet been supplanted by the manufactured assortment.

Of the polymer materials utilized as a part of designing practice plastics are, by a long shot, delivered in most noteworthy sum and record for approximately 90% of U.S. tonnage. It is regular to subdivide plastics into two main groups: thermoplastics and thermosets. Thermoplastics dissolve on warming and are handled in this state by an assortment of expulsion and trim procedures. Polyethylene, polyvinyl chloride, polypropylene, and polystyrene, in a specific order, are the four most generally created thermoplastics. Thermosets (or gums) incorporate phenolic gums, epoxies, unsaturated polyesters, and polyurethanes—substances that can't be softened and remelted like thermoplastics, however irreversibly set. Thermoplastic generation is around six times that of thermosets in tonnage and esteem.

Rubbers or elastomers frame another gathering of polymeric materials. The regular elastic industry in light of latex originated before the presentation of plastics and began before it was perceived that rubbers were polymeric materials. Notwithstanding regular elastic, vital manufactured elastomers incorporate polybutadiene, styrene-butadiene, polyisoprene, and silicone elastic.

In spite of the fact that the properties of particular classes of polymers vary broadly, they are all made out of long particles that have a spine or spine of covalently fortified carbon molecules. Hydrogen particles and hydrocarbon gatherings are commonly circulated along and around the spine. Individual subatomic strands are then held together along the side by powerless van der Waals optional bonds, as in thermoplastics, or by more grounded covalent bonds that crosslink them together, as in thermosets. Feeble polymer bonds break down at a point very little above the room temperature. That is the reason thermoplastics can be promptly dissolved and framed at low temperatures. It is likewise the motivation behind why these materials have a tendency to be mechanically shaky and distort or extend with time at moderately low temperatures. In these regards polymers comprehensively contrast from metals and earthenware production.

Naturally occurring polymers—those derived from plants and animals—have been used for many centuries; these materials include wood, rubber, cotton, wool, leather, and silk. Other natural polymers such as proteins, enzymes, starches, and cellulose are important in biological and physiological processes in plants and animals. Modern scientific research tools have made possible the determination of the molecular structures of this group of materials, and the development of numerous polymers, which are synthesized from small organic molecules. Many of our useful plastics, rubbers, and fibre materials are synthetic polymers. In fact, since the conclusion of World War II, the field of materials has been virtually revolutionized by the advent of synthetic polymers. The synthetics can be produced inexpensively, and their properties may be managed to the degree that many are superior to their natural counterparts. In some applications metal and wood parts have been replaced by plastics, which have satisfactory properties and may be produced at a lower cost. As with metals and ceramics, the properties of polymers are intricately related to the structural elements of the material.

Linear polymers are those in which the repeat units are joined together end to end in single chains. These long chains are flexible and may be thought of as a mass of spaghetti, where each circle represents a repeat unit. For linear polymers, there may be extensive van der Waals and hydrogen bonding between the chains. Some of the common polymers that form with linear structures are polyethylene, poly(vinyl chloride), polystyrene, poly(methyl methacrylate), nylon, and the fluorocarbons.

Polymers may be synthesized in which side-branch chains are connected to the main ones, these are fittingly called **branched polymers.** The branches, considered to be part of the main-chain molecule, may result from side reactions that occur during the synthesis of the polymer. The chain packing efficiency is reduced with the formation of side branches, which results in a lowering of the polymer density. Those polymers that form linear structures may also be branched. For example, high density polyethylene (HDPE) is primarily a linear polymer, while low density polyethylene (LDPE) contains short chain branches.

In **crosslinked polymers,** adjacent linear chains are joined one to another at various positions by covalent bonds. The process of crosslinking is achieved either during synthesis or by a nonreversible chemical reaction. Often, this crosslinking is accomplished by additive atoms or molecules that are covalently bonded to the chains.

Multifunctional monomers forming three or more active covalent bonds, make three-dimensional networks and are termed **network polymers.** Actually, a polymer that is highly crosslinked may also be classified as a network polymer. These materials have distinctive mechanical and thermal properties; the epoxies, polyurethanes, and phenol-formaldehyde belong to this group.

The response of a polymer to mechanical forces at elevated temperatures is related to its dominant molecular structure. In fact, one classification scheme for these materials is according to the behaviour with rising temperature. *Thermoplastics* (or **thermoplastic polymers**) and *thermosets* (or **thermosetting polymers**) are the two subdivisions. Thermoplastics soften when heated (and eventually liquefied) and hardened when cooled – processes that are totally reversible and may be repeated. On a molecular level, as the temperature is raised, secondary bonding forces are diminished (by increased molecular motion) so that the relative movement of adjacent chains is facilitated when a stress is applied. Irreversible degradation results when a molten thermoplastic polymer is raised to a high temperature. In addition, thermoplastics are relatively soft. Most linear polymers and those having some branched structures with flexible chains are thermoplastic. These materials are normally fabricated by the simultaneous application of heat and pressure. Most linear polymers are thermoplastics. Examples of common thermoplastic polymers include polyethylene, polystyrene, poly (ethylene terephthalate), and poly (vinyl chloride).

Thermosetting polymers are network polymers. They become permanently hard during their formation, and do not soften upon heating. Network polymers have covalent crosslinks between adjacent molecular chains. During heat treatments, these bonds anchor the chains together to resist the vibrational and rotational chain motions at high temperatures. Thus, the materials do not soften when heated. Crosslinking is usually extensive, in that 10 to 50% of the chain repeat units are crosslinked. Only heating to excessive temperatures will cause severance of these crosslink bonds and polymer degradation. Thermoset polymers are generally harder and stronger than thermoplastics and have better dimensional stability. Most of the crosslinked and network polymers, which include vulcanized rubbers, epoxies, and phenolics and some polyester resins, are thermosetting.

10.3 POLYMER MORPHOLOGY

10.3.1 Crystalline Growth

We now consider the chain architecture in bulk polymers at several levels of microstructural organization beyond individual molecules. The forms and inner structures observed in optical and electron microscopes characterize polymer morphology. It is the subtle interplay at all of these structural levels that determines polymer properties. Polymers conceal their internal morphology very effectively but much can be learnt by melting a small thermoplastic bead between glass slides. Optical observation of the solidified material between crossed polarizers reveals the presence of spherulites. High in the structural hierarchy of polymers, spherulites are also observed in crystalline minerals, and provide the first clue that some bulk polymers are crystalline. The individual small plate-like lamellar single crystals of polyethylene exhibit more convincing evidence of crystallinity.

X-ray diffraction methods have demonstrated that many polymers are crystalline. It is, however, clear that they possess none of the classic structural perfection of metal crystals whose atoms and unit cells are predictably and beautifully ordered over very long distances. Instead, polymer crystals start with good geometric intentions, and ordered unit cells (Figure 10.2) can be identified within groups of chain molecules arranged in parallel bundles. Long strands of molecules then fold like computer paper whose stack width is much less than the overall bundle length. With repeated folding and spreading something like a shaggy rug composed of polymer is woven over two dimensions as depicted in Figure 10.3a. Through the thickness the ~100-carbon-long crystalline segments are well ordered, but where they repeatedly loop over in up and

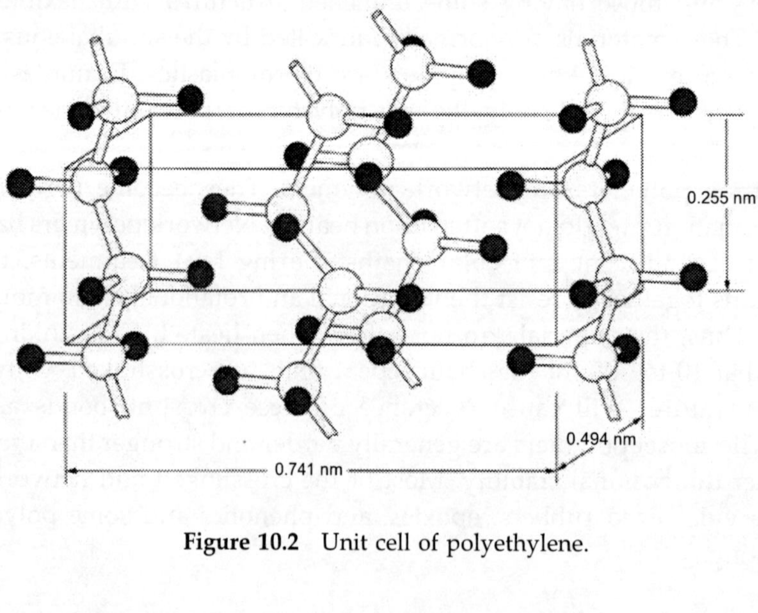

Figure 10.2 Unit cell of polyethylene.

down bends, bundles occasionally tear. This frays the otherwise geometric (quasicrystalline) rug pattern, leaving an amorphous tangle of loose molecular ends that interfaces with the surrounding melt or solution.

The shaggy crystalline platelets often organize themselves in spherulites by first attaching at some central crystalline nucleus (Figure 10.3b). These plates then extend radially outward, executing complex growth mechanisms and patterns that resemble a spherical spark shower at a fireworks display on July 4th. Initial ordered stacking of platelets is short-lived, however, because amorphous polymer melt gets trapped in between. As the spherulite keeps on developing the plates spread or fan separated until in the long run they run into different spherulites. The subsequent grain-like structure is shown by various thermoplastics (e.g., polyethylene, nylon, polystyrene). Strikingly, spherulitic development in polymers parallels that of graphite drops in iron castings.

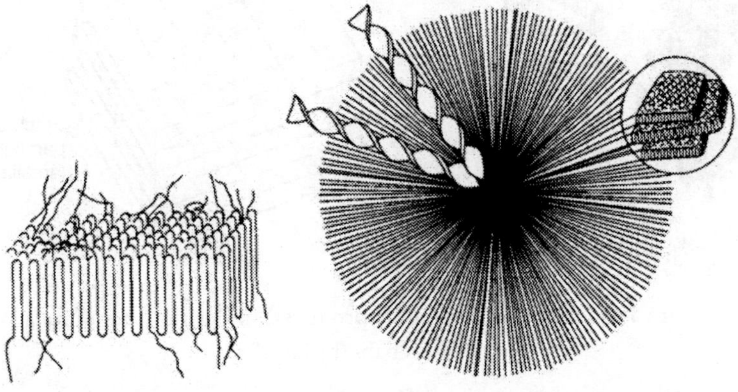

Figure 10.3 (a) Chain-folded polymer single crystal. The continual folding of the linear molecule onto itself creates thin carpet-like plates that extend over micrometre-sized domains. The frayed edges are an amorphous component of the crystalline structure. (b) Model of spherulite structure. Both planar and twisted crystalline plates grow radially outward from the central nucleus.

Polymers normally solidify under softer stream conditions as opposed to under the tranquil conditions that prompt to spherulitic development. In such cases prolonged precious stones adjust along the stream course and a fibrous microstructure comes about. The "seek kebab" structure appeared in Figure 10.4a is regular, where the "seek" is the prolonged gems in the lines, and the kebab, the congested crystalline plates along the vertical spine. A more outrageous arrangement of gems happens if a polymer is pulled or drawn frosty (underneath the liquefy temperature). The stacked lamellae of the spherulites are then persuasively adjusted (Figure 10.4b) into a stretched fibrillar structure. X-beam diffraction designs then regularly uncover differing degrees of crystallinity going, maybe, to 95%. Unavoidable subatomic ensnarements keep exceedingly crystallizable straight polymers from accomplishing more elevated amounts of flawlessness.

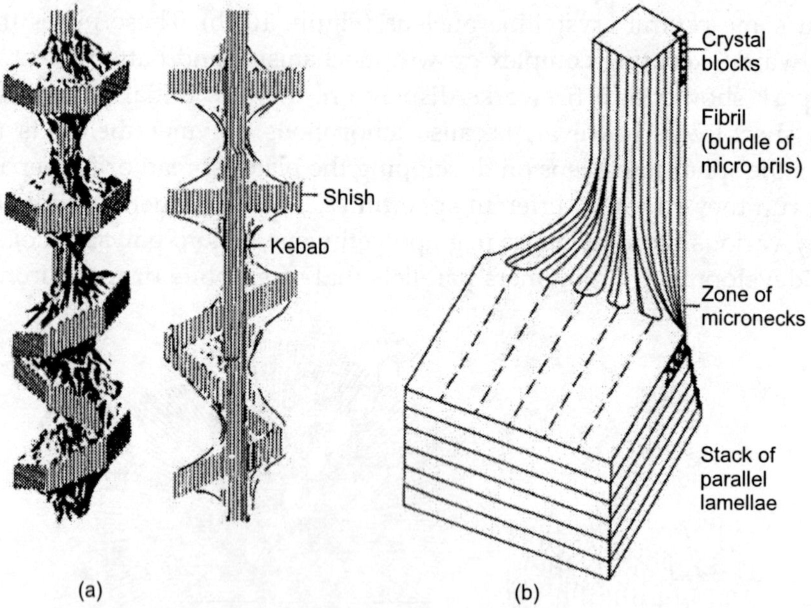

Figure 10.4 Polymer structures produced under flow conditions: (a) Seek kebab. (b) Fibrillar.

Situating and broadening chain precious stones are exceptionally alluring from a designing viewpoint on the grounds that both the modulus of flexibility and the quality of polymers are upgraded. Fibre-fortified or reinforced composites, tyres, and mechanical belts are applications for high-modulus filaments.

10.3.2 Amorphous Growth

We have as of now insinuated the way that polymers accept undefined structures at the atomic level. To put indistinct polymers inside the expansive setting that incorporates the melts from which they and also different shapeless solids (e.g., silica glass, metal glasses) frame, it is helpful to allude to Figure 10.5. This imperative representation shows the volume (V) of the polymer being referred to as a component of temperature. The fluid is fundamentally cluttered however as the dissolve is cooled beneath the softening point (T_M) it doesn't solidify the way a metal does. Metals generally experience an unexpected volume diminish at T_M, and unless cooled in an uncommonly fast manner they set as crystalline solids. A few metals additionally develop cementing. Crystalline polymers copy metals to some degree, but since there is a spread in atomic chain lengths the dissolving point sharpness is obscured.

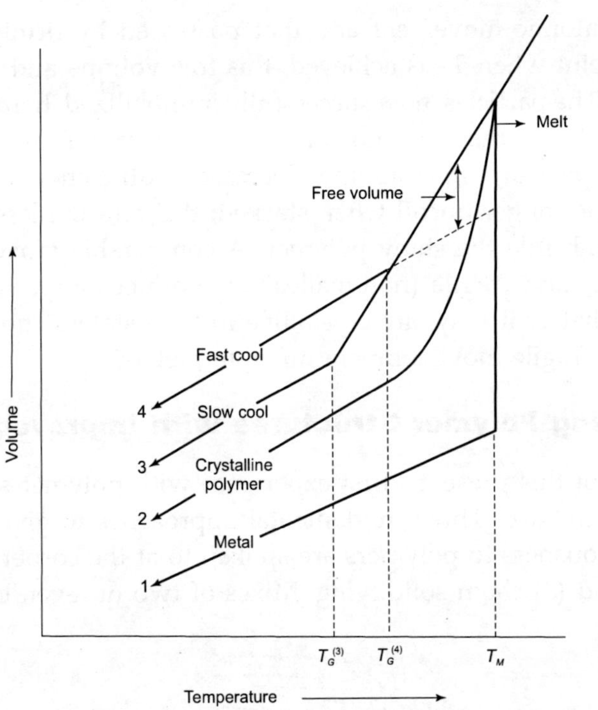

Figure 10.5 Volume change with function of temperature in (1) metals, (2) partially crystalline polymers, (3) amorphous polymers cooled at low rates, and, (4) amorphous polymers cooled at high rates.

By progressively meddling with the stereo-normality of the particles through atacticity, crosslinking, and expansion of arbitrary copolymer branches, any inclination to take shape is completely smothered. The liquefying point vanishes and rather the polymer dissolve hardens as a shapeless glass over a scope of temperatures. Volume withdrawal happens on the grounds that nuclear improvements permit atoms to pack all the more productively. Higher polymer thickness (less volume) is the consequence of lower cooling rates since particles have more opportunity to achieve more tightly pressing. A pretty much direct lessening in volume happens until the glass move temperature (T_G) is come to. Ordinarily, T_G = 0.5 to 0.75 T_M. Properties frequently experience huge property changes in the region of $T = T_G$. Above T_G the indistinct polymer is rubbery and develops promptly; beneath it weak, glasslike conduct is watched.

Amid cooling of the dissolve the atoms of the polymer are allowed to decipher, pivot, expand, curve, turn, etc., with an opportunity relative to the distinction in temperature amongst T and T_G. At raised soften temperatures there is an extreme measure of purported "free-volume" accessible for atoms to execute their movements in. The free volume at any temperature is the contrast between the volume display for unconstrained subatomic movement and that possessed by firmly stuffed unmoving particles. At the point when T_G is achieved, this free volume and the last remnants of the liquefy vanish. The particles, now successfully immobilized, harden as an unbending, strong glass. Beneath T_G there are no further nuclear modifications, and the volume changes with temperature mirrors the decreases sufficiency of warm vibrations. Subatomic chains no longer uncoil when stacked; this and the less deformable carbon spine scheme to embrittle the shiny polymer. A comparable move between promptly stretchable (ductile) and fragile (non-malleable) conduct happens in some designing metals which, be that as it may, are crystalline in both states. Thus, T_G is to polymers what the bendable fragile move temperature is to metals.

10.3.3 Designing Polymer Structures with Improved Properties

A fitting synopsis of this present's first experience with polymer structure is found in the triangle of Figure 10.6. Three fundamental approaches to give mechanical quality and warm imperviousness to polymers are spoken to at the corners: (1) crystallization, (2) crosslinking, and (3) chain solidifying. Mixes of two or, even better, three of these

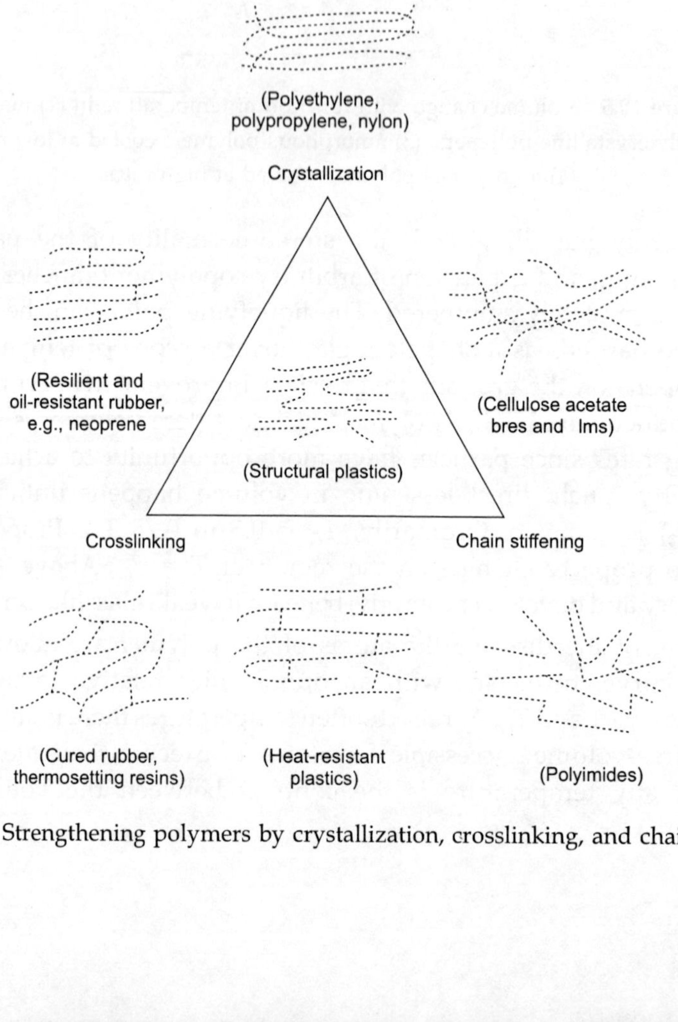

Figure 10.6 Strengthening polymers by crystallization, crosslinking, and chain stiffening.

fortifying components have demonstrated viable in accomplishing valuable polymer properties. For adaptable chain atoms the main choices for fortifying are crystallization and crosslinking (e.g., elastic). Strengthening by chain stiffening is accomplished by hanging bulky side groups that prevent the chain from bending. As an example, the large benzene rings attached to the carbon spine serve to stiffen polystyrene, a polymer that is neither crystalline nor crosslinked. Lastly, in polyethylene terephthalate (PET) the combination of moderately stiff chains and crystallization raises the melting point and strengthens the polymer.

10.3.4 Defects in Polymers

The point defect concept is different in polymers from metals and ceramics as a consequence of the chain-like macromolecules and the nature of the crystalline state for polymers. Point defects similar to those found in metals have been observed in crystalline regions of polymeric materials; these include vacancies and interstitial atoms and ions. Chain ends are considered to be defects because they are chemically dissimilar to normal chain units. Vacancies are also associated with the chain ends. However, additional defects can result from branches in the polymer chain or chain segments that emerge from the crystal. A chain section can leave a polymer crystal and reenter it at another point creating a loop, or can enter a second crystal to act as a tie molecule. Screw dislocations also occur in polymer crystals. Impurity atoms/ions or groups of atoms/ions may be incorporated in the molecular structure as interstitials; they may also be associated with main chains or as short side branches.

10.4 INORGANIC GLASSES

Glass is one of our most versatile materials and also one of the oldest. Obsidian, a common volcanic glass, was widely used for arrowheads, spearheads and knives in the Stone Age. The earliest example of an object completely crafted of glass dates to 7000 BC. Today, 700 different glass compositions are in commercial use. These are fabricated into tens of thousands of products having combinations of properties for countless uses.

10.4.1 Structure and Composition

Silica Glass: If vitreous (glassy) silica (SiO_2) were examined at the atomic level it would appear to be a random three-dimensional network of SiO_4 groups as envisioned in Figure 10.7a. Each group consists of an ionic tetrahedral structure consisting of four O^{2-} ions surrounding a smaller Si^{4+} ion. These tetrahedra interlock in the sense that each O belongs to two adjacent groups and are thus linked to form one giant (polymerized) molecule. In contrast, the regularity of crystalline SiO_2 is schematically depicted in Figure 10.7b. Corresponding X-ray diffraction patterns of *glassy* silica and cristobalite, a crystalline form of SiO_2, are compared in Figure 10.8. Both patterns peak at a common diffraction angle, suggesting approximately the same atomic surroundings, coordination numbers, and interatomic distances. Actually, the average Si-O distance in the glass is 0.162 nm compared with 0.160 nm in crystalline silicates. Other models of glass structure assume it to be little different from that of the melt from which it formed.

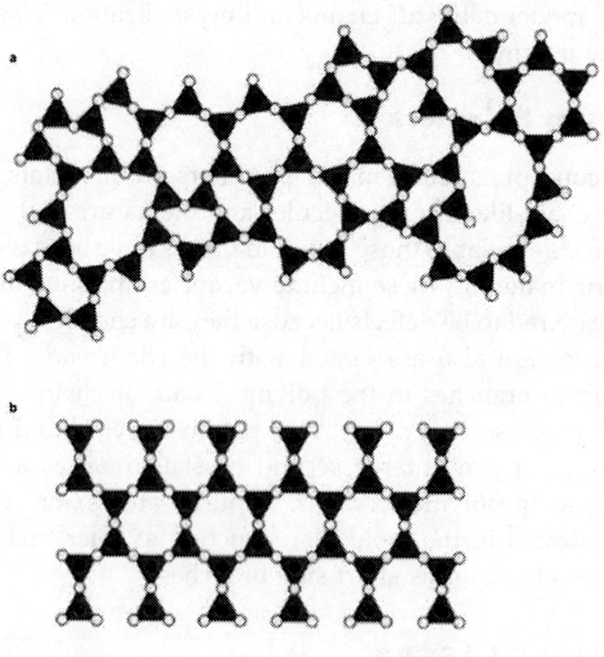

Figure 10.7 Schematic representation of SiO_2 structures: (a) Amorphous SiO_2. (b) Crystalline SiO_2.

Extending what had been learnt about SiO_2, Zachariasen in 1932 postulated a number of conditions that must be fulfilled before an oxide can be a glass former:

1. Each oxygen atom must be linked to no more than two cations.
2. The number of oxygen atoms around each cation must be small, that is, three or four.
3. The oxygen polyhedra must share corners, not faces or edges, to form a three-dimensional network.
4. At least three corners must be shared.

Silica is a classic example of these rules. Glasses based on SiO_2 are nearly universal solvents and virtually all elemental oxides dissolve in them to a considerable degree. Broadly speaking, the various metal oxides indicated in Table 10.1 are classified into one of the three categories: glass formers, intermediates, and modifiers. Interestingly, all of the good glass formers have high bond strengths. They also possess bonds that are estimated to have 50% or more covalent character (e.g.. Si–O, Ge–O, B–O, etc). Strong bonds mean that deformation and flow will be difficult, and that melts of pure glass formers are likely to be viscous. Modifier cations open the network by bonding ionically to the anions. Adding a molecule of Na_2O releases two Na^+ ions, which connect to the oxygens at the tetrahedron corners making them nonbridging. These low-vitality bonds cause the structure to open, in this manner bringing down the thickness and encouraging melt stream or smoothness attributes. This implies the glass can be worked at lower temperatures. Electrical conductivity and warm extension are both expanded by modifier increases. Middle of the road oxides straddle both universes. In spite of the fact that not fit for shaping a glass they expect both system and modifier destinations in the glass. A schematic representation of a glass with modifiers and intermediates is appeared in Figure 10.9.

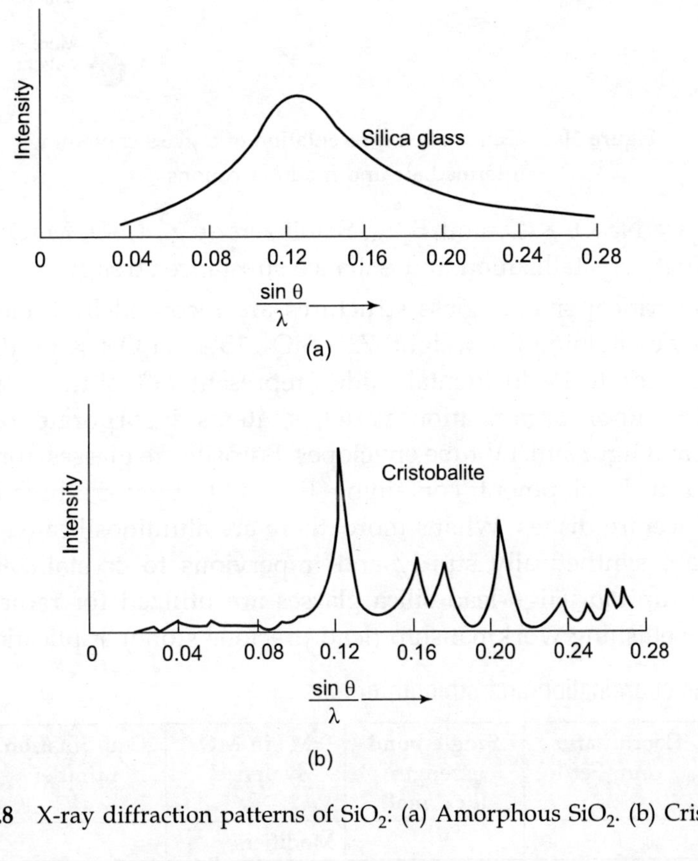

Figure 10.8 X-ray diffraction patterns of SiO_2: (a) Amorphous SiO_2. (b) Cristobalite.

Another grouping of oxides stresses their practical nature. In this plan the normal oxides are isolated into glass formers (as before), fluxes, and stabilizers. Fluxes are added to bring down the liquefying and working temperatures by diminishing the thickness.

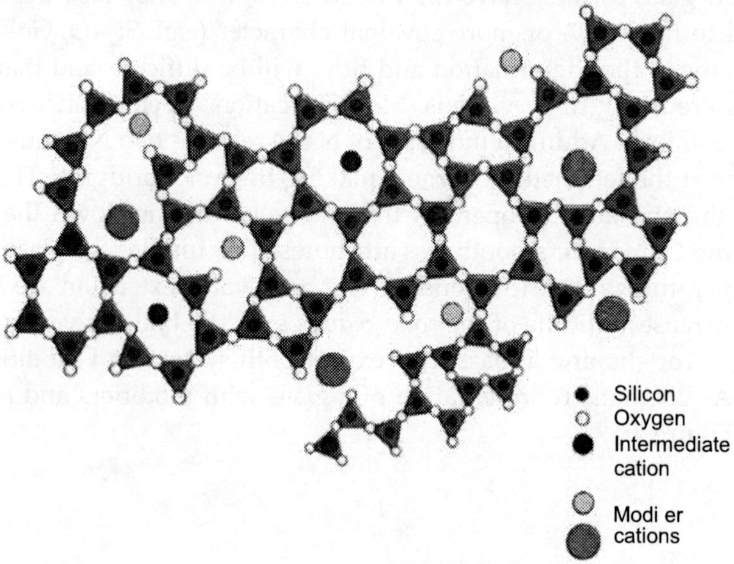

Figure 10.9 Schematic representation of a glass containing intermediate and modifier cations.

Normal fluxes are Na_2O, K_2O, and B_2O_3. Stabilizers (e.g., CaO, MgO, and Al_2O_3) are added to anticipate crystallization and enhance substance strength.

Some of the critical silicate glass structures are recorded in Table 10.2. Pop lime glasses, normally containing (by weight) 72% SiO_2, 15% $Na_2O + K_2O$, 10% CaO + MgO, 2% Al_2O_3, in addition to 1% incidental oxides, represent 90% of the tonnage of all glass delivered. With minor organization varieties, items incorporate reinforced glass, compartments, and light and TV tube envelopes. Borosilicate glasses, for example, Pyrex display low warm development; containing 13% B_2O_3, they are broadly utilized as a part of research centre dishes. What's more, there are aluminosilicate and lead glasses. The previous are synthetically sturdy and impervious to crystallization and raised temperature corruption. High-lead silica glasses are utilized for radiation protecting, optical, and embellishing workmanship (lead precious stone) applications.

Table 10.1 Role coordination and strength of ions

M* in MO (valence)	Coordination number	Single-bond strength (kcal/mol)	M* in MO (valence)	Coordination number	Single-bond strength (kcal/mol)
Glass former			Modifiers		
B (3)	3	119	Y (3)	8	50
Si (4)	4	106	Sn (4)	6	46
Ge (4)	4	108	In (3)	6	43

Al (3)	4	101-79	Mg (2)	6	37
P (5)	4	111-88	Li (1)	4	36
V (5)	4	112-90	Pb (2)	4	36
As (5)	4	87-70	Zn (2)	4	36
Sb (5)	4	85-68	Ba (2)	8	33
Zr (4)	6	81	Ca (2)	8	32
			Cd (2)	4	30
			Na (1)	6	20
			K (1)		13
Intermediates					
Ti (4)	6	73			
Zn (2)	2	72			
Cd (2)	2	60			
Al (3)	6	67-53			
Zr (4)	8	61			

Table 10.2 Composition and uses of silica glasses

Glass*	SiO_2	Al_2O_2	B_2O_3	Na_2O	K_2O	CaO	PbO	MgO	Applications
0010	63	1		7	7		22		Lamp tubing
0080	73	1		17		5		4	Lamp bulbs
1720	62	17	5	1		8		7	Ignition tubes
7070	71	1	26	1	1				Low-loss electrical
7740	81	2	13	4					General
7900	96	0.3	3						High temperature
8363	5	3	10				82		Radiation shielding
Window	72	1		14		10		3	
Fibre glass	54.5	14.5	8.5	0.5		22			Composites
Borosilicate	81	2.5	12	4.5					Low expansion, chemical industry

Non-Silicate Glass: There is a considerable number of non-silicate glasses. Polymers, the most imperative of these, have as of now been examined finally. Formless silicon and germanium thin films are promptly stored, even at moderately raised temperatures. Glass-framing oxides instead of SiO_2 are the premise of a few business sytheses. They incorporate a 83% B_2O_3 glass for high X-beam transmission applications and a 80% GeO_2

optical glass. A fraction of these open in the melt and, with moderately rapid cooling, can be supercooled to form elemental glasses. These group VI elements are the base of the chalcogenide glasses which also contain assorted metals. Some of them behave like semiconductors but commercial applications have not yet been found.

10.4.2 Solidification of Glassy Melts

Definitions of glass stress that they are products of melts cooled to a rigid condition without crystallizing or that they have attained such a high degree of viscosity as to be rigid for all practical purposes. Be that as it may, glass is not just a supercooled liquid, as can be delineated in the officially commonplace volume (V)- temperature (T) plot of Figure 10.11. A supercooled liquid is the straight augmentation of the fluid warm development line extrapolated to low temperature. Close to T T_G, the glass change temperature, there is a twist in the bend and the consistency achieves estimations of 1011 to 1012 Pa-s. The slant, dV/dT, no longer takes after the (supercooled) liquid line yet rather parallels the warm development conduct for the smooth strong. It is by and large imagined that the incline change is because of motor elements or constrained by drowsy nuclear movement in the dissolve. The thickness turns out to be large to the point that basic change can't be identified in the time size of estimation. Dissolves cooled at a lower rate have more opportunity to unwind and the supercooled fluid perseveres to lower temperature, bringing about a denser glass. More fast cooling raises T_G and a rough glass change extent between these two T_G qualities can be connected with a specific glass.

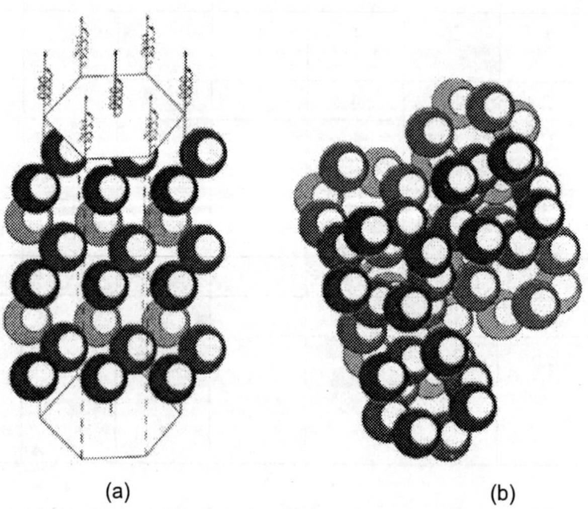

Figure 10.10 Crystalline forms of selenium and tellurium: (a) hexagonal structure with spiral chains (b) puckered ring structure.

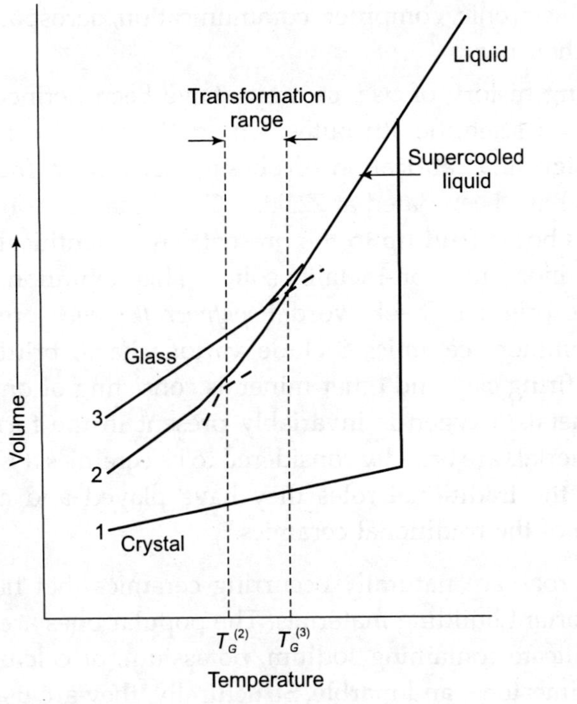

Figure 10.11 Volume-temperature relationships in (1) a crystal, (2) a slowly cooled glass, and (3) a rapidly cooled glass.

10.5 CERAMICS

10.5.1 Traditional Ceramics

They are inorganic and nonmetallic materials. Most ceramics are compounds between metallic and nonmetallic elements for which the interatomic bonds are either totally ionic, or predominantly ionic but having some covalent character. The term "ceramic" comes from the Greek word *keramikos,* which means "burnt stuff," indicating that desirable properties of these materials are normally achieved through a high-temperature heat treatment process called firing.

About the last 60 or so years, the most important materials in this class were termed the "traditional ceramics," those for which the primary raw material is clay; products considered to be traditional ceramics are china, porcelain, bricks, tiles, and, in addition, glasses and high-temperature ceramics. Of late, significant progress has been made in understanding the fundamental character of these materials and of the phenomena that occur in them that are responsible for their unique properties. Consequently, a new generation of these materials has evolved, and the term "ceramic" has taken on a much broader meaning. To one degree or another, these new materials have a rather dramatic effect on our lives; electronic, computer, communication, aerospace, and a host of other industries rely on their use.

With such a long history of use, ceramics have been defined in terms that stress both their artistic and scientific attributes. One definition of ceramics is "the art that deals with the design and fabrication of objects made from fired clay." The earliest known clay figures have been dated at 22,000 BC and may be considered the root of an artistic tree that has borne fruit up to the present day. Scientifically, ceramics are most broadly defined as inorganic non-metallic solids. This definition considerably extends the meaning of the original Greek word *Kspafjiocr {keramos),* referring to pottery or earthenware. In common, ceramics include various hard, brittle, corrosion-resistant materials made by firing clay and other minerals consisting of one or more metals and one or more nonmetals; oxygen is invariably present in the form of oxides. Because many inorganic materials are broadly considered to be ceramics it is instructive to classify them according to the traditional roles they have played and continue to play. The following are some of the traditional ceramics.

Stone: Stone and rock are naturally occurring ceramics that have always been and continue to be important building materials. The popular ones are granite [a mixture of feldspar (aluminosilicate containing sodium, potassium, or calcium), quartz, and mica (aluminosilicate)], limestone, and marble. Structurally, they are usually agglomerates of crystallites held together by an intergranular glassy network. The latter two stones can support compressive loads of 700 kg/cm^2, or 10,000 lb on every square inch of area, and granite twice as much. Their brittle nature facilitates quarrying and polishing, while their low coefficients of thermal expansion (~8×10^{-6} C^{-1}), relatively low density (~2.5-2.7 g/cm^3), and low tendency to absorb water are beneficial attributes in structural applications.

Clay Products: Ceramics have traditionally meant fired whiteware and structural clay products. The ingredient common to all of these is clay, which varies widely in chemical, mineralogical, and physical characteristics. Clay basically consists of electrically neutral aluminosilicate layers in the form of tiny crystalline platelets that readily slide over each other. In combination with water they form plastic masses that are easily shaped. Whiteware includes porcelain, sanitaryware, electrical insulators, and dishes. They are made from mixtures of clay, feldspar, and flint. Porcelain, an important member of this group, can be ideally thought of as a complex mixture of K_2O, SiO_2, and Al_2O_3 that is fired at very high temperature (~1400°C), yielding a translucent product. Terminated porcelain structures commonly contain a shiny antacid network in which little precious stones of quartz and Mullite ($3\,Al_2O_3 \cdot 2SiO_2$) are inserted. Auxiliary earth items or clay products incorporate the bricks, tiles, and sewer pipes usually utilized in the development business. Different cases of comparative arrangements incorporate stoneware and pottery, for example, terracotta. Muds of various places are utilized as a part of these items and terminating temperatures are lower than for whiteware. For both enhancing purposes and to make terminated mud impenetrable to water infiltration, surface coatings (glasses) are for the most part connected.

Glasses: Previously in this chapter we have given the impression that glasses and ceramics are distinct materials. Silica glasses are, however, generally classified as a subgroup of ceramics. And like glass, ceramics can even sometimes be made optically transparent. One interesting category of glass has unambiguous attributes of ceramics; these are the glass-ceramics, and corning cookware is perhaps the most visible example. They are formed as a glass but, in the ensuing processing treatments, are converted to a fine-grained crystalline ceramic that is dense, is heat resistant, and has a high thermal shock resistance.

Refractories: These ceramics largely provide the thermal insulation, crucibles, and hardware in all kinds of casting operations, as well as elevated temperature processing and heat treatment furnaces. Whether they line the immense blast furnaces used in steel making or line moulds for casting turbine blades, they must withstand the ravages of direct contact with molten metals and glasses (e.g., slags) without decomposing or cracking. For these applications, refractories are commonly sold in brick form but they can also be purchased in rolls, powder, and loose form. Refractory oxides used for these general purposes most commonly include fireclays containing alumina – silicates with varying $Al_2O_3 - SiO_2$ proportions, magnesium oxide-rich compositions, relatively pure silica, and zirconia (ZrO_2). A problem of concern in refractories is their porosity, which may exceed 20% of the volume. High-density ceramics are stronger and are not as susceptible to chemical attack as the more porous varieties. Specialized refractories processed from pure Al_2O_3, ZrO_2, BeO, and other compounds can be made relatively pore free for special applications.

Abrasives: Abrasive ceramic materials are critical in the grinding, lapping, and polishing operations that enable the production of parts requiring high-dimensional tolerances. The sharp abrasive grains must necessarily be harder and tougher than the material being cut so that they do not round off or fracture during contact. Materials used as abrasives include silicon carbide, tungsten carbide, alumina, silica, and industrial diamond dust. They are most commonly bonded to grinding wheels or to paper and cloth (e.g., sandpaper), but are also used in loose form as well as embedded in pastes and waxes.

10.5.2 New Ceramics

In the previous quite a few years there has been a quickly developing enthusiasm for what are called new, fine, progressed, or building artistic materials. The capacities, properties, and utilizations of these ceramics production are helpfully shown in a dense arrangement in Figure 10.12. A look at the applications instantly uncovers the tremendous financial capability of these cutting edge materials. Some of these engineering ceramics are not new. Be that as it may, numerous others are, similar to the applications and the strategies used to prepare and describe them. It is enlightening to look at the ordinary and new pottery. The customary and new fired materials don't for the most part rival each other in the commercial centre; rather, they coincide in generally free applications. Albeit a portion of the subtle elements are saved for later dialog, the accompanying contrasts are promptly evident.

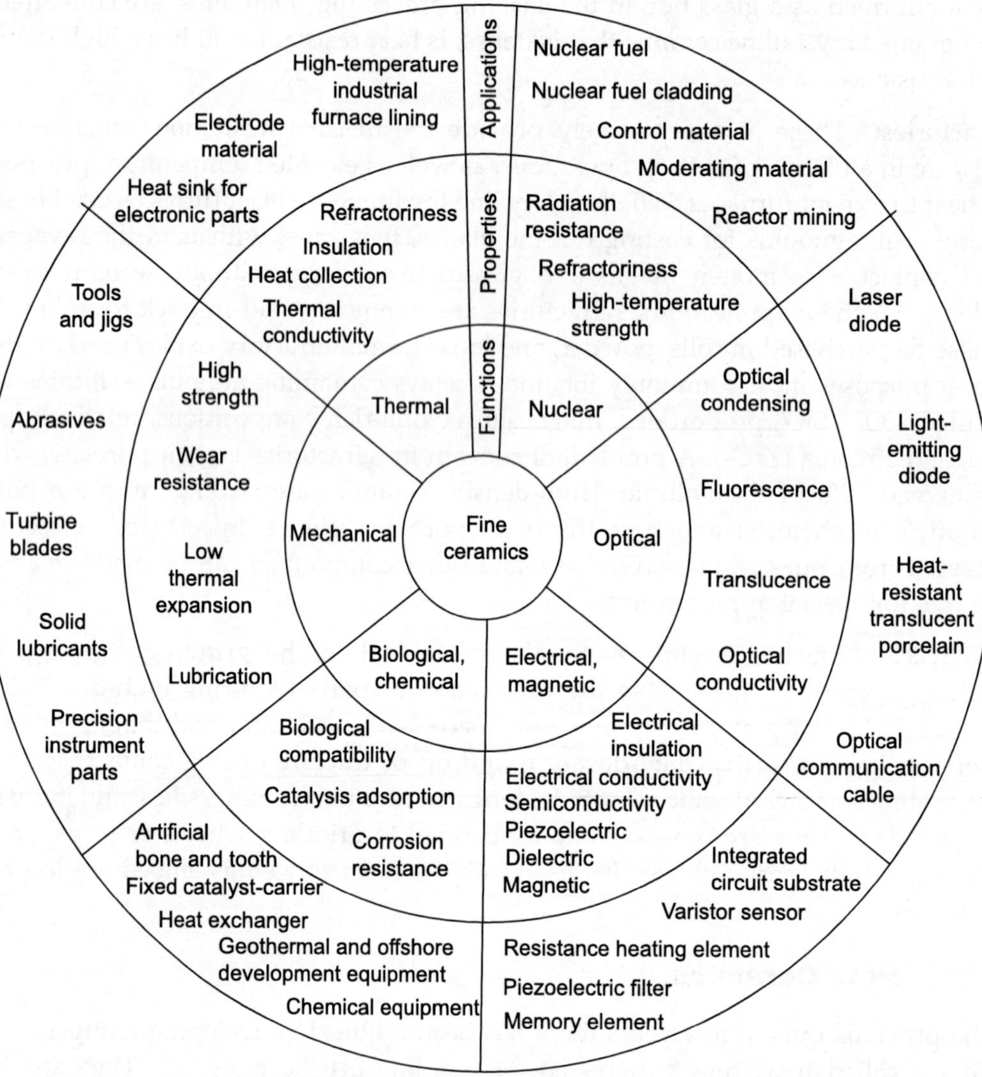

Figure 10.12 Functions, properties, and applications of the new ceramics.

Materials: Conventional ceramics are very heterogeneous physical and chemical mixtures of oxide compounds and glasses, containing relatively large impurity concentrations. The new ceramics are frequently composed of single compounds such as carbides (e.g., SiC, TiC) and nitrides (e.g., Si_3N_4, TiN, BN) whose stoichiometry, purity, and grain size must be well controlled. Therefore, conventional ceramics rely on the time-tested mines, sources of clays, and quarries. Fine ceramics, on the other hand, are often synthesized in laboratory-like environments from pure or well-controlled precursor raw materials.

Forming and Processing: Traditional methods of producing ceramicware like casting, pressing, and throwing clay on a potter's wheel are still widely practised. And firing in large, poorly temperature-controlled kilns is also common. At the other extreme many of the new ceramics must be in the form of single crystals which are carefully grown from the melt. Hot-pressing powders in well-controlled furnaces to achieve high-density consolidation are other techniques used to fabricate the new ceramic materials. Not only are bulk ceramics of interest, but there is increasing demand for thin films to coat assorted substrates. Methods like chemical vapour deposition that are spinoffs from microelectronic processing have been adapted to deposit ceramic films and coatings from suitable vapours.

Structures: Conventional ceramics are inhomogeneous even at levels that can sometimes be seen with the naked eye. But the new ceramics require electron microscopes and other modern analytical tools to assess structural and chemical variations. Depending on the method of preparation, ceramics can be either crystalline or glassy. This is especially true of thin films and coatings. But, for the most part, the new ceramic materials are crystalline, and in some applications (e.g., lasers, optics) single crystals are essential. For this reason the next section primarily, though not exclusively, emphasizes crystalline ceramics. Consideration of ionic groupings is a good place to start.

10.6 STRUCTURE OF CERAMICS

10.6.1 Crystalline Ceramics

The varied and complex structures of ceramics derive from the fact that two or more atoms (ions) of different size are usually involved and the condition that charge neutrality must be maintained. An extensive treatment of these structures is not possible and only a few will actually be presented in figures. Instead an extensive compilation of ceramic oxide structures is provided in Table 10.3.

Table 10.3 Ionic structures grouped according to anionic packing

Anion packing	Coordination number of M and O	Cation sites	Structure name	Examples
Cubic close-packed	6:6 MO	All octahedral	Rocksalt	NaCl, MgO, FeO, NiO, CaO, SrO, MnO, BaO
Cubic close-packed	4:4 MO	Half tetrahedral	Zinc blende	ZnS, BeO, SiC
Cubic close-packed	4:8 M_2O	All tetrahedral	Antifluorite	Li_2O, Na_2O, K_2O, Rb_2O
Cubic close-packed	12:6:6 ABO_3	One-fourth octahedral (B)	Perovskite	$CoTiO_3$, $SrTiO_3$, $SrSnO_3$, $SrZrO_3$
Simple cubic	8:8 MO	All cubic	CsCl	CsCl, CsBr, CsI
Simple cubic	8:4 MO_2	Half cubic	Fluorite	ThO_2, CeO_2, PrO_2, UO_2, ZrO_2, HfO_2
Hexagonal close-packed	4:4 MO	Half tetrahedral	Wurtzite	ZnS, ZnO, SiC
Distorted cubic	6:3 MO_2	Half octahedral	Rutile	TiO_2, GeO_2, SnO_2, PbO_2, VO_2, NbO_2
Hexagonal close-packed	6:4 M_2O_3	Two-thirds octahedral	Corundum	Al_2O_3, Fe_2O_3, Cr_2O_3, Ti_2O_3

Rocksalt: The position of cations and anions is well known. On account of MgO, for instance, O particles are considered to populate FCC Bravais grid focuses. The (interstitial) Mg particles are octahedrally organized so that six anions encompass every cation and the other way around.

Zinc Blend: This structure, exhibited by GaAs. On account of BeO, oxygen atoms populate all the FCC Bravais grid focuses. Be, however, possesses just 50% of the tetrahedral destinations. Four O particles encompass each Be particle, and the other way around. Different materials displaying this structure incorporate ZnS and SiC.

Perovskite: Three ions are involved in these structures, whose idealized cubic unit cell is shown in Figure 10.13. In $CaTiO_3$ the central ion is Ca^{2+} (the larger of the two cations), and it lies in 12-fold coordination with O^{2-} ions and in 8-fold coordination with Ti^{4+}. Meanwhile, each Ti^{4+} ion is octahedrally coordinated to 6 O^{2-} ions. It is instructive to account for both the stoichiometry and the charge balance in this unit cell. The one Ca^{2+} ion is wholly contained within the cell. Each of the corner Ti^{4+} ions is shared by 8 identical cells so there is effectively only $8 \times 1/8 = 1$ Ti atom. Lastly, each of the 12 O^{2-}

ions is shared by 4 cells, and therefore contributes a total of 3 O atoms. The sum of the positive charge (2+) + (4 +) = 6+ balances the negative charge (3) × (2 –) = 6–. Furthermore, the centres of the positive and negative charges coincide, unlike the situation in $BaTiO_3$, a slightly distorted perovskite.

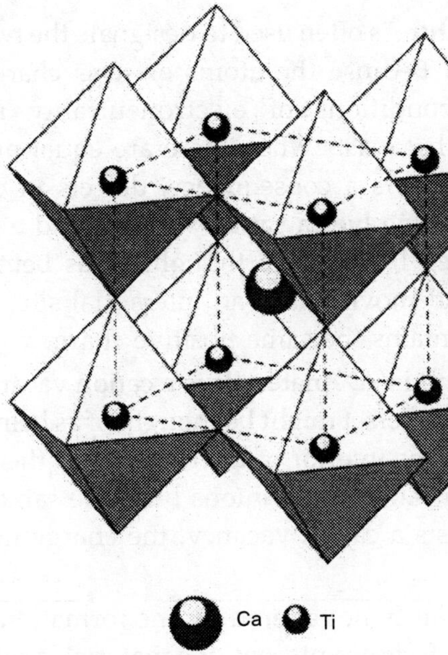

Figure 10.13 Perovskite structure of $CaTiO_3$.

10.6.2 Silicates

After carbon, silicon forms more compounds than any other element. The earth's crust is largely composed of silicon-oxygen compounds, or silicates and many of them find engineering uses in glass, cement, brick, and electrical insulators. They all have structures that are based on the fundamental tetrahedral SiO_4~ building block, the orthosilicate anion. By linking to one another across the bridging oxygen atoms one-, two-, and three-dimensional subunits can be constructed. These connect to one another through intervening cations such as Al^{3+}, Mg^{2+}, Ca^{2+} and Fe^{2+} to create a diverse array of rather complex crystalline structures.

10.7 IMPERFECTIONS IN CERAMICS

10.7.1 Atomic Point Defects

Atomic defects involving host atoms may exist in ceramic compounds. As with metals, both vacancies and interstitials are possible; however, since ceramic materials contain ions of at least two kinds, defects for each ion type may occur. For example, in NaCl, Na interstitials and vacancies and Cl interstitials and vacancies may exist. It is highly improbable that there would be appreciable concentrations of anion interstitials. The anion is relatively large, and to fit into a small interstitial position, substantial strains on the surrounding ions must be introduced.

The expression **defect structure** is often used to designate the types and concentrations of atomic defects in ceramics. Because the atoms exist as charged ions, when defect structures are considered, conditions of electroneutrality must be maintained. **Electroneutrality** is the state that exists when there are equal number of positive and negative charges from the ions. As a consequence, defects in ceramics do not occur alone. One such type of defect involves a cation-vacancy and a cation-interstitial pair. This is called a **Frenkel defect.** It might be thought of as being formed by a cation leaving its normal position and moving into an interstitial site. There is no change in charge because the cation maintains the same positive charge as an interstitial.

Another type of defect found in AX materials is a cation vacancy-anion vacancy pair known as a **Schottky defect**. This defect might be thought of as being created by removing one cation and one anion from the interior of the crystal and then placing them both at an external surface. Since both cations and anions have the same charge, and since for every anion vacancy there exists a cation vacancy, the charge neutrality of the crystal is maintained.

The ratio of cations to anions is not altered by the formation of either a Frenkel or a Schottky defect. If no other defects are present, the material is said to be stoichiometric. **Stoichiometry** may be defined as a state for ionic compounds wherein there is the exact ratio of cations to anions as predicted by the chemical formula. For example, NaCl is stoichiometric if the ratio of ions to ions is exactly 1:1. A ceramic compound is *nonstoichiometric* if there is any deviation from this exact ratio.

Nonstoichiometry may occur for some ceramic materials in which two valence (or ionic) states exist for one of the ion types. Iron oxide is one such material, for iron can be present in both Fe^{2+} and Fe^{3+} states; the number of each of these ion types depends on temperature and the ambient oxygen pressure. The formation of an Fe^{3+} ion disrupts the electroneutrality of the crystal by introducing an excess +1 charge, which must be offset by some type of defect. This may be accomplished by the formation of one Fe^{3+} vacancy (or the removal of two positive charges) for every two Fe^{3+} ions that are formed. The crystal is no longer stoichiometric because there is one more O ion than Fe ion; however, the crystal remains electrically neutral. This phenomenon is fairly common in iron oxide, and, in fact, its chemical formula is often written as $Fe1\text{-}xO$ (where x is some small and variable fraction substantially less than unity) to indicate a condition of nonstoichiometry with a deficiency of Fe.

10.7.2 Impurities in Ceramics

Impurity atoms can form solid solutions in ceramic materials much as they do in metals. Solid solutions of both substitutional and interstitial types are possible. For an interstitial, the ionic radius of the impurity must be relatively small in comparison to the anion. Since there are both anions and cations, a substitutional impurity will substitute for the host ion to which it is most similar in an electrical sense: if the impurity atom normally forms a cation in a ceramic material, it most probably will substitute for a host cation. For example, in sodium chloride, impurity Ca^{2+} and O^{2-} ions would most likely substitute for Na^+ and Cl^- ions, respectively. To achieve any appreciable solid solubility of substituting impurity atoms, the ionic size and charge must be nearly the same as those of one of the host ions. For an impurity ion having a charge different from the host ion for which it substitutes, the crystal must compensate for this difference in charge so that electroneutrality is maintained with the solid. One way this is accomplished is by the formation of lattice defects—vacancies or interstitials of both ion types, as discussed above.

Example 10.1 If electroneutrality is to be preserved, what point defects are possible in NaCl when a Ca^{2+} substitutes for an Na^+ ion? How many of these defects exist for every Ca^{2+} ion?

Solution: Replacement of an Na^+ by a Ca^{2+} ion introduces one extra positive charge. Electroneutrality is maintained when either a single positive charge is eliminated or another single negative charge is added. Removal of a positive charge is accomplished by the formation of one Na^+ vacancy. Alternatively, a Cl^- interstitial will supply an additional negative charge, negating the effect of each Ca^{2+} ion. However, as mentioned above, the formation of this defect is highly unlikely.

10.8 CEMENT AND CONCRETE

10.8.1 Introduction

Concrete and reinforced concrete are two of the most important civil engineering materials. Their tonnage consumed far exceeds that of steel, wood, and polymers combined. They are the essential ingredients in some of the largest structures such as high-rise buildings and airport runways. Many large concrete structures have to restrain water from rivers as in dams, whereas others must withstand ocean forces and marine environments. As an important example of the latter consider their use in ocean platforms for oil drilling.

10.8.2 Cement

Concrete is composed of cement, aggregates (sand, gravel, crushed rock), and water. The water is important because it makes the concrete mouldable and is needed for the hydration reactions that harden the cement. As the cement hardens it binds the aggregates and gives concrete the stone-like quality of bearing high compressive loads. Cement, a generic term for concrete binder, is the key ingredient. The most important binder for concrete is Portland cement and it will be discussed exclusively. Portland cement is produced from an initial mixture of 75% limestone ($CaCO_3$), 25% clay, assorted aluminosilicates, and iron and alkali oxides. This mixture is ground and fed into a rotary kiln (a large cylindrical rotating furnace) together with powdered coal. Progressive reactions at temperatures extending to 1800 K break down clays, decompose the limestone to yield quicklime or CaO, and then fuse them to produce clinker (pellets) 5-10 mm in size. After cooling, the latter is mixed with 3-5 wt% gypsum ($CaSO_4$) and ground to powder form that is Portland cement.

The final powder size influences the setting rate; finer sizes mean a higher specific surface area and faster hydration and setting rates. Portland cement is composed of the following compounds:

Calcium oxide (lime) CaO = C

Silicon oxide (silica) SiO_2 = S

Aluminum oxide (alumina) Al_2O_3 = A

Water H_2O = H

Some 75 per cent by weight is made out of tricalcium silicate (C_3S) and dicalcium silicate (C_2S). Diverse extents of these and different fixings are mixed to create unique concretes that either set gradually or all the more quickly, free less warmth of hydration, or are expected to oppose debasement by water containing sulphates and different mixes. Wet bond is workable just until it sets, and this period sets up the pot life of the blend. The average time reliance of warmth development and solidifying of bond is shown in Figure 10.14. After a substantial introductory burst of heat there are much littler auxiliary pinnacles that harmonize with the beginning of solidification. Amid the setting time frame bond has no quality, however with cementing, quality ascents.

Chemical responses that are in charge of heat advancement and solidifying are perplexing and not so much caught on. Cement does not solidify by a drying reaction, but rather, then again, requires water; shockingly, the responses proceed for a considerable length of time. A few imperative hydration responses happen at various rates:

$$C_3A + 6H \rightarrow C_3AH_6 + \text{heat (hours)}$$

$$2C_3S + 6H \rightarrow C_3S_2H_3 + 3CH + \text{heat (days)}$$

$$2C_2S + 4H \rightarrow C_3S_2H_3 + CH + \text{heat (months)}$$

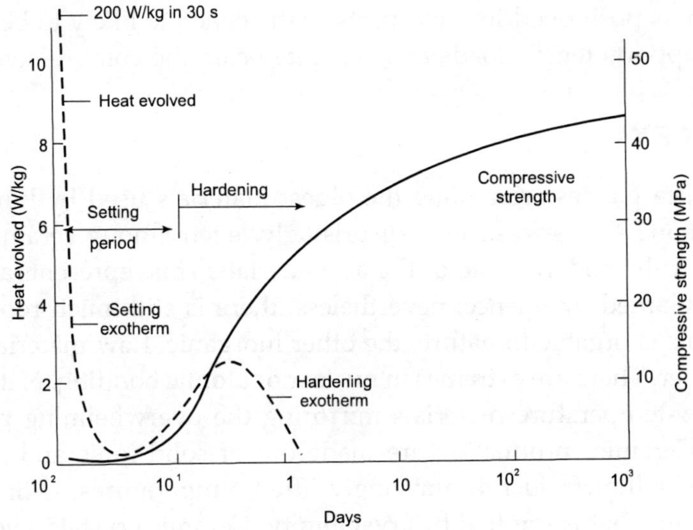

Figure 10.14 Time-dependent heat evolution and strength change during cement hydration.

The first reaction causes the cement to set; the second and third reactions cause hardening. Some 70% of the structure is taken up by $(CaO)_3(SiO_2)_2(H_2O)_3$ or $C_3S_2H_3$ which is hydrated $(CaO)_3(SiO_2)_2$. Despite the long reaction times, the hydrated tricalcium sulphate product is in the form of small particles (less than 1 μm in size) and is known as tobomorite gel. Gels are swollen polymer-like networks of high viscosity, and they coat the surfaces of the unreacted cement grains, slowing hydration reactions of the latter. Hardening and strengthening occur with the growth and intertwining of these spine-like gels. The increasing silicate content with time suggests the corner-to-corner joining of $(SiO_4)^{4-}$ tetrahedra:

$$2(SiO_4)^{4-} \rightarrow Si_2O_7^{6-} + O^{2-}$$

10.8.3 Concrete

Concrete is the first of a number of synthetic composites or engineered materials mixtures discussed in the book. In addition to cement, concrete contains aggregate consisting of sand (less than 2 mm in diameter) and larger particles of gravel and crushed rock. A high proportion of fine aggregate is required for smooth finishes. The typical ratio of aggregate to cement is 5 : 1 , whereas that of sand to gravel is about 3 : 2. Aggregates contain particles with a distribution of sizes that enable smaller ones to fit neatly into the spaces between larger ones. Cement then flows into the remaining spaces. For many applications steel bars are embedded in the concrete to reinforce it. Concrete, like other ceramics, is weak when pulled in tension but strong when compressed. Steel, which is strong in tension, is positioned in those parts of the concrete likely to be so stressed. In that way steel supports tensile loads and concrete bears the compressive loads.

10.9 SUMMARY

Ceramics and silica glasses are among the oldest materials used by humans, whereas polymers are among the newest. But, surprisingly, even among ceramics the new or fine varieties are truly modern, state-of-the-art materials. This represents a vivid example of art being supplanted by science; nevertheless, there is still much room for art. One group of materials is organic in nature, the other inorganic. Raw materials and sources for them thus differ. There are extremes in nuclear or atomic bonding. Natural or organic polymers are low-temperature materials mirroring the overwhelming part of van der Waals holding. Ceramics production are made out of solid ionic and covalent bonds that disintegrate or liquefy just at amazingly lifted temperatures. Ultimately, there is the colossal inlet in what is implied by crystallinity. Ceramic crystals, even those found in nature, are standards of wonderfully requested structures; in reality the investigation of crystallography was initially given to unfurling their inner nuclear courses of action. Crystalline polymers, then again, show an exceptionally flawed crystallinity and structures are scattered or even shapeless at the edges of precious crystals.

Most polymeric materials are composed of large molecular chains with side groups of various atoms (O, Cl, etc.) or organic groups such as methyl, ethyl, or phenyl groups. These macromolecules are composed of repeat units, smaller structural entities, which are repeated along the chain. Repeat units for some of the chemically simple polymers [i.e., polyethylene, polytetrafluoroethylene, poly(vinyl chloride), and polypropylene] were presented.

With regard to behaviour at elevated temperatures, polymers are classified as either thermoplastic or thermosetting. The former have linear and branched structures; they soften when heated and harden when cooled. In contrast, thermosets, once having hardened, will not soften upon heating; their structures are crosslinked and network.

Although the point defect state concept in polymers is different from metals and ceramics, vacancies, interstitial atoms, and impurity atoms/ions and groups of atoms/ions as interstitials have been found to exist in crystalline regions. Other defects include chains ends, dangling and loose chains, as well as dislocations.

Both crystalline and noncrystalline states are possible for ceramics. The crystal structures of those materials for which the atomic bonding is predominantly ionic are determined by the charge magnitude and the radius of each kind of ion. Some of the simpler crystal structures are described in terms of unit cells; several of these were discussed (rock salt, caesium chloride, zinc blende, diamond cubic, graphite, fluorite, perovskite, and spinel structures).

With regard to atomic point defects, interstitials and vacancies for each anion and cation type are possible. These imperfections often occur in pairs as Frenkel and Schottky defects to ensure that crystal electroneutrality is maintained. Addition of impurity atoms may result in the formation of substitutional or interstitial solid solutions. Any charge imbalance created by the impurity ions may be compensated by the generation of host ion vacancies or interstitials.

At room temperature, virtually all ceramics are brittle. Microcracks, the presence of which is difficult to control, result in amplification of applied tensile stresses and account for relatively low fracture strengths (flexural strengths). This amplification does not occur with compressive loads, and, consequently, ceramics are stronger in compression. Fractographic analysis of the fracture surface of a ceramic material may reveal the location and source of the crack-producing flaw, as well as the magnitude of the fracture stress. Representative strengths of ceramic materials are determined by performing transverse bending tests to fracture.

EXERCISE

1. What is the degree of polymerization of a polystyrene sample that has a molecular weight of 129,000?
2. If the degree of polymerization of polyvinyl chloride is 729, what is the molecular weight of the polymer?
3. Plasticizers in polymers behave like modifiers in silica glasses. Speculate the behaviour of plasticizers.
4. List compositional and structural differences between the new and structural ceramic materials?
5. Given its density is 6.02 Mg/m^3. What is the lattice parameter of $BaTiO_3$?
6. Explain why clay minerals like pyrophyllite cleave readily?
7. Approximately calculate the density of $YBa_2Cu_3O_7$ which is a high-temperature superconductor.
8. Differentiate between cement and concrete.
9. Differentiate between glasses and ceramics.
10. Write about any two of the traditional ceramics.
11. Explain the structure and composition of silica glass.
12. What is polymer morphology? Summarize any two aspects of it.
13. Make comparisons of thermoplastic and thermosetting polymers (a) on the basis of mechanical characteristics upon heating, and (b) according to possible molecular structures.

14. Is it possible to grind up and reuse phenol-formaldehyde? Why or why not?
15. Is it possible to grind up and reuse polypropylene? Why or why not?
16. Would you expect Frenkel defects for anions to exist in ionic ceramics in relatively large concentrations? Why or why not?
17. In your own words, briefly define the term "stoichiometric."
18. Briefly explain why there may be significant scatter in the fracture strength for some given ceramic material.
19. Specify the reason why fracture strength increases with decreasing specimen size.
20. Cite one reason why ceramic materials are, in general, harder yet more brittle than metals.

11

Processing and Forming Operations of Materials

Learning Objectives

After going through this chapter the reader will be able to:

- introduce solidification processes of metals;
- understand mechanical forming operations;
- cite the processing of polymers and glasses; and
- throw light on forming of glass.

11.1 INTRODUCTION

Engineering materials are useful only if they can be processed in large quantities and fabricated in varied shapes, while meeting composition and property specifications. **Primary processing** is concerned with large-scale production of basic materials like steel and polymer precursors. This output is then sent either to **secondary processors** for alloying or blending or directly to the manufacturing industries (e.g., auto and plastic packaging companies), which additionally shape and assemble them into final products. Many variations on this theme exist and some prime producers also manufacture intermediate finished products for further assembly, for example, I-beams for construction and huge rotors for steam turbines. Some companies and industries derive relatively little benefit, but others involved with semiconductors and electronics reap huge added value profits from additional processing.

Here a comparative approach is adopted, stressing similarities in processing methods for different materials, for example, die casting of metals and injection moulding of polymers, and powder pressing and sintering of metal and ceramic powders. At the same time, different competitive routes to production of the same component, for example, by forging, casting, or powder metallurgy, are compared. Qualitative process descriptions, engineering modeling of their features, product design, and elimination of processing defects are some of the additional issues addressed.

11.2 SOLIDIFICATION OF METALS

Casting is a viable process because liquid metals flow readily, have a high density, and are good conductors of heat. Fluidity, or the ability to flow, is inversely proportional to viscosity. Liquid metals are unique in having very low viscosities and therefore, flow readily. Coupled with high densities, liquid metals easily fill moulds, whereas less dense nonmetallic fluxes, drosses, and oxide slags float to the surface where they can be skimmed off. High heat transfer rates make it possible to complete solidification rapidly, making casting production economical.

11.2.1 Expandable Mould Casting

Lost Wax Method: Lost wax casting employs expandable moulds that are used just once because they are destroyed in the process of making a single impression. Except for new metal compositions and improvements in waxes and mould materials, tooling, furnaces, and other aspects, the basic process steps have not changed since antiquity. First the sculptor created an exact wax replica or pattern of what the final bronze was envisioned to look like. Then wax was removed completely. This was done through a series of steps:

(i) Wax rods are initially attached to the wax pattern to create a type of plumbing system which directs the flow of metal from the **pouring cup,** through a **sprue** and into gates that metre the molten metal flow into the casting. Provision must also be made to attach a wax **riser.** When filled, the riser provides a reservoir of molten metal that can flow into and fill casting cavities caused by metal shrinkage during the transformaton of liquid to solid bronze.

(ii) The wax and the gating system are either immersed in plaster-like, water slurry of refractory oxide powder (e.g., SiO_2, CaO, ZrO_2, plus binders) or stuccoed over with this mould material until completely surrounded. To replicate wax copies or additional castings split or multipiece permanent moulds with parting lines are used.

(iii) After drying, the whole shape gathering is modified and set in a heater. The wax melts and depletes. In the process the shape solidifies and is equipped to withstand the stun of hot metal contact.

(iv) Molten metal then fills the hollow pit made by the lost wax. After cementing, the shape is broken/separated to free the throwing. The strong gating and riser poles are sawed off and the metal surface experiences a long completing operation to improve the excellence of the bronze and the creative nature of the last figure.

Sand Casting: In the light of a weight rule, more sand castings are delivered than some other sort. Given iron a role as well as aluminium motor pieces, cast press machine apparatus bases, steel hardware segments of various types, stainless-steel pump lodgings, plumbing installations of all sizes, and colossal bronze ship propellers are cases of some sand throwing applications and the metals utilized. Wooden or plastic examples of the sought shapes together with connected entryways and risers are initially installed into an extraordinary throwing sand contained inside equal parts of a split form. This permits the example and its joined doors and risers to be effectively evacuated to make the depression, which is thus filled by liquid metal, after the parts have been rejoined. In a vital variation of sand giving referred to a role as shell form throwing, a thin shell (0.5-1 cm thick) of a sand-tar blend is utilized as the shape. In this frame split moulds can be put away; when required they are cinched together, supported by sand, and after that loaded with liquid metal. Another extra form handle utilizes effortlessly moulded and delivers low thickness polystyrene examples of the articles to be thrown. After they are implanted in sand, liquid metal is poured in, quickly vaporizing the polystyrene. At last, metal totally dislodges the first polystyrene design. Applications for such full shape castings have included aluminium barrel heads, brake segments, and manifolds for vehicles

11.2.2 Permanent Mould Casting

Die Casting: Moulds are utilized again and again as a part of lasting mould throwing forms. The most broadly honed of these strategies is known as kick the bucket throwing. Passes on are critical assembling devices made of hard, extreme compound steels that empower parts made of metals, polymers, and earthenware production to be moulded and recreated. They are not just utilized as moulds to contain liquid metals and polymers, but also find broad applications in mechanical framing operations like expulsion, wire drawing, sheet metal shaping, and the squeezing of metal and fired powders. Since the procedure chamber and kick the bucket steels can't withstand long hoisted temperature contact with liquid metals, bite the dust throwing is restricted transcendently to the low-liquefying point metals like zinc and aluminium amalgams. In the generally utilized hot chamber handle, shown in Figure 11.1, liquid metal is initially pressurized to anywhere in the range of 14 to 35 MPa (2-5 ksi). It is then infused into the clasped, split pass on shape, by means of a gooseneck, and held under weight until it cements. The "bite the dust" is then opened, the sofidified part catapulted and the cycle rehashed. Generation rates rely on the amalgam utilized and shape and weight of the throwing. Roughly 900 shots or infusions of zinc every hour are conceivable.

A variation of "bite the dust" throwing is the cool chamber handle. Higher-dissolving point metals like aluminium, magnesium, and some copper amalgams can be thrown along these lines by first spooning out metal into the shot chamber and afterwards quickly smashing it into the bite the dust depression. Weights utilized range from 20 to 70 MPa (3-10 ksi), however, can be significantly higher. Parts made by pass on throwing incorporate lawnmower motor and engine lodgings, carburettors, hand instrument lodgings and parts, machine segments, entryway handles, and zipper teeth. Most "kick the bucket" castings weigh somewhere around 0.1 and 25 kg.

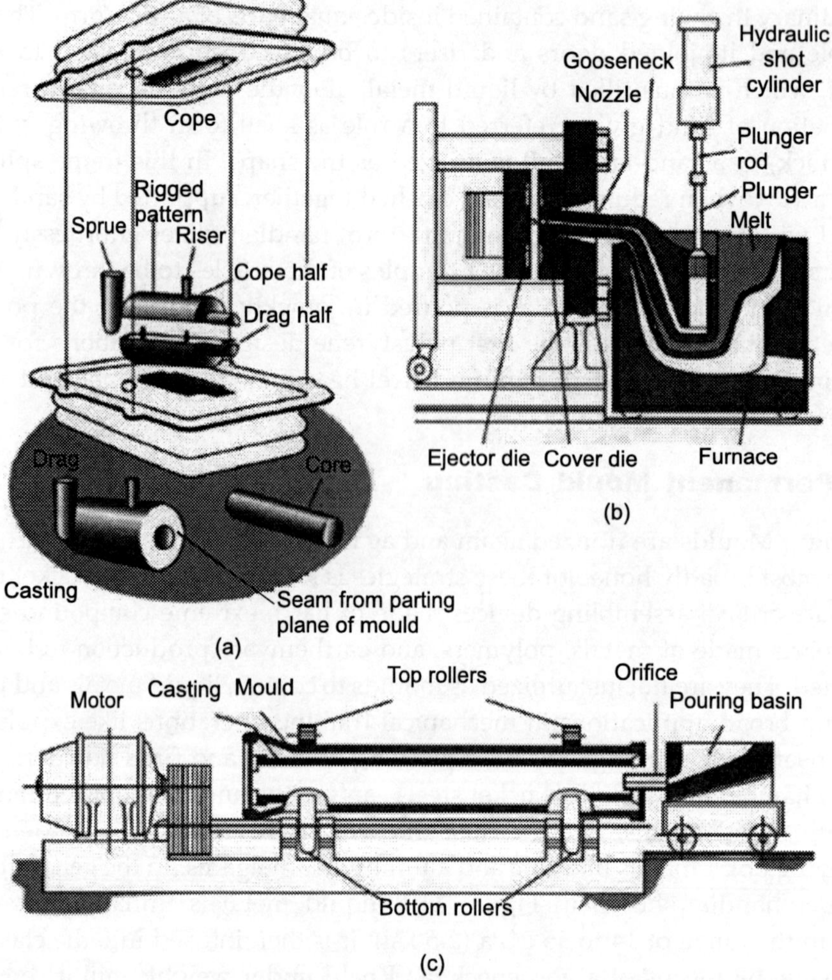

Figure 11.1 Casting methods. (a) Sand casting. Fronn H. J. Heine, *Casting Design and Applications* (Fall, 1989). (b) Hot chamber die casting. (c) Centrifugal casting of pipes.

Centrifugal Casting: As the name suggests, centrifugal casting involves rotation to increase the inertial force that drives molten metal to fill the mould cavities. By spinning permanent moulds at −1000 rpm, sounder pipes, cylinder liners, and hollow cylindrical parts of all metals can be cast while retaining good dimensional tolerances. Investment cast gold rings are often rotated about the vertical sprue axis during pouring to better fill intricately shaped moulds. Now that a number of expandable and permanent mould casting processes have been introduced, it is instructive to compare their relative merits. This is done in a self-explanatory way in Table 11.1.

Table 11.1 General characteristics of casting processes

Process	Metals cast	Weight (kg)		Dimensional accuracy*	Shape complexity*	Porosity*
		min	max			
Sand	All	0.05	No limit	3	1-2	4
Shell	All	0.05	100 +	2	2-3	4
Investment	All (High melting point)	0.005	100+	1	1	3
Permanent mold	All	0.5	300	1	3-4	2-3
Die	Nonferrous (Al, Mg, Zn, Cu)	< 0.05	50	1	3-4	1-2
Centrifugal	All	–	5000+	3	3-4	1-2

11.3 MECHANICAL FORMING OPERATIONS

Unlike the casting of liquids that readily flow to fill moulds by gravity, mechanical forming operations redistribute solid matter under the influence of applied forces. Examples of common bulk mechanical forming operations are schematically depicted in Figure 11.2. Included are forging, rolling, extrusion, and drawing processes that one associates with the primary forming of metals.

Mechanical framing of metals regularly supplies them with predominant mechanical properties. The customary course of metals preparing regularly incorporates ingot throwing after by either (hot) moving, expulsion, or manufacturing at hoisted temperatures. This causes a cast structure that is dendritic, artificially inhomogeneous, and loaded with shrinkage voids or porosity to be separated by the consolidated impacts of push and high-temperature dissemination. The earlier throwing of sensible quality yet low pliability is changed over into a much harder and homogeneous material. At that point the metal is much of the time moulded by low-temperature framing, for instance, frosty rolling, where the essential quality, malleability, and uniform grain structure are presented for definite assembling operations, for instance, profound shaping of sheet metal.

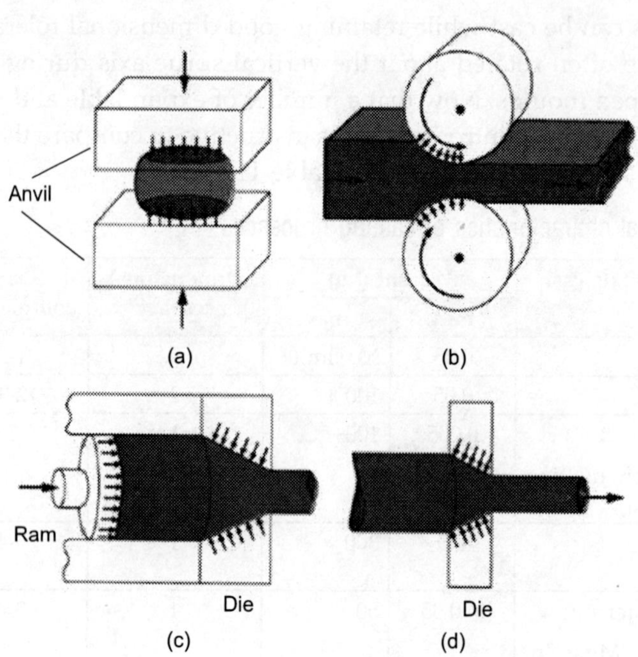

Figure 11.2 Mechanical forming operations.

11.3.1 Forging

Consider the forging of a rectangular block of metal compressed between a pair of flat, parallel anvils as shown in Figure 11.3. In practice, hammer-like impact loads are usually applied. If the yield stress is σ_o we wish to know the compressive load required to deform the block. The presence of friction at the work—anvil interface means that tangential shear stresses exist such that $\tau/\sigma y = f$, where, f is the coefficient of friction. It also means that compressive stresses in the x direction, generated by friction forces, are distributed throughout the workpiece. The effect is cumulative. At the edges $x = \pm a'$ the frictional forces are zero, but they build symmetrically to a maximum value at the centre of the forging. A so-called **friction hill,** which becomes higher and steeper with increasing f, develops. The larger f is, the greater is the applied force necessary to plastically deform the block. Undesirable barrel-shaped forgings are a manifestation of work-die friction.

$$\sigma_y = -\sigma_o \exp[2f(a - x)/h)]$$

Similarly, σ_o, the compressive stress in the x direction, also reflects symmetrical friction hill. The two stresses are related by mentioned equations and their variation with x is indicated in Figure 11.3b. Because the total area under the o-y stress distribution is proportional to the total forging load, every effort is made to reduce f and σ_o. And since elevated temperatures reduce the yield stress, forging and other bulk metal working operations such as rolling and extrusion are carried out hot.

Another simple formula can be used after a series expansions is as follows:

$$\sigma_y = -\sigma_o \exp\left[1 + 2f(a - x)/h\right]$$

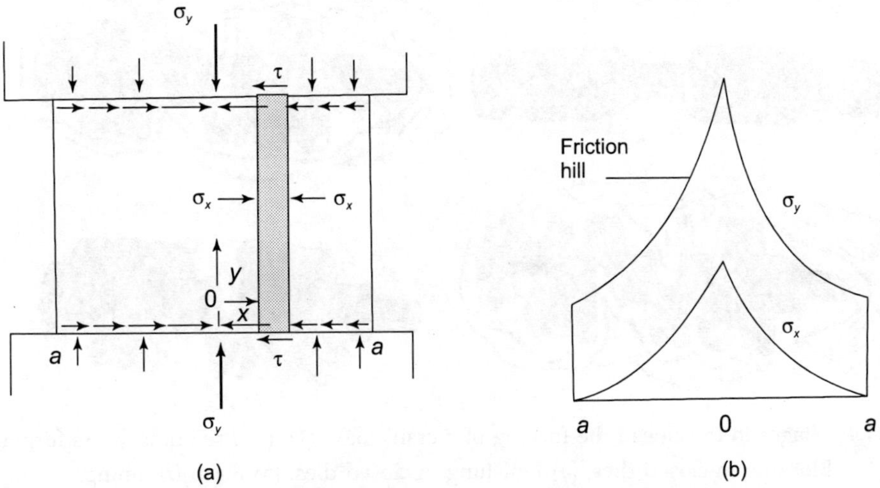

Figure 11.3 (a) Forging geometry in a rectangular block. Stresses acting on the interior shaded element are indicated. (b) Distribution of compressive stresses, σ_x and σ_y, versus distance.

11.3.2 Forging: Practices and Products

At the start a qualification amongst open and close die forging ought to be noted. Open die the bucket fashioning is the sort to which the past segment applies. Substantial shapes, for example, steam turbine rotors and military guns are manufactured between open bite the dust blacksmith's irons of straightforward shapes. A smithy's mallet and iron block would freely estimated open die forging; in any case, numerous extra completed items like vehicle crankshafts, torques, hand instruments, turbine circles, railroad wheels, and riggings have, at some time, been fashioned inside formed kicks the bucket that nearby around the metal inside. On account of the crankshaft shown in Figure 11.4, the steel pole is first generally manufactured to swell the metal in the right places.

At that point it is dynamically manufactured between half-way closed blocker die which shape it promote until finishing dies can be utilized. After a few strikes a dedicated impression of both pass on appearances is made. The crankshaft emerges but with a thin flare or flash of steel that spreads outwards from the die parting line. After the flash is ground away the finished forged shaft emerges. Since the flash is thin (h small), in the above equations indicates that the forging load rapidly mounts. Simultaneously, the flash is sufficiently strengthened so that it prevents metal in the closed die from extruding out. Friction promotes metal filling of the cavity, and without it closed die forging would not be feasible.

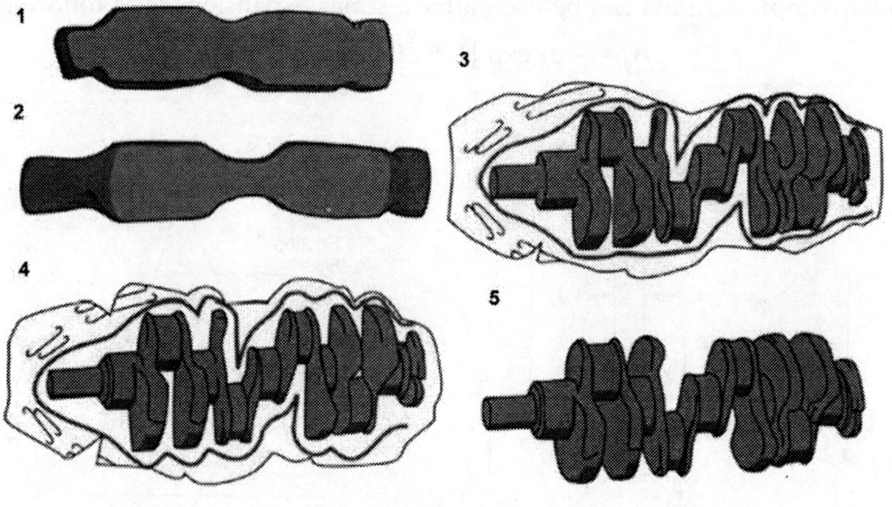

Figure 11.4 Stages in the closed die forging of a crankshaft. (1), (2) Preliminary roll forging. (3) Blocking in closed dies. (4) Finishing in closed dies. (5) Flash trimming.

11.3.3 Rolling

Rolling of metals, a popular mechanical forming process, results in a reduction of the initial thickness. Work in the form of slabs or sheets is fed in at the left and the thinned material exits the rolls at higher speed. If the process is frozen in time, rolling is not unlike forging. The difference is that an asymmetrical stress distribution (friction hill) arises in the deformed metal because of the more complicated curved roll (anvil) – work geometry. The sheer volume of steel that passes between rolling mills dwarfs, by far, that of other metals. Irrespective of metal, however, the process is carried out at hot, warm, and cold temperatures.

Hot refers to temperatures of > = $0.6\,T_M$ where T_M is the melting point in degrees kelvin. Recrystallization of the worked matrix can occur at such temperatures. Between $0.3\ T_M$ and $0.6\ T_M$ the warm working regime is operative. Cold rolling is usually done at temperatures lower than $0.3\ T_M$ and most frequently at room temperature. Normalizing to the melting point of the material means that temperatures for cold working of tungsten are comparable to those required for hot rolling steel. As the rolling temperature drops, the thickness of the stock that can be rolled must necessarily decrease. Also, thinner stock is rolled at higher roll velocities and reaches delivery speeds of over a mile a minute in the case of sheet steel or aluminium foil.

If the initial stock has an equiaxed grain structure, cold rolling will elongate the grains and string out the grain boundaries in a distinctive manner. The dislocation density grows enormously and the metal becomes harder. As shown in Figure 11.5, strength properties increase significantly, but simultaneously, ductility measured by percentage reduction in area suffers with increasing amount of cold work. Surprisingly, these microstructural and mechanical property trends can be reversed by annealing; a metal approximating the original one prior to rolling can be regenerated through heating to elevated temperatures.

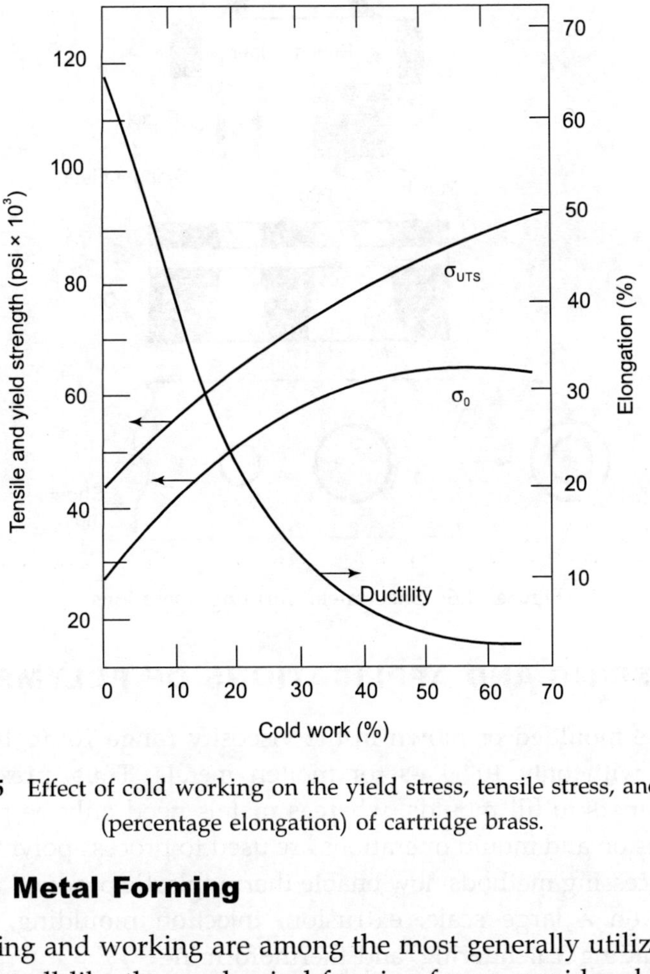

Figure 11.5 Effect of cold working on the yield stress, tensile stress, and ductility (percentage elongation) of cartridge brass.

11.3.4 Sheet Metal Forming

Sheet metal forming and working are among the most generally utilized procedures in assembling. Not at all like the mechanical framing forms considered above where the proportion of the volume to surface territory is expansive, the switch is valid in sheet shaping. Car bodies, flying machine outlines, real home apparatus lodgings, and building installations are a portion of the fields where expansive territory sheet items are utilized. Be that as it may, there are endless applications for littler sheet metal parts, for instance, washers, lage jars, kitchenware, and utensils. Punches and bites the dust are the essential moulding and cutting instruments utilized as a part of sheet framing. These instruments are mounted in presses that give the important powers to oblige workpieces that must be sheared, essentially bowed, bent, profoundly recessed, or inspired with an example in help (authoring). Two or three cases will delineate sheet shaping extremes. To start with consider the creation of washers, maybe the most broadly delivered sheet metal item.

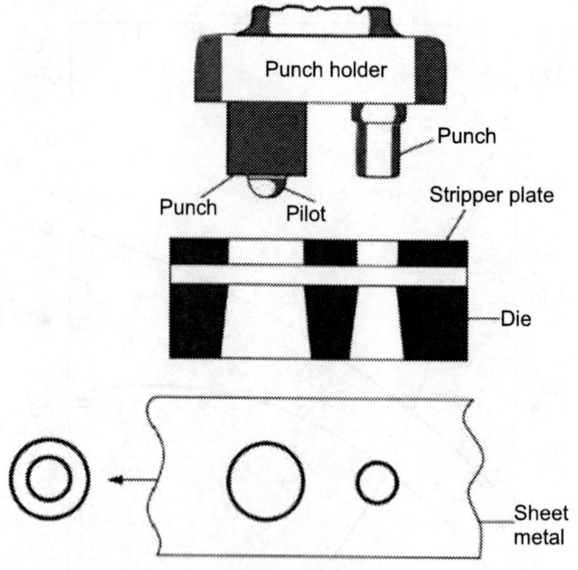

Figure 11.6 Sheet metal forming operations.

11.4 PROCESSING AND APPLICATIONS OF POLYMERS

Polymer melts are moulded or blown in the viscosity range 10^3 to 10^5 Pa-s (10^4-10^6 P), which compares with only 10^{-4} Pa-s for molten metals. Thus, pressure is generally required for polymers to fill moulds, whereas metals need only be poured to be cast. A variety of extrusion and mould operations are used to process polymers. In particular, the following processing methods now enable thermoplastic products of complex shape to be produced on a large scale: extrusion, injection moulding, blow moulding, compression moulding, calendering, and thermoforming.

11.4.1 Extrusion

Extrusion of polymers resembles extrusion of metals. Lengths of tubular, sheet, and rod products possessing both simple and complex cross sections emerge from dies after pushing polymer feedstock through them. The details differ because metals do not melt during extrusion. Batch processing (one billet of metal at a time), unlike continuous extrusion of polymers, is another difference. In this sense polymer extrusion also resembles continuous casting of metals. (In continuous casting, metal is continually poured from a large, replenishable melt reservoir into an open, chilled mould, where it solidifies. Later, downstream, the solidified moving billet is cut into convenient lengths.) Thermoplastic materials (resins) in pellet form are fed in at one end of the extruder as shown in Figure 11.7. Polymer feeds often contain additives, the most important of which are *lubricants,* which assist processing; *fillers and plasticizers* to modify mechanical properties; *flame retardants;* and *stabilizers* (antioxidants) to enhance degradation resistance. The polymer feed is conveyed down the extruder barrel by one (or more) long, rotating (and reciprocating) screw(s) where it is compressed. Through contact with the heated walls and the mechanical action of the screw, polymer **melting** occurs. In some systems polymer is pushed along by a piston. Next, melt **pumping** or **metering** under pressure through a tapered region and into a shaped die enables the product to exit. The extrudate must have a sufficiently high viscosity when it leaves the die to prevent it from deforming mechanically or even collapsing in an uncontrolled way. Therefore, water or air sprays are required to cool the product. The cooling rate must be controlled, however, because it can influence the crystallinity and also has some effect on mechanical properties. Many rollers can be produced by using these dies.

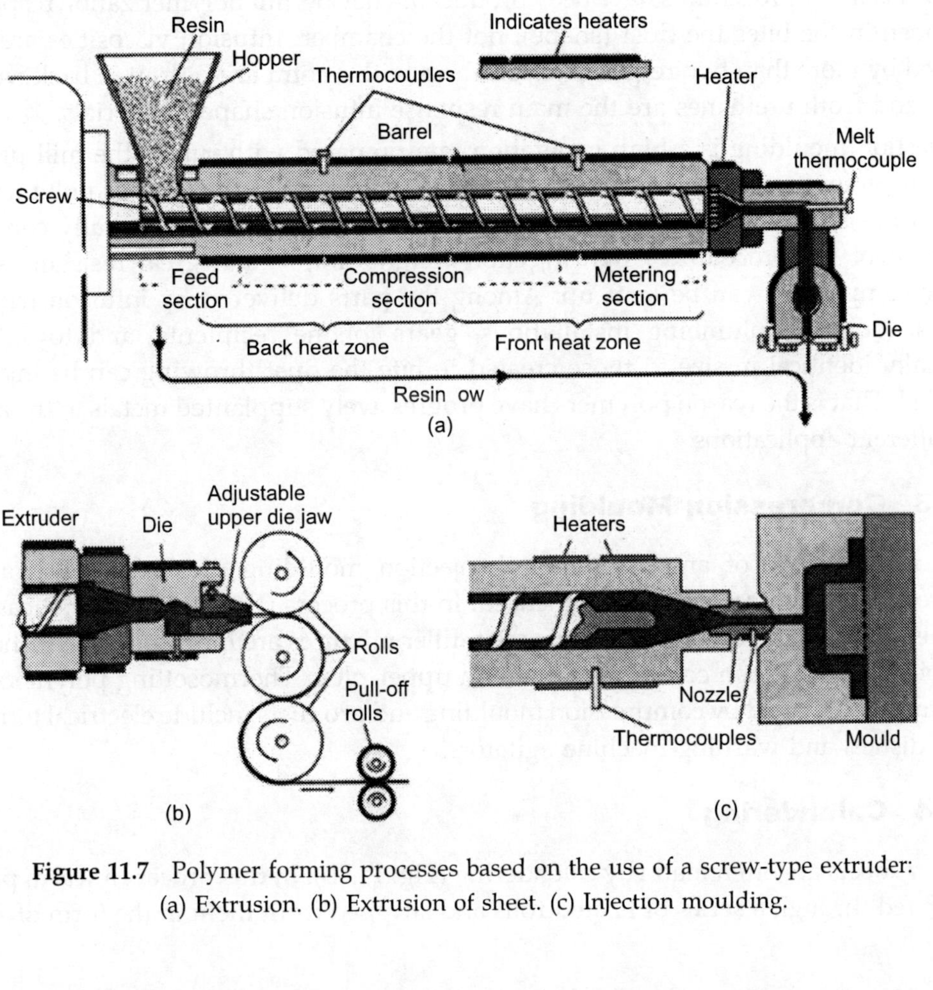

Figure 11.7 Polymer forming processes based on the use of a screw-type extruder: (a) Extrusion. (b) Extrusion of sheet. (c) Injection moulding.

11.4.2 Injection Moulding

To amplify the parallels with metal preparing, infusion embellishment or injection moulding of formed polymer parts nearly takes after "kick the bucket" throwing. The procedure shown in Figure 11.7c begins in an indistinguishable path from an expulsion. Rather than being constrained through a "bite the dust" into the unpressurized encompassing, the liquefy is infused under weight into a split bite the dust pit. Infusion trim is typically completed at 1.5 T_G, where T_G is the glass move temperature of the polymer. Weights in the form are commonly kept up at to some degree larger amounts than those utilized in kick the bucket throwing, and extend from 35 to 140 MPa (5-20 ksi). After the part cools underneath T_G (under weight) the bite the dust opens and the part is launched out. The chamber is basically a synthetic reactor that permits the blended and warmed fixings to mix, polymerize, and crosslink amid concurrent moulding. Strands and particles can likewise be presented, empowering the generation of composites. As the sizes of parts increment so do the extents of the infusion weight, and the bracing power expected to keep the bites the dust together. To address both issues *reaction injection moulding* procedures have been produced. They permit polymerization responses to happen in the bites the dust (shape), not the chamber. Infusion viscosities are then lessened by more than two requests of greatness with a comparing drop in back weight. Strong and froth urethanes are the main response infusion shaped materials.

Injection moulding is a high-generation rate prepared with run of the mill process durations going from seconds to a few minutes relying upon the kind of polymer and the part measure. Poor warmth exchange properties of polymers essentially constrain the speed of the procedure. High injection weights imply that great resistances and surface completions can be kept up. Among the parts delivered by infusion trim are holders, lodgings, plumbing installations, gears, phone recipients, and toys. Parts practically identical in size to those created in bite the dust throwing can be injection moulded. That is the reason polymers have progressively supplanted metals in the above and different applications.

11.4.3 Compression Moulding

Where the expense of an extruder and injection moulding dies is not warranted, compression moulding of parts is practised. In this process (Figure 11.8a) premeasured volumes of polymer powder or viscous resin-filler mixtures are introduced into a heated multipiece die and then compressed with an upper plug. Thermosetting polymers and elastomers are shaped by compression moulding and products include electrical terminal strips, dishes, and washing machine agitators.

11.4.4 Calendering

Polymer sheets are often made by calendering (Fig. 11.8b). In this process a warm plastic mass is fed through a series of heated rolls and stripped from them in the form of sheet.

Because thermoplastics have a high strain hardening coefficient they undergo large uniform deformations, making sheet production possible.

11.4.5 Thermoforming

Thermoforming, shown in Figure 11.8c, parallels methods used to form sheet metals. Sheet polymer (e.g., formed by calendering) is preheated and then laid over a mould having the desired shape. The applied air pressure is normally sufficient to make the sheet flow plastically and conformally cover the mould interior. Parts made this way include advertising signs, panels for shower stalls, appliance housings, and refrigerator linings.

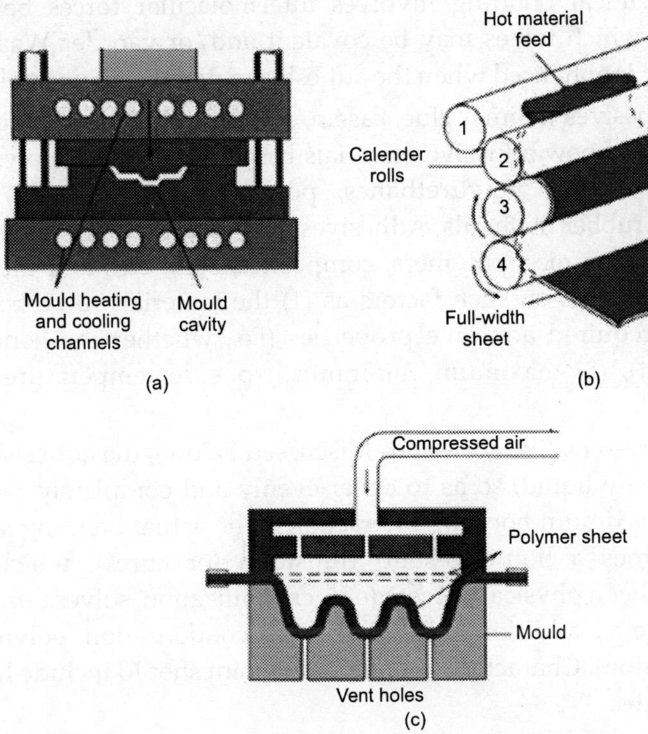

Figure 11.8 (a) Compression mould. (b) Calendering. (c) Thermoforming.

11.4.6 Applications of Polymers

Coatings: Coatings are frequently applied to the surfaces of materials to serve one or more of the following functions: (1) to protect the item from the environment that may produce corrosive or deteriorative reactions; (2) to improve the item's appearance; and (3) to provide electrical insulation. Many of the ingredients in coating materials are polymers, the majority of which are organic in origin. These organic coatings fall into several categories, as follows: paint, varnish, enamel, lacquer, and shellac. Many common coatings are latexes. A latex is a stable suspension of small insoluble polymer particles dispersed in water. These materials have become increasingly popular because they don't contain large quantities of organic solvents that are emitted into the environment, that is, they have low volatile organic compound (VOC) emissions. VOCs react in the atmosphere to produce smog. Large users of coatings such as automobile manufacturers continue to reduce their VOC emissions to comply with environmental regulations.

Adhesives: An **adhesive** is a substance used to bond together the surfaces of two solid materials (termed "adherends"). There are two types of bonding mechanisms: mechanical and chemical. For mechanical there is actual penetration of the adhesive into surface pores and crevices. Chemical bonding involves intermolecular forces between the adhesive and adherend, which forces may be covalent and/or van der Waals; degree of van der Waals bonding is enhanced when the adhesive material contains polar groups.

Although natural adhesives (animal glue, casein, starch, and resin) are still used for many applications, a host of new adhesive materials based on synthetic polymers have been developed; these include polyurethanes, polysiloxanes (silicones), epoxies, polyimides, acrylics, and rubber materials. Adhesives may be used to join a large variety of materials, viz., metals, ceramics, polymers, composites, skin, etc., and the choice of which adhesive to use depends on such factors as (1) the materials to be bonded and their porosities; (2) the required adhesive properties (i.e., whether the bond is to be temporary or permanent); (3) maximum/minimum exposure temperatures; and (4) processing conditions.

For all but the pressure-sensitive adhesives (discussed below), the adhesive material is applied as a low-viscosity liquid, so as to cover evenly and completely the adhered surfaces, and allow for maximum bonding interactions. The actual bonding joint forms as the adhesive undergoes a liquid-to-solid transition (or cures), which may be accomplished through either a physical process (e.g., crystallization, solvent evaporation) or a chemical process [e.g., addition polymerization, condensation polymerization (Section 15.20), vulcanization]. Characteristics of a sound joint should include high shear, peel, and fracture strengths.

Adhesive bonding offers some advantages over other joining technologies (e.g., riveting, bolting, and welding) including lighter weight, the ability to join dissimilar materials and thin components, better fatigue resistance, and lower manufacturing costs. Furthermore, it is the technology of choice when exact positioning of components as well as processing speed are essential. The chief drawback of adhesive joints is service temperature limitation; polymers maintain their mechanical integrity only at relatively low temperatures, and strength decreases rapidly with increasing temperature. The maximum temperature possible for continuous use for some of the newly-developed polymers is 300°C. Adhesive joints are found in a large number of applications especially in the aerospace, automotive, and construction industries, in packaging, and some household goods.

A special class of this group of materials is the pressure-sensitive adhesives (or self-adhesive materials), such as those found on self-stick tapes, labels, and postage stamps. These materials are designed to adhere to just about any surface by making contact and with the application of slight pressure. Unlike the adhesives described above, bonding action does not result from a physical transformation or a chemical reaction. Rather, these materials contain polymer tackifying resins; during detachment of the two bonding surfaces, small fibrils form that are attached to the surfaces and tend to hold them together. Polymers used for pressure-sensitive adhesives include the acrylics, styrenic block copolymers, and natural rubber.

Films: Polymeric materials have found widespread use in the form of thin films. Films having thicknesses between 0.025 and 0.125 mm (0.001 and 0.005 in) are fabricated and used extensively as bags for packaging food products and other merchandise, as textile products, and a host of other uses. Important characteristics of the materials produced and used as films include low density, a high degree of flexibility, high tensile and tear strengths, resistance to attack by moisture and other chemicals, and low permeability to some gases, especially water vapour. Some of the polymers that meet these criteria and are manufactured in film form are polyethylene, polypropylene, cellophane, and cellulose acetate.

Foams: **Foams** are plastic materials that contain a relatively high volume percentage of small pores and trapped gas bubbles. Both thermoplastic and thermosetting materials are used as foams; these include polyurethane, rubber, polystyrene, and poly (vinyl chloride). Foams are commonly used as cushions in automobiles and furniture as well as in packaging and thermal insulation. The foaming process is often carried out by incorporating into the batch of material a blowing agent that, upon heating, decomposes with the liberation of a gas. Gas bubbles are generated throughout the fluid mass, which remains in the solid upon cooling and give rise to a sponge-like structure. The same effect is produced by dissolving an inert gas into a molten polymer under high pressure. When the pressure is rapidly reduced, the gas comes out of solution and forms bubbles and pores that remain in the solid as it cools.

11.4.7 Additives for Polymers

Most of the properties of polymers discussed earlier in this chapter are intrinsic ones, that is, characteristic of or fundamental to the specific polymer. Some of these properties are related to and controlled by the molecular structure. Many times, however, it is necessary to modify the mechanical, chemical, and physical properties to a much greater degree than is possible by the simple alteration of this fundamental molecular structure. Foreign substances called *additives* are intentionally introduced to enhance or modify many of these properties, and thus render a polymer more serviceable. Typical additives include filler materials, plasticizers, stabilizers, colorants, and flame retardants.

Fillers: **Filler** materials are most often added to polymers to improve tensile and compressive strengths, abrasion resistance, toughness, dimensional and thermal stability, and other properties. Materials used as particulate fillers include wood flour (finely powdered sawdust), silica flour and sand, glass, clay, talc, limestone, and even some synthetic polymers. Particle sizes range all the way from 10 nm to macroscopic dimensions. Polymers that contain fillers may also be classified as composite materials. Often the fillers are inexpensive materials that replace some volume of the more expensive polymer, reducing the cost of the final product.

Plasticizers: The flexibility, ductility, and toughness of polymers may be improved with the aid of additives called plasticizers. Their presence also produces reductions in hardness and stiffness. Plasticizers are generally liquids having low vapour pressures and low molecular weights. The small plasticizer molecules occupy positions between the large polymer chains, effectively increasing the interchain distance with a reduction in the secondary intermolecular bonding. Plasticizers are commonly used in polymers that are intrinsically brittle at room temperature, such as poly(vinyl chloride) and some of the acetate copolymers. The plasticizer lowers the glass transition temperature, so that at ambient conditions the polymers may be used in applications requiring some degree of pliability and ductility. These applications include thin sheets or films, tubing, raincoats, and curtains.

Stabilizers: Some polymeric materials, under normal environmental conditions, are subject to rapid deterioration, generally in terms of mechanical integrity. Additives that counteract deteriorative processes are called **stabilizers.**

One common form of deterioration results from exposure to light [in particular ultraviolet (UV) radiation]. Ultraviolet radiation interacts with and causes a severance of some of the covalent bonds along the molecular chains, which may also result in some crosslinking. There are two primary approaches to UV stabilization. The first is to add a UV absorbent material, often as a thin layer at the surface. This essentially acts as a sunscreen and blocks the UV radiation before it can penetrate into and damage the polymer. The second approach is to add materials that react with the bonds broken by UV radiation before they can participate in other reactions that lead to additional polymer damage.

Another important type of deterioration is oxidation. It is a consequence of the chemical interaction between oxygen [as either diatomic oxygen (O_2) or ozone (O_3)] and the polymer molecules. Stabilizers that protect against oxidation either consume oxygen before it reaches the polymer, and/or they prevent the occurrence of oxidation reactions that would further damage the material.

Colourants: **Colourants** impart a specific colour to a polymer; they may be added in the form of dyes or pigments. The molecules in a dye actually dissolve in the polymer. Pigments are filler materials that do not dissolve, but remain as a separate phase;

normally they have a small particle size and a refractive index near to that of the parent polymer. Others may impart opacity as well as colour to the polymer.

Flame Retardants: The flammability of polymeric materials is a major concern, especially in the manufacture of textiles and children's toys. Most polymers are flammable in their pure form; exceptions include those containing significant contents of chlorine and/or fluorine, such as poly(vinyl chloride) and polytetrafluoroethylene. The flammability resistance of the remaining combustible polymers may be enhanced by additives called **flame retardants.** These retardants may function by interfering with the combustion process through the gas phase, or by initiating a different combustion reaction that generates less heat, thereby reducing the temperature; this causes a slowing or cessation of burning.

11.5 FORMING OF GLASS

As with all materials that have been worked for a long time, many methods have evolved and developed to form glass. The important ones discussed below include pressing, blowing, casting, rolling, and float moulding. These processing techniques are schematically depicted in Figure 11.9.

Glass is produced by heating the raw materials to an elevated temperature above which melting occurs. Most commercial glasses are of the silica-soda-lime variety; the silica is usually supplied as common quartz sand, whereas Na_2O and CaO are added as soda ash (Na_2CO_3) and limestone ($CaCO_3$). For most applications, especially when optical transparency is important, it is essential that the glass product is homogeneous and pore free. Homogeneity is achieved by complete melting and mixing of the raw ingredients. Porosity results from small gas bubbles that are produced; these must be absorbed into the melt or otherwise eliminated, which requires proper adjustment of the viscosity of the molten material.

Four different forming methods are used to fabricate glass products: pressing, blowing, drawing, and fibre forming. Pressing is used in the fabrication of relatively thick-walled pieces such as plates and dishes. The glass piece is formed by pressure application in a graphite-coated cast iron mould having the desired shape; the mould is ordinarily heated to ensure an even surface.

Although some glass blowing is done by hand, especially for art objects, the process has been completely automated for the production of glass jars, bottles, and light bulbs. From a raw gob of glass, a *parison,* or temporary shape, is formed by mechanical pressing in a mould. This piece is inserted into a finishing or blow mould and forced to conform to the mould contours by the pressure created from a blast of air.

Drawing is used to form long glass pieces such as sheet, rod, tubing, and fibres, which have a constant cross section. Flatness and the surface finish may be improved significantly by floating the sheet on a bath of molten tin at an elevated temperature;

the piece is slowly cooled and subsequently heat treated by annealing. Continuous glass fibres are formed in a rather sophisticated drawing operation. The molten glass is contained in a platinum heating chamber. Fibres are formed by drawing the molten glass through many small orifices at the chamber base. The glass viscosity, which is critical, is controlled by chamber and orifice temperatures.

11.5.1 Pressing

An example of pressing involves the forming of glassware by compressing it in a mould with a plunger. The process (Figure 11.9a) resembles closed die forging of metals and compression mould of polymers. Parts with diameters of 30 cm, depths of 10 cm, and weights of up to 15 kg are pressed using pressures of 0.69 MPa (100 psi).

11.5.2 Blowing

Hand blowing of glass is still practised was in antiquity. Glass is gathered on the end of an iron blowpipe and, with lung power or the use of compressed air, the glass blower shapes the "gathers." Artistic glassware is hand blown while the glass is continually rotated. The largest pieces reach weights of 15 kg, lengths of almost 2 m, and diameters of 1 m. Examples of machine-blown ware include glass containers and light bulb envelops. Gobs of glass are delivered to the blank mould (1, 2), where they are preformed by either blowing or pressing. After the blank is rotated, it is then blown in the blow mould (3, 4) to finish the operation.

11.5.3 Casting

In casting, glass is poured into or on moulds, tables, or rolls. One of the largest pieces ever cast is the Mount Palomar reflector, a cylinder measuring 5.08 m (200 in) in diameter and 0.457 m thick. Other products routinely cast are radiation-shielding window blocks and television and cathode-ray tubes. The latter are centrifugally cast, and as the molten gobs are spun the glass flows up to create a uniform wall thickness (Figure 11.9c). There is a 10^5-fold difference in viscosity between molten glasses and metals. The implication of this with respect to casting can be appreciated by comparing how honey and water pour.

11.5.4 Rolling and Float Moulding

Both rolling and float moulding are utilized to create reinforced glass. In persistent procedures, crude materials are encouraged towards one side of substantial level heaters (with limits of 1000 tonnes), where progressive softening and refining of glass happen. At the flip side the glass is bolstered into a couple of cooled rollers and the rising strip of strong glass is then passed on rollers through a strengthening lehr. Fortified glass about 4 m wide and 1 cm thick can be moved at rates more prominent than 6 m/min.

The surfaces of moved reinforced glass must be further ground and cleaned. The progressive Pilkington coast shaping procedure (Figure 11.9d) is generally used to make fortified glass. In this procedure delicate glass is drifted on the level surface of a liquid tin composite that controls the lower glass surface temperature. The upper surface of the glass is created by gravity. As the temperature is decreased the glass solidifies and is passed on to be toughened. This procedure produces bending free fortified glass practically identical in evenness to that delivered by moving glass. A downside of the procedure is consolidation of Sn into the glass.

11.6 PROCESSING OF CERAMICS

11.6.1 Forming Methods

With millennia of experience as background, many processes and variants of processes have been used to make vitreous (glass containing) ceramic products. They generally involve a number of common steps including preparing and mixing the raw ingredients (sands, clays) and binders (in a water base), forming or shaping the green body (e.g., on a potters wheel), drying it, firing the body to densify it, and finishing the product by grinding, cutting, polishing, and glazing. Shaping the green body is facilitated because the hydroplastic nature of the clay particles enables them to slide easily over one another. Upon firing, the glassy (vitreous) phase melts and coats the higher-melting clay particles, bonding them together. Most of the new ceramic material processes are similar in spirit, but often there is no glassy phase. They start with powders that are generally purer. Additives to the composition may include binders (for green strength), lubricants (to reduce friction and release the body from moulds), wetting and water retention agents, deflocculants (to control pH, electrostatic charge and particle dispersion), and sintering aids. Green bodies of both traditional and new ceramics are formed by a number of methods that are enumerated below.

Slip Casting: First, a slip consisting of a suspension of clay or ceramic powder in water is prepared. This is cast into an absorbent mould as indicated in Figure 11.10a. After sufficient water loss, the partially wet solid body is removed and dried further prior to firing. Vases as well as large and complex parts like Si_3N_4 turbine rotors have been reproduced this way.

Pressing: The exceptionally arranged dry powder is set in a kick the bucket and squeezed under high uniaxial powers to make a green conservative (Figure 11.10b). The procedure is particularly similar to squeezing of metal powders in powder metallurgy bites the dust. Just moderately little parts can be made along these lines. A change over applying uniaxial compressive burdens is to minimal the powder isostatically. Here the part is encased in an adaptable hermetically sealed elastic sack and inundated in a chamber loaded with pressure driven liquid. Hydrostatic (uniform over all headings)

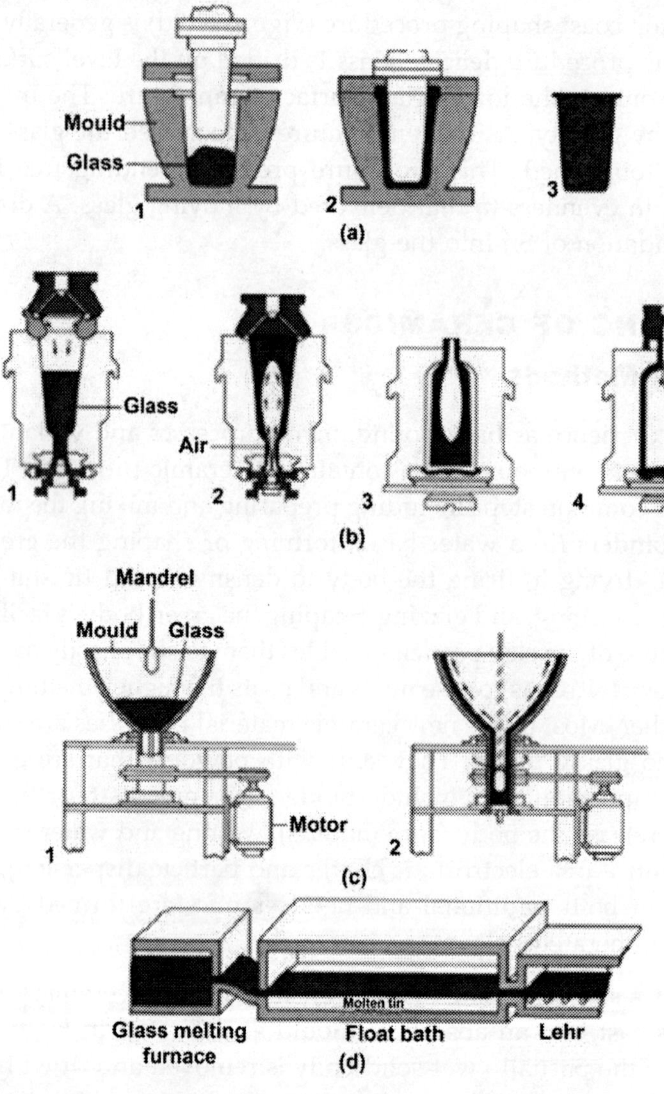

Figure 11.9 Processes to form and shape glass: (a) Pressing a glass container. (b) Blowing a bottle. (c) Centrifugal casting of a cathode-ray tube. (d) Plate glass manufacture by the Pilkington method.

weight is connected amid (cool) compaction, dispensing with the intricate weight dissemination all through the pass on because of shape divider rubbing. Electrical protectors, bushings, attractive parts, and start connect bodies are delivered to along these lines for resulting sintering.

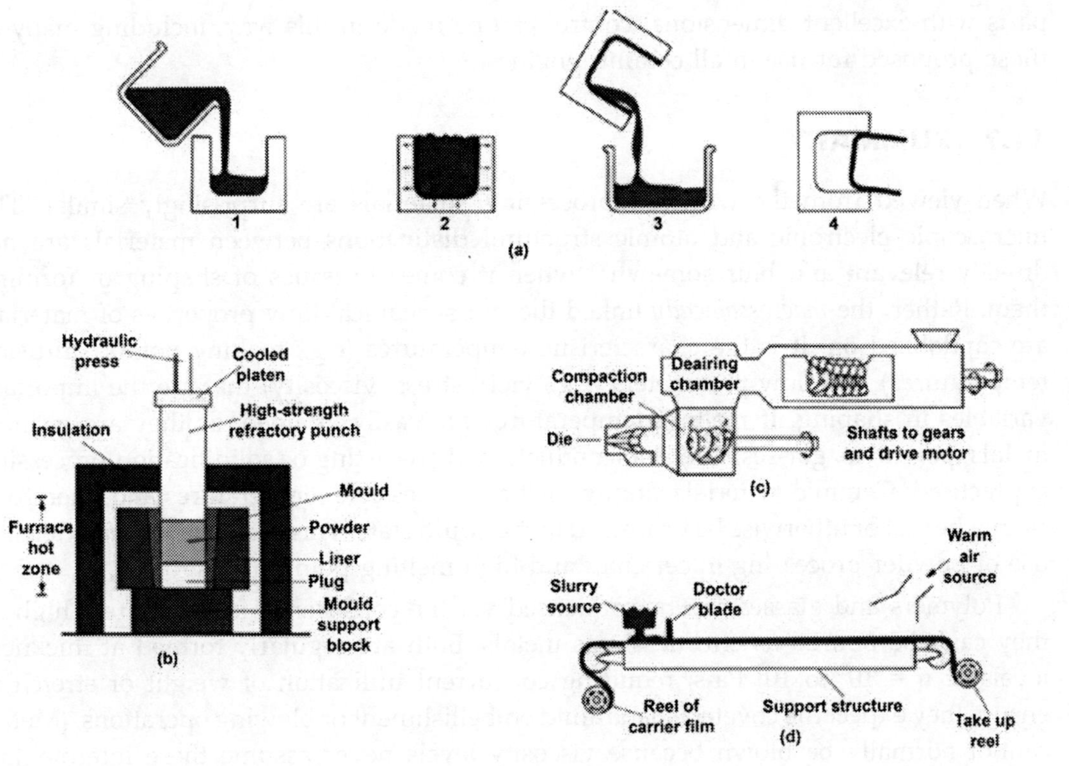

Figure 11.10 Several techniques used to process ceramics. (a) Slip or drain casting: (1) Pouring the clay/ceramic suspension. (2) Water absorption by plaster mould until a solid shell forms. (3) Pouring out the excess suspension. (4) Removal of body for firing. (b) Pressing. (c) Extrusion. (d) Tape forming.

Extrusion: Extrusion of green bodies parallels the extrusion of polymers in that powder and screws (augers) are involved. Polymers, however, are extruded in the molten state whereas ceramic green body extrusions are created at room temperatures from pliable, powder-binder mixtures. This material is extruded through a die as shown schematically in Figure 11.10c, and then sintered. Long tubes, rods, and honeycomb structures used to shield thermocouples or for heat exchanger applications are made in this way.

Tape Forming: Tape forming, shown in Figure 11.10d, is used to make thin ceramic parts like alumina substrates for integrated circuit chips and special capacitors. A thin layer of slip is laid on a flat carrier which can be a paper sheet or polymer film. A doctor blade controls the slip thickness, resulting in a tape that can be stamped into small shapes. These can be stacked or coiled for subsequent sintering.

Injection Moulding: In injection moulding, a hybrid process, ceramic powder is mixed with thermosetting (organic) polymer binder and injected under pressure into a die or mould. The bodies are then ejected and sintered while new ones are replicated. Complex parts with excellent dimensional control can be made in this way, including many of those proposed for use in all ceramic engines.

11.7 SUMMARY

When viewed from the world of processing, materials are surprisingly similar. The microscopic electronic and atomic structural distinctions between materials are not directly relevant and blur somewhat when it comes to issues of shaping or forming them. Rather, the *macroscopically* linked thermomechanical flow properties of materials are capitalized on. It is the characteristic temperatures (e.g., melting points, diffusion temperatures) and flow parameters (e.g., yield stress, viscosity) that are the important variables in shaping. If melting temperatures are easily attained, as they are in most metals, polymers, glasses, and semiconductors, then casting or solidification processing is practised. Ceramic materials simply melt at inaccessible temperatures and, therefore, cannot be cast or otherwise be processed in the liquid state. This accounts for the extensive use of powder processing in ceramics and high-melting point metals.

Polymers and glasses can be condensed yet the consistency is moderately high so they can't be poured as effortlessly as metals. Both are regularly formed at thickness levels of $\eta = 10^3$ to 10^5 Pa-s, requiring concurrent utilization of weight or stretch to ensure they expect the coveted shape amid embellishment or blowing operations. (Metals cannot normally be blown because viscosity levels never assume these intermediate levels.) The main reason that polymers appreciate handling focal points with respect to both metals and glasses is the low temperatures. For instance, plastic jugs are ordinarily blown at ~100°C contrasted and ~1000°C for glass bottles. What's more, polymer framing burdens are regularly lower than those for metals. Having accentuated refinements amongst polymers and metals, it is not out of question that we bring up similitudes, for instance, between injection moulding and die casting, between compression moulding and closed die forging, and expulsion in both materials.

The final sections of this chapter treated synthesis and fabrication techniques for polymeric materials. Synthesis of large molecular weight polymers is attained by polymerization, of which there are two types: addition and condensation. The properties of polymers may be further modified by using additives; these include fillers, plasticizers, stabilizers, colourants, and flame retardants.

Ceramics production dependably requires a two-stage handling treatment: the first to frame or shape them into a green item or minimal, and the second to flame or sinter them. The trouble of these means relies on the earthenware being referred to. Customary ceramics are by and large less demanding to handle than the new earthenware production in light of the fact that the low-dissolving shiny stage ties and combine particles on terminating. For new ceramics, hot pressing and sintering routes must often be followed in the quest to achieve fully dense products. High melting points and compressive strengths, coupled with low ductility, limit conventional deformation processing of ceramics.

Clay is the principal component of the whiteware and structural clay products. Other ingredients may be added, such as feldspar and quartz, which influence the changes that occur during firing. Many of our modern technologies utilize and will continue to utilize advanced ceramics because of their unique mechanical, chemical, electrical, magnetic, and optical properties and property combinations. The following advanced ceramic materials were discussed briefly: piezoelectric ceramics, microelectromechanical systems (MEMS), and ceramic ball bearings.

Since glasses are formed at elevated temperatures, the temperature–viscosity behaviour is an important consideration. Melting, working, softening, annealing, and strain points represent temperatures that correspond to specific viscosity values. Knowledge of these points is important in the fabrication and processing of a glass of given composition. Four of the more common glass-forming techniques—pressing, blowing, drawing, and fibre forming—were discussed briefly. After fabrication, glasses may be annealed and/or tempered to improve mechanical characteristics.

EXERCISE

1. Metal castings of a sphere and a cube of the same metal have the same surface area. What is the ratio of the solidification times for the sphere and the cube?
2. Castings of a sphere and a cube of the same metal have the same volume. What is the ratio of the solidification times for the sphere and the cube?
3. Design the dimensions of a cylindrical riser with height twice the diameter that would be suitable to cast the bowling pin. Assume the riser solidification time is 25% longer than that of the casting. The pin has a volume of 1000 cm^2 and with surface area of 700 cm^2.
4. Find the extrusion ratio if a material is reduced in area by 50% by extrusion.
5. List some possible ways of producing different powder sizes during atomization processes in metals.
6. Polymers not basically shaped into parts by the powder metallurgy processes employed for metals and ceramics. Why?
7. What pressure is required to extrude a bar of stainless steel at 1000°C from an initial diameter of 10 cm to 2.5 cm?
8. Why are thermosetting polymers suitable for injection moulding processes?
9. Write about any two of the traditional ceramics.
10. Provide the reason behind hardness of fused silica more than that of shape of soda lime glass.
11. Ceramics undergo larger shrinkage than powdered parts. Why?

12. Distinguish the applications. Porosity in ceramics is usually undesirable but there are applications where they are desired.
13. Thick ceramicware more prone to cracking than the thin ones during its processing. Explain why?
14. Briefly explain how each of the following influences the tensile modulus of a semicrystalline polymer and why:
    (a) Molecular weight
    (b) Degree of crystallinity
    (c) Deformation by drawing
15. What is the distinction between dye and pigment colourants?
16. Write a short note on the polymer additives. What is the significance of using additives?
17. Cite four factors that determine which fabrication technique is used to form polymeric materials.
18. Contrast compression, injection, and transfer moulding techniques that are used to form plastic materials.
19. Cite the two desirable characteristics of glasses.
20. What is the distinction between glass transition temperature and melting temperature?
21. Briefly describe what happens when a glass piece is thermally tempered.
22. Explain why residual thermal stresses are introduced into a glass piece when it is cooled.
23. Are thermal stresses introduced upon heating? Why or why not?
24. How does the processing of polymers differ from that of ceramics?
25. Briefly explain the processing of glass/glass forming methodology.

12
Engineering Materials

Learning Objectives

After going through this chapter the reader will be able to:

- have an idea about some of the engineering materials; and
- understand aspects associated with alloys, ferrous materials, dielectric materials, ferromagnetic materials, nanomaterials, shape memory alloys, smart materials, etc.

12.1 INTRODUCTION

Engineering materials play a vital role in many engineering applications. This chapter deals with the most useful engineering materials like alloys, ferrous materials, dielectric materials, ferromagnetic materials, nanomaterials, shape memory alloys, smart materials, etc. Classification of some engineering materials and their importance and applications are also elaborated.

12.2 ALLOYS

Metallic materials prepared by mixing two or more molten metals are called alloys. They are used for many purposes, such as construction, and are central to the transportation and electrical industries. Some common alloys are given in Table 12.1.

Table 12.1 Composition of typical alloys

Alloy	Mass percentage composition
Brass	up to 40% zinc in copper
Bronze	a metal other than zinc or nickel in copper (casting bronze: 10% Sn and 5% Pb)
Cupronickel	nickel in copper (coinage cupronickel: 25% Ni)
Pewter	6% antimony and 1, 5% copper in tin
Solder	tin and lead
Stainless steel	more than 12% chromium in iron

In homogeneous amalgams or alloys, atoms of the distinctive components are dispersed consistently. Cases incorporate metal, bronze, and the coinage alloys. Heterogeneous combinations or alloys, for example, tin-lead weld and the mercury alloys at times used to fill teeth, comprise a blend of crystalline stages with various sytheses. The structures of compounds are more muddled than those of unadulterated metals since they are worked from particles of at least two components with various nuclear radii. Since the metallic radii of the group components are all comparative, they can shape a broad scope of amalgams with each other with little contortion of the first precious stone structure, e.g., copper-zinc amalgam utilized for some "copper" coins. Since zinc atoms are nearly of same size as copper atoms and have similar electronic properties (they belong to neighbouring groups), they can take the place of some of the copper atoms in the crystal. An alloy in which atoms of one metal are substituted for atoms of another metal is called a substitutional alloys.

Elemental components that can shape substitutional alloys have particles with nuclear radii that contrast by close to around 15%. Since there are slight contrasts in size and electronic structure, the less inexhaustible particles in substitutional compound mutilate the state of the cross section of the more bounteous atoms of the host metal and obstruct the stream of electrons. Since the cross section is misshaped, it is harder for one plane of particles to slip past another. In this way, in spite of the fact that a substitutional alloy has brought down electrical and heat conductivity lower than immaculate component, it is harder and more grounded.

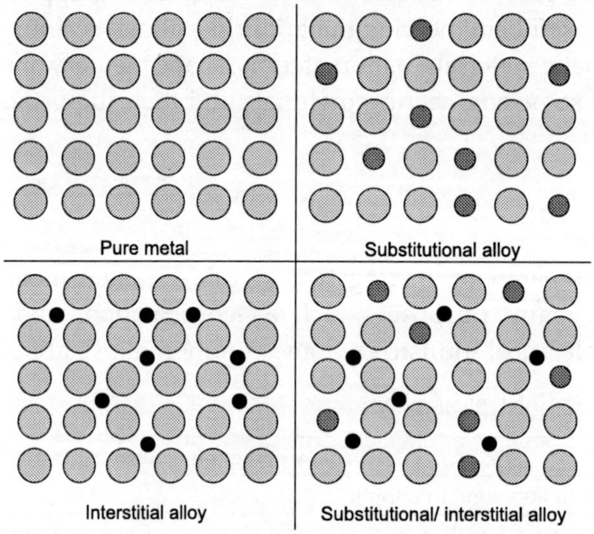

Figure 12.1 Alloys.

Steel is a combination of around 2% or less carbon in iron. Carbon atoms are much less than iron particles, thus they can't substitute for iron in the precious stone grid. In reality, they are small to the point that they can fit into the interstices (the openings) in the iron cross section. The subsequent material is called an interstitial amalgam. For two components to shape an interstitial compound, the nuclear sweep of the solute component must be not be exactly around 60% of the nuclear range of the host metal. The interstitial particles meddle with the electrical conductivity and with the development of the atoms shaping the cross section. This limited movement makes the composite harder and more grounded than the unadulterated host metal would be.

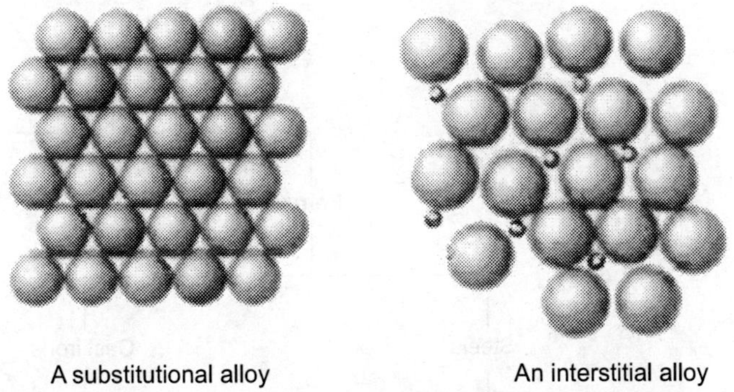

Figure 12.2 Substitutional and interstitial alloys.

Some alloys are gentler than the segment metal. The nearness of enormous bismuth molecules relaxes a metal and lowers its softening point, much as melons would destabilize a pile of oranges since they simply don't fit together well. A low-dissolving point combination of lead, tin, and bismuth is utilized to control water sprinklers utilized as a part of certain fire-stifling frameworks. The warmth of the fire liquefies the composites, which initiates the sprinklers before the fire can spread.

Alloys of metals have a tendency to be more grounded and bring down electrical conductivity more than the unadulterated metals. In substitutional alloys, molecules of the solute metal replace a few particles of metal of comparative nuclear span. In interstitial composites, atoms of the solute component fit into the interstices in a cross section framed by molecules of a metal with a bigger nuclear range. On account of combination hardening, the information produced by heat examination are exceptionally helpful for a subjective or quantitative portrayal of the dynamic cementing and solidification systems.

12.3 FERROUS MATERIALS

Ferrous alloys, those of which iron is the prime constituent, are created in bigger amounts than whatever other metal sort. They are particularly critical as building development materials. Their wide utilization is due to three variables: (1) press containing mixes exist in copious amounts inside the earth's outside layer; (2) metallic iron and steel composites might be delivered utilizing moderately prudent extraction, refining, alloying, and creation strategies; and (3) ferrous combinations are to a great degree flexible, in that they might be custom-made to have an extensive variety of mechanical and physical properties. The chief disservice of numerous ferrous composites is their powerlessness to erosion. This area talks about pieces, microstructures, and properties of various distinctive classes of steels and cast irons.

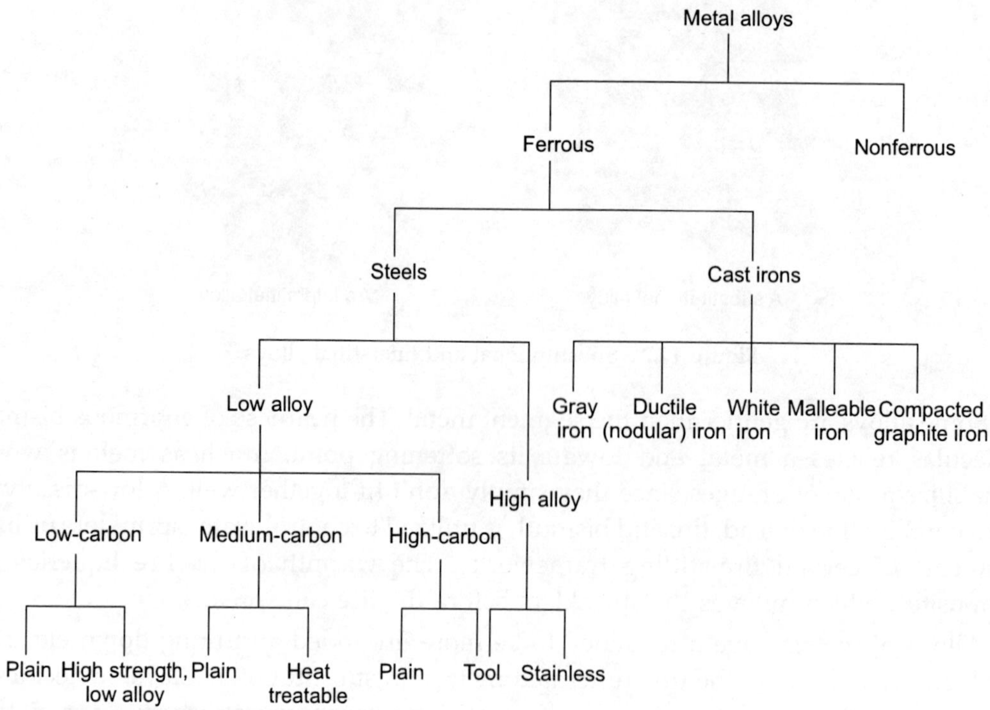

Figure 12.3 Classification of metal alloys.

Steels are iron-carbon combination alloys that may contain calculable centralizations of other alloying components; there are a great many compounds that have diverse organizations as well as warmth medications and heat treatments. The mechanical properties are touchy to the substance of carbon, which is typically under 1.0 wt%. A portion of the more basic steels are characterized by focus—to be specific, into low-, medium-, and high carbon sorts. Subclasses likewise exist inside every gathering as per the grouping of other alloying components. Plain carbon steels contain just lingering groupings of debasements other than carbon and a little manganese. For compound steels, additional alloying components are deliberately included.

The medium-carbon steels have carbon focuses between around 0.25 and 0.60 wt%. These combinations might be warmth treated by austenitizing, extinguishing, and after that hardening to enhance their mechanical properties. They are frequently used in the tempered condition, having microstructures of tempered martensite. The plain medium-carbon steels have low hardenabilities and can be effectively warm treated just in thin areas and with exceptionally quick extinguishing rates. Increases of chromium, nickel, and molybdenum enhance the limit of these amalgams to be warmth treated offering ascend to an assortment of strength-ductility blends. These warmth treated amalgams are more grounded than the low-carbon steels. Applications incorporate railroad haggles, riggings, crankshafts, and other machine parts and high-quality auxiliary segments requiring a blend of high quality, wear resistance, and durability.

Stainless steels are highly resistant to corrosion (rusting) in a variety of environments, especially the ambient atmosphere. Their predominant alloying element is chromium; a concentration of at least 11 wt%. Corrosion resistance may also be enhanced by nickel and molybdenum addition. Stainless steels are divided into three classes on the basis of the predominant phase constituent of the microstructure—martensitic, ferritic, or austenitic. Table 11.4 lists several stainless steels, by class, along with composition, typical mechanical properties, and applications. A wide range of mechanical properties combined with excellent resistance to corrosion make stainless steels versatile in their applicability.

Martensitic stainless steels are fit for being warmth treated in a manner that martensite is the prime microconstituent. Increase of alloying components in huge fixations creates sensational adjustments in the iron-iron carbide stage chart. For austenitic stainless steels, the austenite (or) stage field is reached out to room temperature. Ferritic stainless steels are made out of the ferrite (BCC) phase. Austenitic and ferritic stainless steels are solidified and reinforced by icy work since they are not warm treatable. The austenitic stainless steels are the most consumption safe on account of the high chromium substance furthermore the nickel augmentations; and they are delivered in the biggest amounts. Both martensitic and ferritic stainless steels are attractive.

Nonexclusively, cast irons are a class of ferrous composites with carbon substance above 2.14 wt%; practically speaking, most cast irons contain somewhere around 3.0 and 4.5 wt% of carbon. A re-evaluation of the iron-iron carbide stage graph (Figure 9.24) uncovers that compounds inside this structure turn out to be totally fluid at temperatures between 1150 and 1300°C, which is extensively lower than for steels. Along these lines, they are effortlessly softened and amiable to blowing. Moreover, some cast irons are extremely weak, and blowing is the most advantageous creation or fabrication method.

12.4 DIELECTRIC MATERIALS

Various ceramics and polymers are used as protectors. Large number of the ceramics, including glass, porcelain, steatite, and mica, have dielectric constants inside the scope of 6 to 10. These materials show a high level of dimensional solidness and mechanical quality. Average applications incorporate electrical cable and electrical protection, switch bases, and light containers. The titania (TiO_2) and titanate pottery, for example, barium titanate ($BaTiO_3$), can be made to have to a great degree high dielectric constants, which render them particularly valuable for some capacitor applications.

Piezoelectric materials are used in transducers, which are gadgets that change over electrical vitality into mechanical strains, or the other way around. Some other well-known applications that utilize piezoelectrics include phonograph cartridges, amplifiers, speakers, perceptible cautions, and ultrasonic imaging. In a phonograph cartridge, as the stylus crosses the depressions on a record, a weight variety is forced on a piezoelectric material situated in the cartridge, which is then changed into an electric flag and is opened up before setting off to the speaker.

12.5 FERROMAGNETIC MATERIALS

All materials can be grouped, as far as their magnetic property is concerned, into one of five classes depending on their attractive power. The two most normal sorts of attraction are diamagnetism and paramagnetism, which represent the attractive magnetic properties of the majority of the usual class of components and elements at room temperature.

The two most common types of magnetism are diamagnetism and paramagnetism, which account for most of the periodic table of elements at room temperature. These elements are usually referred to as non-magnetic, whereas those which are referred to as magnetic are actually classified as ferromagnetic. The only other type of magnetism observed in pure elements at room temperature is antiferromagnetism. Finally, magnetic materials can also be classified as ferrimagnetic although this is not observed in any pure element but can only be found in compounds, such as the mixed oxides, known as ferrites, from which ferrimagnetism derives its name.

Ferromagnetism is conceivable when atoms are masterminded in a grid and the nuclear attractive minutes can associate to adjust parallel to each other. This impact is clarified in established hypothesis by the nearness of a subatomic field inside the ferromagnetic material, which was initially hypothesized by Weiss in 1907. This field is adequate to charge the material to immersion. In quantum mechanics, the Heisenberg model of ferromagnetism depicts the parallel arrangement of attractive minutes as far as a trade cooperation between neighbouring minutes. Weiss hypothesized the nearness of attractive areas inside the material, which are districts where the nuclear attractive minutes are adjusted. The development of these areas decides how the material reacts to an attractive field and as a result a component of connected attractive field is powerless. Along these lines, ferromagnetic materials are generally thought about as far as immersion (magnetisation when all spaces are adjusted) as opposed to susceptibility. In the intermittent table of components just Fe, Co and Ni are ferromagnetic at or more space temperature. When ferromagnetic materials are warmed, the warm fomentation of the molecules implies that the level of arrangement of the nuclear attractive minutes diminishes and henceforth the immersion magnetisation additionally diminishes. In the end the warm disturbance turns out to be great to the point that the material gets to be paramagnetic; the temperature of this move is the Curie temperature, T_C (Fe: T_C = 770°C, Co: T_C = 1131°C and Ni: T_C = 358°C). Above T_C the susceptibility fluctuates as indicated by the Curie-Weiss law.

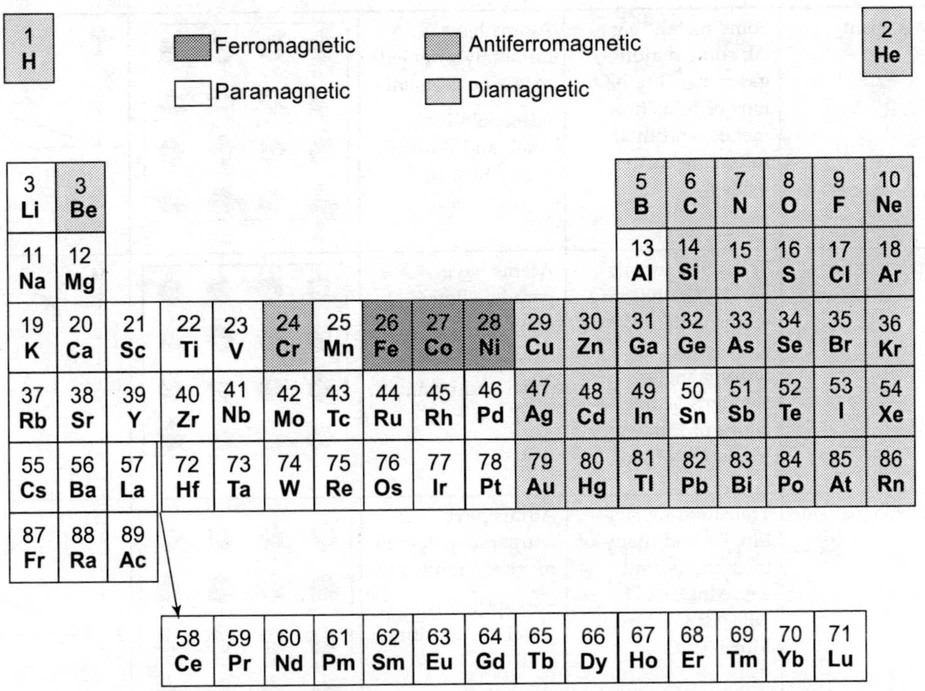

Figure 12.4 Ferromagnetic materials.

In the intermittent periodic table the main component displaying antiferromagnetism at room temperature is chromium. Antiferromagnetic materials are fundamentally the same as ferromagnetic materials, however, the trade connection between neighbouring particles prompts the counter parallel arrangement of the nuclear attractive minutes. In this manner, the attractive field counteracts and the material seems to carry on in an indistinguishable path from a paramagnetic material. Like ferromagnetic materials these materials get to be paramagnetic over a move temperature, known as the Néel temperature.

Type	Example	Atomic/Magnetic Behaviour		
Diamagnetism	Inert gases; many metals, e.g., Au, Cu, Hg; non-metallic elements, e.g., B, Si, P, S; many ions, e.g., Na^+, Cl^- & their salts; diatomic molecules, e.g. H_2, N_2–, H_2O; most organic compounds	Atoms have no magnetic moment. Susceptibility is small and negative, -10^{-8} to -10^{-5}	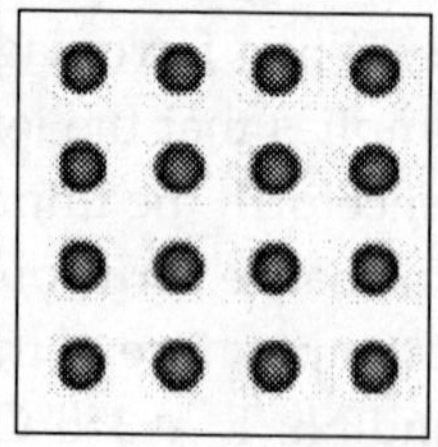	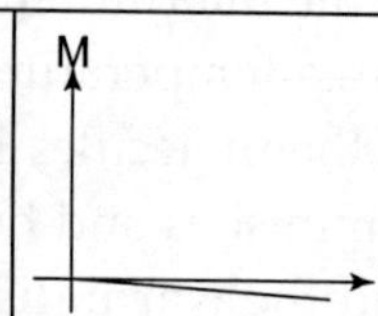
Paramagnetism	Some metals, e.g., Al; some diatomic gases, e.g., O_2, NO; ions of transition metals, and their salts; rare earth oxides.	Atoms have randomly oriented magnetic moments. Susceptibility is small and positive, $+10^{-5}$ to $+10^{-3}$	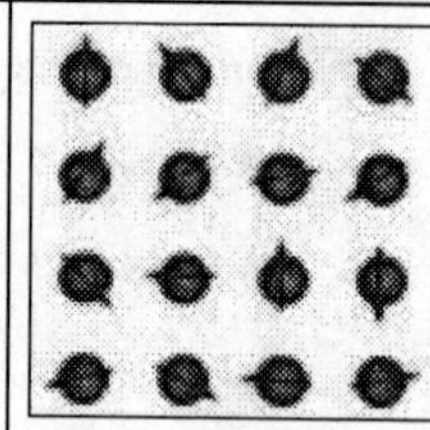	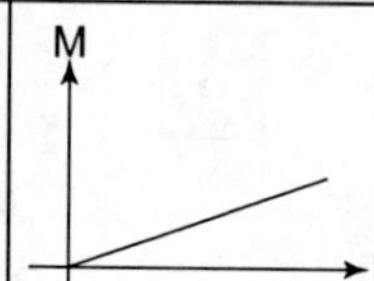
Ferromagnetism	Transition metals Fe, H, Co, Ni; rare earths with 64<Z<69; alloys of ferromagnetic elements; some alloys of Mn, e.g., MnBi, Cu_2MnAl.	Atoms have parallel aligned magnetic moments. Susceptibility is small and positive, $+10^{-5}$ to $+10^{-3}$	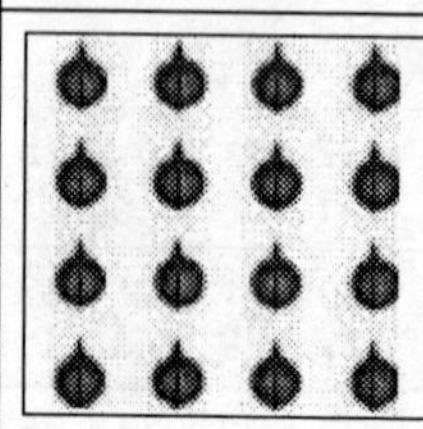	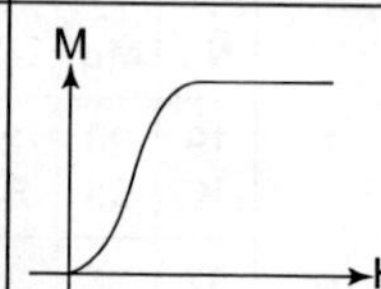
Antiferromagnetism	Transition metals Mn, Cr and many of their compound, e.g., MnO, CoO, NiO, Cr_2O_3, MnS, MnSe, CuC_{12}.	Atoms have antiparallel aligned magnetic moments. Susceptility is small and positive, $+10^{-5}$ to $+10^{-3}$	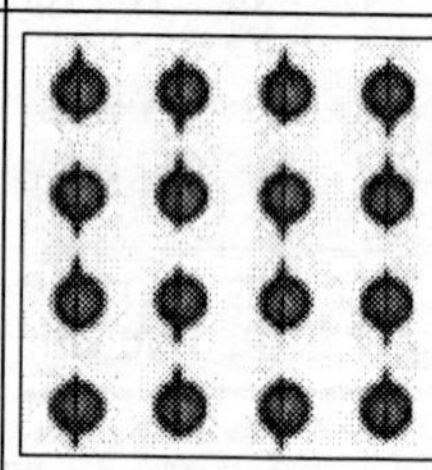	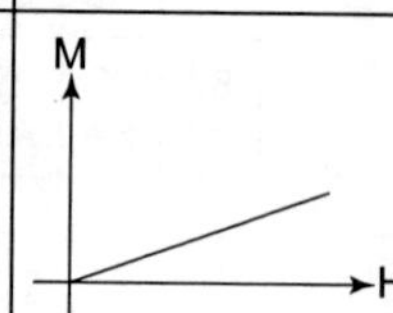
Ferrimagnetism	Fe_3O_4 (magnetite); γ-Fe_2O_3 (magnemite); mixed oxides of iron and other elements such as Sr ferrite.	Atoms have mixed parallel and anti-parallel aligned magnetic moments. Susceptibility is large (below T_c)	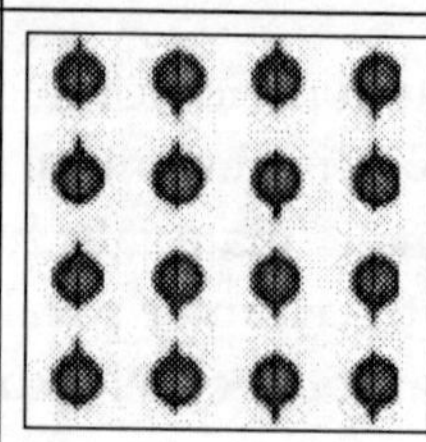	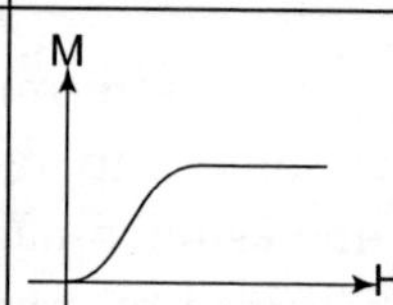

Figure 12.5 Atomic behaviour of magnetic materials.

Ferrimagnetism is seen in mixes and compounds, which have more perplexing precious crystal structures than immaculate components. Inside these materials the trade communications prompt parallel arrangement of molecules in a portion of the precious stone destinations and hostile to parallel arrangement of others. The material separates into attractive areas, much the same as a ferromagnetic material and the attractive conduct is likewise fundamentally the same as, despite the fact that ferrimagnetic materials for the most part have brought down immersion magnetisations. For instance in Barium ferrite ($BaO \cdot 6Fe_2O_3$) the unit cell contains 64 particles of which the barium and oxygen particles have no attractive minute, 16 Fe^{3+} particles have minutes adjusted parallel and 8 Fe^{3+} adjusted antiparallel giving a net magnetisation parallel to the connected field, yet with a moderately low extent as just 1/8 of the particles add to the magnetisation of the materials.

12.6 NANOMATERIALS

Nanomaterials are cornerstones of nanoscience and nanotechnology. Nanostructure science and technology is a broad and interdisciplinary area of research and development activity that has been growing explosively worldwide in the past few years. It has the potential for revolutionizing the ways in which materials and products are created and the range and nature of functionalities that can be accessed. It is already having a significant commercial impact, which will assuredly increase in the future.

Nanoscale materials are defined as a set of substances where at least one dimension is less than approximately 100 nanometres. A nanometre is one millionth of a millimetre – approximately 100,000 times smaller than the diameter of a human hair. Nanomaterials are of interest because at this scale unique optical, magnetic, electrical, and other properties emerge. These emergent properties have the potential for great impacts in electronics, medicine, and other fields.

Engineered nanomaterials are resources designed at the molecular (nanometre) level to take advantage of their small size and novel properties which are generally not seen in their conventional, bulk counterparts. The two main reasons why materials at the nanoscale can have different properties are increased relative surface area and new quantum effects. Nanomaterials have a much greater surface area to volume ratio than their conventional forms, which can lead to greater chemical reactivity and affect their strength. Also at the nanoscale, quantum effects can become much more important in determining the materials properties and characteristics, leading to novel optical, electrical and magnetic behaviours.

Nanomaterials are already in commercial use, with some having been available for several years or decades. The range of commercial products available now is huge, including stain-resistant and wrinkle-free textiles, cosmetics, sunscreens, electronics, paints and varnishes. Nanocoatings and nanocomposites are finding use in diverse consumer products, such as windows, sports equipment, bicycles and automobiles. There are novel UV-blocking coatings on glass bottles which protect beverages from damage by sunlight, and long-lasting tennis balls using butylrubber/nano-clay composites. Nanoscale titanium dioxide, for instance, is finding applications in cosmetics, sun-block creams and self-cleaning windows, and nanoscale silica is used as filler in a range of products, including cosmetics and dental fillings.

Nanomaterials have extremely small size which having at least one dimension 100 nm or less. Nanomaterials can be nanoscale in one dimension (e.g., surface films), two dimensions (e.g., strands or fibres), or three dimensions (e.g., particles). They can exist in single, fused, aggregated or agglomerated forms with spherical, tubular, and irregular shapes. Common types of nanomaterials include nanotubes, dendrimers, quantum dots and fullerenes. Nanomaterials have applications in the field of nanotechnology, and display different physical/chemical characteristics from normal chemicals (i.e., silver nano, carbon nanotube, fullerene, photocatalyst, carbon nano, silica).

12.6.1 Importance of Nanomaterials

These materials have created a high interest in recent years by virtue of their unusual mechanical, electrical, optical and magnetic properties. Some examples are given below:

(i) Nanophase ceramics are of particular interest because they are more ductile at elevated temperatures as compared to the coarse-grained ceramics.

(ii) Nanostructured semiconductors are known to show various non-linear optical properties. Semiconductor Q-particles also show quantum confinement effects which may lead to special properties, like the luminescence in silicon powders and silicon germanium quantum dots as infrared optoelectronic devices. Nanostructured semiconductors are used as window layers in solar cells.

(iii) Nanosized metallic powders have been used for the production of gas tight materials, dense parts and porous coatings. Cold welding properties combined with the ductility make them suitable for metal-metal bonding especially in the electronic industry.

(iv) Single nanosized magnetic particles are mono-domains and one expects that also in magnetic nanophase materials the grains correspond with domains, while boundaries on the contrary to disordered walls. Very small particles have special atomic structures with discrete electronic states, which give rise to special properties in addition to the superparamagnetism behaviour. Magnetic nanocomposites have been used for mechanical force transfer (ferrofluids), for high density information storage and magnetic refrigeration.

(v) Nanostructured metal clusters and colloids of mono- or plurimetallic composition have a special impact in catalytic applications. They may serve as precursors for new type of heterogeneous catalysts (Cortex-catalysts) and have been shown to offer substantial advantages concerning activity, selectivity and lifetime in

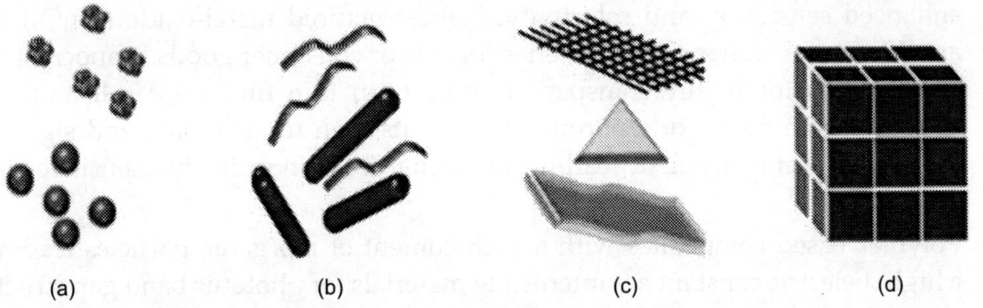

Figure 12.6 Classification of nanomaterials (a) 0D spheres and clusters, (b) 1D nanofibres, wires, and rods, (c) 2D films, plates, and networks, (d) 3D nanomaterials.

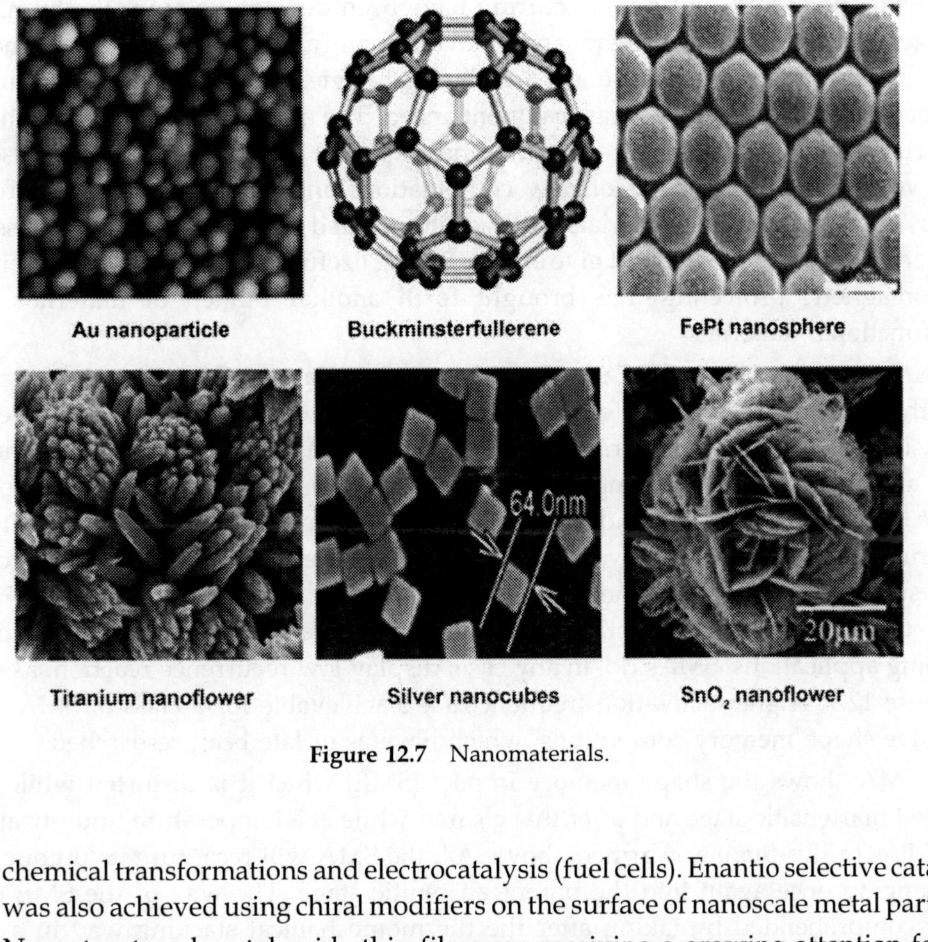

Figure 12.7 Nanomaterials.

chemical transformations and electrocatalysis (fuel cells). Enantio selective catalysis was also achieved using chiral modifiers on the surface of nanoscale metal particles.

(vi) Nanostructured metal-oxide thin films are receiving a growing attention for the realization of gas sensors (NOx, CO, CO_2, CH_4 and aromatic hydrocarbons) with enhanced sensitivity and selectivity. Nanostructured metal-oxide (MnO_2) finds application for rechargeable batteries for cars or consumer goods. Nanocrystalline silicon films for highly transparent contacts in thin film solar cell and nano-structured titanium oxide porous films for its high transmission and significant surface area enhancement leading to strong absorption in dye sensitized solar cells.

(vii) Polymer based composites with a high content of inorganic particles leading to a high dielectric constant are interesting materials for photonic band gap structure.

12.7 SHAPE MEMORY ALLOYS

For quite a long time, metals have assumed a vital part as auxiliary materials. Methods of alloying, purifying, and manufacturing have been developing since the bronze and iron ages. With headways in science and innovation, and a more profound comprehension of the impacts of microstructure and handling strategies on the materials conduct, the field of materials science has drastically enhanced. The ability to build diverse material properties (mechanical, warm, electrical, and so forth.) For an assortment of uses has empowered the advancement of new combinations and composites. The interest for lighter, more grounded materials with custom fitted properties that address both stringent basic prerequisites and give extra building usefulness (e.g., detecting, incitation, electromagnetic protecting) has brought forth another branch of materials called multifunctional materials.

Shape memory alloys (SMAs) are an interesting class of shape memory materials with the capacity to recuperate their shape when the temperature is expanded. An expansion in temperature can bring about shape recuperation even under high connected loads along these lines bringing about high activation vitality densities as shown in Figure 12.8. Furthermore, under particular conditions, SMAs can assimilate and disseminate mechanical vitality by experiencing a reversible hysteretic shape change when subjected to connected mechanical cyclic stacking. These attributes of SMAs have made them famous for detecting and activation, affect assimilation and vibration damping applications. SMAs do, in any case, display low recurrence reaction, as shown in Figure 12.9. Higher activation frequencies are achievable for a class of SMAs called attractive shape memory compounds, which have as of late been researched.

A SMA shows the shape memory impact (SME) when it is distorted while in the twinned martensitic stage and after that cleared while at a temperature underneath 'A_s'. When it is in this manner warmed above 'A_f', the SMA will recapture its unique shape by changing once again into the parent austenitic stage. The way of the SME can be better comprehended by taking after the thermomechanical stacking way in a joined push strain-temperature space.

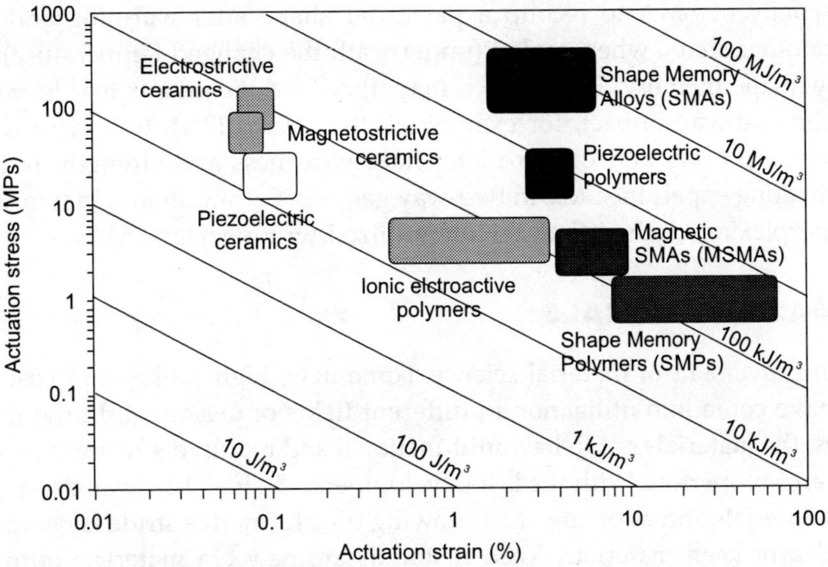

Figure 12.8 Actuation energy density diagram indicating typical ranges of actuation stress, actuation strain, and the actuation energy densities of different active materials that exhibit direct coupling.

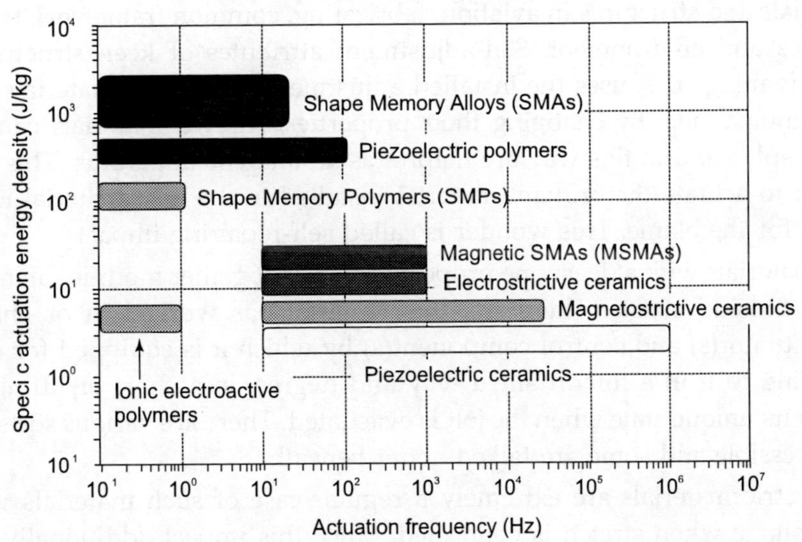

Figure 12.9 Actuation frequency diagram comparing the actuation frequency ranges of different active materials that exhibit direct coupling.

The capacity of SMA to recoup a particular shape after warming and after that arrival to another shape when cooled (underneath the changed temperature) is known as two-way shape memory. Be that as it may, there are constraints that lessen the ease of use of the two-way impact, for example, littler strains (2%), to a great degree low cooling change powers and obscure long haul weariness and strength. Indeed, even slight overheating expels the SME in two-way gadgets. Setting shapes in two-way SMAs is a more perplexing method than the one utilized with oneway SMAs.

12.8 SMART MATERIALS

With the improvement of material science, some new, high calibre and cost-proficient materials have come into utilization in different fields of designing. In the most recent ten decades, the materials got to be multifunctional and required streamlining of various portrayals and properties. With the last development, the idea has been driving towards composite materials and as of late, the following transformative stride is being pondered with the idea of keen materials. Keen materials are new era materials outperforming the traditional basic and utilitarian materials. These materials have versatile abilities to outside jolts, for example, burdens or environment, with inalienable insight.

Smartness depicts self-flexibility, self-detecting, memory and numerous functionalities of the materials or structures. These attributes give various conceivable applications to these materials and structures in aviation, fabricating, common framework frameworks, biomechanics and environment. Self-adjustment attributes of keen structures are an awesome advantage that uses the installed adjustment of brilliant materials like shape memory combinations. By changing their properties, shrewd materials can recognize blames and splits and in this way are helpful as an analytic apparatus. This trademark can be used to actuate the brilliant material installed in the host material legitimately to make up for the blame. This wonder is called self-repairing impact.

These materials with at least one property can be altogether modified in a controlled manner by outside boosts. Therefore, this material has worked in or characteristic sensor(s), actuator(s) and control component(s) by which it is equipped for detecting a boost, reacting to it in a foreordained way and degree, in a short or fitting time and returning to its unique state when the jolt is evacuated. There are various sorts of shrewd material accessible and some are talked about beneath.

Piezoelectric materials are extremely a regular case of such materials where they deliver a voltage when stretch is connected. Since this impact additionally applies in the turn around way, a voltage over the example will create stretch inside the specimen. Reasonably planned structures produced using these materials can in this manner be made that twist, grow or contract when a voltage is connected. They can likewise be utilized as a part of optical-GPS beacons, attractive heads, speck framework printers, PC consoles, high-recurrence stereo speakers, accelerometers, miniaturized scale telephones, weight sensors, transducers and igniters for gas flame broils.

Thermo-responsive materials, either shape memory alloys or shape memory polymers, are materials that can hold distinctive shapes at different temperatures. They can be twisted and come back to their unique shape by warming. All the while, they produce an inciting power. Shape memory combination, for example, nitinol, an amalgam of nickel and titanium, which has an erosion resistance like stainless steel, making it especially helpful for biomechanical applications. Such sorts of materials can be utilized as a part of espresso pot indoor regulator, super versatile exhibition outlines, stents for veins, while shape memory polymer can recapture its unique shape when warmed. These are by and large utilized as a part of biodegradable surgical sutures that will naturally fix to the right pressure furthermore in self-repairing auto bodies that will recuperate shape on tender warming after an imprint.

Magneto-prohibitive materials like piezoelectrics, react to just attractive fields as opposed to electric. They are normally utilized as a part of low-recurrence, high-control sonar transducers, engines and water powered actuators, alongside the shape-memory amalgam. Nitinol, magneto-prohibitive materials are viewed as promising contender for accomplishing dynamic damping of vibrations. PH-delicate materials are the materials that change their shading as an after-effect of evolving sharpness. This can be appropriate for paints that can change shading to demonstrate erosion in the metal underneath them. Chromogenic frameworks change shading because of electrical, optical or warm changes. These incorporate electrochromic materials, which change their shading or haziness on the use of a voltage. Extremely basic case of such sort is liquid gem shows.

Photochromic materials are materials which change shades because of light. These can be seen in light delicate scenes that obscure when presented to splendid daylight. There are additionally sure paints, for example, the monochromic and photochromic paints, which change shading on warming and on exposure to light individually.

12.8.1 Applications of Smart Materials

Smart materials offer an extensive variety of utilizations because of their fluctuating reaction to outer boosts. The diverse regions of use can be: aviation, structural designing applications and mechatronics to give some examples. The extent of utilization of keen material incorporates taking care of designing issues with unfeasible productivity and gives a chance to formation of new items. Imperative element identified with savvy materials and structures is that it incorporates all fields of science and designing. To the extent the specialized utilization of smart materials is concerned, it includes composite materials inserted with fibre optics, actuators, sensors, Micro Electro Mechanical Systems (MEMSs), vibration control, sound control, shape control, item well-being or lifetime observing, cure checking, dynamic and detached controls, self-repair (mending), simulated organs, novel demonstrating gadgets, planned magnets, damping aeroelastic solidness and push disseminations. Shrewd structures are found in autos, space frameworks, settled and rotating wing air ships, maritime vessels, common structures, machine apparatuses, diversion and restorative gadgets. The sort of "smartness" shown by these materials is by and large modified by material organization, unique handling, presentation of deformities or by altering the small scale structure, in order to adjust to the different levels of boosts in a controlled manner. Like smart structures, the terms 'shrewd and "clever" are utilized conversely for savvy materials. Takagi (1990) characterized keen materials as the materials which react to natural changes at the most ideal conditions and show their own capacities as indicated by the earth. The criticism capacities inside the material are consolidated with properties and elements of the materials.

Smart frameworks likewise sense their surroundings and react, yet are not built from a solitary material. They may consolidate brilliant materials, yet can likewise be developed utilizing conventional innovation. Pacemakers are a savvy framework intended to react to an unpredictable heart rate with an electrical motivation that manages it. Brilliant framework alludes to a gadget which can detect changes in its surroundings and can make an optimal reaction by changing its material properties, geometry, mechanical or electromagnetic reaction. Both the sensor and the actuator capacities with their fitting criticism must be appropriately incorporated. The framework could lose its application or could be perilous if the reaction is too moderate or too quick.

Smart materials and frameworks have an extensive variety of uses. Interest in innovative work is driven by elements, for example, enactment, lessening waste and interest for higher personal satisfaction, for example, structures, spans, pipelines, boats and flying machine must be emphatically outlined and routinely investigated to avert 'wear and tear'. Review is costly and tedious, while planning to forestall harm can trade off execution. With some current materials.

12.9 SUMMARY

In this chapter the various types of engineering materials are considered and their properties, importance and applications are also listed. These engineering materials are of so many uses. This chapter began with a discussion of the various types of metal alloys. With regard to composition, metals and alloys are classified as either ferrous or nonferrous. Ferrous alloys (steels and cast irons) are those in which iron is the prime constituents. Smart materials are also of many applications in present scenario. The adjective "smart" implies that these materials are able to sense changes in their environments and then respond to these changes in predetermined manners traits that are also found in living organisms. In addition, this "smart" concept is being extended to rather sophisticated systems that consist of both smart and traditional materials. Hence, this chapter provides a summary of engineering materials and their applications in modern world.

EXERCISE

1. Briefly explain the importance of engineering materials.
2. Summarize the properties of ferromagnetic materials.
3. List applications of nanomaterials.
4. What are smart materials? Discuss in detail.
5. How can ferromagnetic and antiferromagnetic materials be differentiated?
6. List applications of smart materials.
7. What are magnetic materials? Draw their magnetic behaviour.
8. What are various applications of dielectric materials?
9. Write a short note on classification of metal alloys.
10. What is an alloy? What are the advantages of using an alloy in various engineering applications.
11. With a neat actuation energy density diagram indicating typical ranges of actuation stress, actuation strain, and the actuation energy densities of different active materials that exhibit direct coupling explain the phenomenon in SMA.
12. Draw the actuation frequency diagram comparing the actuation frequency ranges of different active materials that exhibit direct coupling in smart memory alloys and explain.
13. Compare the atomic behaviour of magnetic materials.
14. What is an alloy? Write the composition of any five typical metal alloys with their applications.
15. Compare the properties and applications of smart materials and shape memory alloys with examples.

Multiple-Choice Questions

1. % C in medium carbon steels ranges from ________.

   (a) 0.3 – 0.4 (b) 0.3 – 0.5

   (c) 0.3 – 0.6 (d) None

2. Stainless steel is so-called because of its ________.

   (a) High strength (b) High corrosion resistance

   (c) High ductility (d) Brittleness

3. In white cast irons, carbon is present as ________.

   (a) Graphite flakes (b) Graphite nodules

   (c) Cementite (d) Carbon does not exist

4. Refractory metal
   (a) Ag (b) W
   (c) Pt (d) Ni
5. Not a noble metal
   (a) Cu (b) Ag
   (c) Au (d) Pt
6. Noble metal
   (a) Al (b) Ag
   (c) Mo (d) W
7. Usual casting method for making dental crowns
   (a) Sand casting (b) Die casting
   (c) Continuous casting (d) Investment casting
8. Prime structural disadvantage of P/M products
   (a) Low density (b) Porosity
   (c) High damping capacity (d) None
9. _______is not an important heat treatment process parameter
   (a) Heating rate (b) Temperature
   (c) Cooling rate (d) Atmosphere
10. Final structure of austempered steel is
    (a) Pearlite (b) Ferrite + graphite
    (c) Bainite (d) Martensite
11. The word 'ceramic' means _________.
    (a) Soft material (b) Hard material
    (c) Burnt material (d) Dry material
12. Not a characteristic of ceramic material
    (a) High temperature stability (b) High mechanical strength
    (c) Low elongation (d) Low hardness
13. Major ingredient of traditional ceramics is
    (a) Silica (b) Clay
    (c) Feldspar (d) All

14. Not a major contributor of engineering ceramics
(a) SiC (b) SiO_2
(c) Si_3N_4 (d) Al_2O_3

15. The following ceramic product is mostly used as a pigment in paints
(a) TiO_2 (b) SiO_2
(c) UO_2 (d) ZrO_2

16. Most commercial glasses consist of
(a) Lime (b) Soda
(c) Silica (d) All

17. Hot isostatic pressing is not a viable option if the chief criterion is
(a) Strength without grain growth (b) Lost cost
(c) Zero porosity (d) Processing refractory ceramics

18. During sintering densification is not due to
(a) Atomic diffusion (b) Surface diffusion
(c) Bulk diffusion (d) Grain growth

19. The word 'polymer' meant for material made from ________.
(a) Single entity (b) Two entities
(c) Multiple entities (d) Any entity

20. One of the characteristic properties of polymer material is ________.
(a) High temperature stability (b) High mechanical strength
(c) High elongation (d) Low hardness

21. Polymers are ________ in nature.
(a) Organic (b) Inorganic
(c) Both (a) and (b) (d) None

22. These polymers cannot be recycled:
(a) Thermoplasts (b) Thermosets
(c) Elastomers (d) All polymers

23. In general, strongest polymer group is ________.
(a) Thermoplasts (b) Thermosets
(c) Elastomers (d) All polymers

24. These polymers consist of coil-like polymer chains:
    (a) Thermoplasts (b) Thermosets
    (c) Elastomers (d) All polymers
25. Strong covalent bonds exist between polymer chains in _________.
    (a) Thermoplasts (b) Thermosets
    (c) Elastomers (d) All polymers
26. Following is unique to polymeric materials:
    (a) Elasticity (b) Viscoelasticity
    (c) Plasticity (d) None
27. Elastic deformation in polymers is due to _________.
    (a) Slight adjustment of molecular chains
    (b) Slippage of molecular chains
    (c) Straightening of molecular chains
    (d) None of the above
28. Kevlar is commercial name for _________.
    (a) Glass fibres (b) Carbon fibres
    (c) Aramid fibres (d) Cermets
29. Composite materials are classified on the basis of
    (a) Type of matrix (b) Size-and-shape of reinforcement
    (c) Both (d) None
30. Major load carrier in dispersion-strengthened composites is
    (a) Matrix (b) Fibre
    (c) Both (d) Can't define
31. Usually softer constituent of a composite is
    (a) Matrix (b) Reinforcement
    (c) Both are of equal strength (d) Can't define
32. Usually stronger constituent of a composite is
    (a) Matrix (b) Reinforcement
    (c) Both are of equal strength (d) Can't define
33. Last constituent to fail in fibre reinforced composites
    (a) Matrix (b) Fibre
    (c) Both fail at the same time (d) Can't define

34. Size range of dispersoids used in dispersion strengthened composites is
    (a) 0.01-0.1 μm (b) 0.01-0.1 nm
    (c) 0.01-0.1 mm (d) None
35. Rule-of-mixture provides _________ bounds for mechanical properties of particulate composites.
    (a) Lower (b) Upper
    (c) Both (d) None
36. Al-alloys for engine/automobile parts are reinforced to increase their
    (a) Strength (b) Wear resistance
    (c) Elastic modulus (d) Density
37. Mechanical properties of fibre-reinforced composites depend on
    (a) Properties of constituents
    (b) Interface strength
    (c) Fibre length, orientation, and volume fraction
    (d) All the above
38. Longitudinal strength of fibre reinforced composite is mainly influenced by
    (a) Fibre strength (b) Fibre orientation
    (c) Fibre volume fraction (d) Fibre length
39. The following material can be used for filling in sandwich structures
    (a) Polymers (b) Cement
    (c) Wood (d) All
40. Not an example of laminar composite
    (a) Wood (b) Bimetallic
    (c) Coatings/Paints (d) Claddings
41. Solar cell works on
    (a) Laser technology (b) Photo-conduction
    (c) Thermal emission (d) Tyndall effect
42. Optical fibre operates on the principle of
    (a) Total internal reflectance (b) Tyndall effect
    (c) Photoelectric effect (d) Laser technology

43. Usually, _________ is the initial stage of a component manufacturing.
    (a) Design (b) Conception
    (c) Material selection (d) Testing
44. Important waste management techniques is
    (a) Disposal-biodegradable (b) Recycling
    (c) Landfills (d) All
45. Thermosetting materials are
    (a) the plastics which can be softened even after they have set and hardened.
    (b) the plastics which require heat and pressure to mould them into shape.
    (c) both (a) and (b)
    (d) None of the above
46. Which plastic materials contain strong crosslinkings in their molecular structure?
    (a) Thermoplastic materials (b) Thermosetting materials
    (c) Both (a) and (b) (d) None of the above
47. Which of the following characteristics does not affect the properties of plastics?
    (a) Their molecular structure (b) Their degree of polymerization
    (c) Both (a) and (b) (d) None of the above
48. The process of combining two or more distinct polymer molecules to form a new product with different characteristics is called
    (a) Binding (b) Stabilizing
    (c) Filling (d) Blending
49. Why are the plasticizers added with polymers?
    (a) To hold other constituents of plastic together
    (b) To reduce the cost and enhance the strength and hardness of plastics
    (c) To improve flexibility and to reduce temperature and pressure required for moulding of plastics
    (d) None of the above
50. The materials such as lead and barium which are added with polymers to minimize the effect of heat, sunlight and ozone are called
    (a) Binders (b) Blenders
    (c) Stabilizers (d) Fillers

Answers

1. (c)	**2.** (b)	**3.** (c)	**4.** (b)	**5.** (a)
6. (b)	**7.** (d)	**8.** (b)	**9.** (a)	**10.** (c)
11. (c)	**12.** (d)	**13.** (d)	**14.** (b)	**15.** (a)
16. (d)	**17.** (b)	**18.** (b)	**19.** (c)	**20.** (c)
21. (c)	**22.** (b)	**23.** (b)	**24.** (c)	**25.** (b)
26. (b)	**27.** (a)	**28.** (c)	**29.** (c)	**30.** (a)
31. (a)	**32.** (b)	**33.** (a)	**34.** (a)	**35.** (c)
36. (b)	**37.** (d)	**38.** (a)	**39.** (d)	**40.** (a)
41. (b)	**42.** (a)	**43.** (b)	**44.** (d)	**45.** (b)
46. (b)	**47.** (d)	**48.** (d)	**49.** (c)	**50.** (c)

SECTION V

Economic, Environmental and Social Aspects of Engineering Materials

13

Economic, Environmental and Social Issues of Material Usage

Learning Objectives

After going through this chapter the reader will be able to:

- List the environmental and societal considerations.
- Understand the issues related to recycling.
- Cite the lifecycle analysis and its use in design.

13.1 INTRODUCTION

Engineering materials are essential in day-to-day life due to their adaptable auxiliary and physical properties. Determination of the engineering materials depends on these properties. Be that as it may, decision of material has suggestions for the duration of the lifecycle of an item, impacting numerous parts of monetary and ecological execution. As it were, extraordinary material issues should have been managed amid material choice if the item is to be industrially fruitful and aggressive in the market. These unique contemplations are identified with their cost, creation cost, ecological and social issues like contamination, transfer, reusing, and so forth. This chapter deals with these issues of material selection.

Engineering profession manages use of logical and innovative advances to outline and produce segments and frameworks that perform dependably and attractively. In any case, there are financial aspects as the main thrust behind it. Financial aspects of engineering a component relies on upon three variables: segment outline, material use, and assembling costs. These three variables are between related as in maybe a couple may impact the decision of others.

At first, any component need to experience origination and after that design stage. This incorporates era of idea about the component. Later on design stage deals with its size, shape, and setup which will impact its execution in the administration. Usually, engineers deal with not a single component, but with complex assembly of components/a system. Therefore, every part should be intended for more optimal design of the component. This may a few times go about as limitation to ideal plan of a segment. Consequently, plan of a segment is typically an iterative procedure. Less the number of iterations, lower will be the cost of the component/system.

Plan and design stage is trailed by material choice. Material is chosen relying upon its properties, which are reasonable to fill the need amid the administration. Other than the properties, less expensive materials are favoured. Along these lines, more often than not a group of materials is chosen that fulfils the plan limitations, then correlations are made on the premise of cost per unit. This cost additionally incorporates unavoidable material wastage amid assembling stage.

After outline and material determination, it is up to assembling technique to decrease the cost. Generally, fabricating incorporates both essential stage and auxiliary stage. The cost contemplations incorporates the capital on tooling, upkeep cost, work, repair expenses, and material wastage. Increasingly the quantity of assembling stages, higher will be the item cost. Examination, get together, and last bundling will add-on to the item cost.

In engineering practice there are other important criteria that must be considered in the development of a marketable product. Some of these are economic in nature, which, to some degree, are unrelated to scientific principles and engineering practice, and yet are significant if a product is to be competitive in the commercial market. Other criteria that should be addressed involve environmental and societal issues such as pollution, disposal, recycling, and energy. This final chapter offers relatively brief overviews of economic, environmental, and societal considerations that are important in engineering practice.

Some fraction of the cost of a component is associated with its design. In this context, component design is the specification of size, shape, and configuration, which will affect in-service component performance. For example, if mechanical forces are present, then stress analyses may be required. Detailed drawings of the component must be prepared; computers are normally employed, using software that has been generated for this specific function.

It is often the case that a single component is part of a complex device or system consisting of a large number of components (i.e., the television, automobile, DVD player/recorder, etc.). Thus, design must take into consideration each component's contribution to the efficient operation of the complete system. Component design is a highly iterative process that involves many compromises and trade-offs. The engineer should keep in mind that an optimal component design may not be possible due to system constraints.

13.2 ENVIRONMENTAL AND SOCIETAL CONSIDERATIONS

It goes without saying that engineering practice involves utilizing scientific principles to design components and systems that perform reliably and satisfactorily. Another critical driving force in engineering practice is that of economics. Simply stated, the company or institution must realize a profit from the products that it manufactures and sells. The engineer might design the perfect component; however, as manufactured, it must be offered for sale at a price that is attractive to the consumer and, in addition, return a suitable profit to the company. Only a brief overview of important economic considerations as they apply to the materials engineer will be provided. The student may want to consult references provided at the end of this chapter that address engineering economics in detail.

There are three factors over which the materials engineer has control and that affect the cost of a product; they are (1) component design, (2) the material(s) used, and (3) the manufacturing technique(s) that are employed. These factors are interrelated in that component design may affect which material is used, and both component design and the material used will influence the choice of manufacturing technique(s).

Assembling of an item has affect on environment furthermore on society from various perspectives. This is on the grounds that assets required to create an item originates from various parts of the world, and the modern riches and thriving are shared by increasingly the one locale. Alongside these, inconvenient impacts of industrialization additionally spread its wings to different parts of the world.

A material used to create an item experiences a number of stages. These incorporate extraction of crude materials from normal assets through generation, during its service, lastly its transfer. It is here and there known as support to-grave lifecycle of a material. A schematic lifecycle show has been displayed in the Figure 13.1.

As portrayed in Fig. 13.1, crude materials are initially extricated from regular gritty assets through penetrating, mining, and so forth. Later on these are subjected to decontamination, refining to change over them into metals, earthenware production, elastic, fuel, and so on. These essential items are further prepared to get built materials like metallic compound, glass, plastics, semi-transmitters, and so on. Presently, the built materials are mould, warm treated to make parts which are amassed into items, gadgets that are prepared for use by society. Amid the administration, items get to be old, out mould, separate, or may not fill the need proficiently. So they are disposed of. Upon this, they go as waste into landfills, or a few times reused and a few parts might be reused. This contends the lifecycle.

Natural assets are, sadly, restricted. Notwithstanding, some are renewable while greater part are definitely not! As society and hence populace develops, assets per head get to be lesser and lesser. Thus, there is an incredible and in reality quick consideration must be paid to the better and viable utilization of gritty assets. It will be far and away

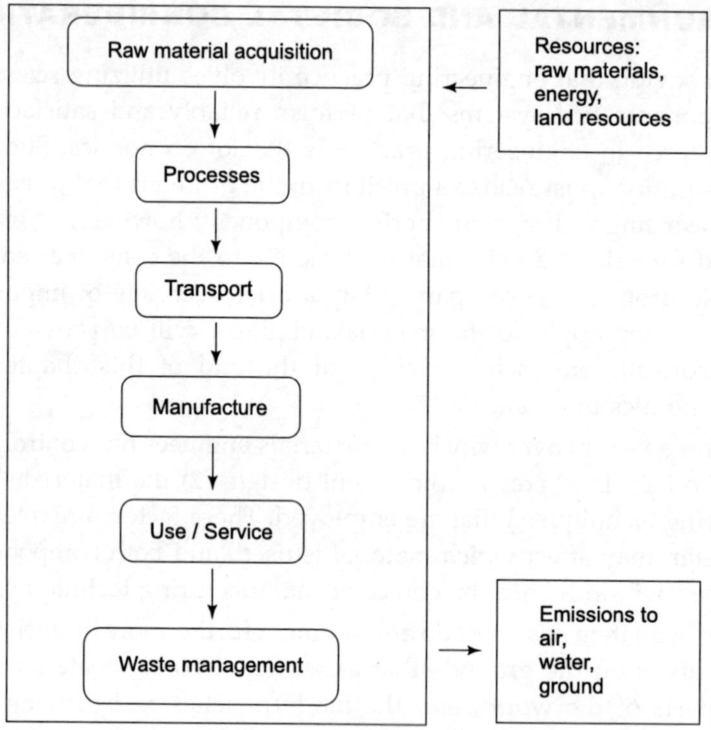

Figure 13.1 Lifecycle model.

superior in the event that we quit utilizing the assets, which is obviously unreasonable. Notwithstanding this, vitality is utilized at each phase of material cycle, which itself is an asset. That implies that vitality supply is restricted, and vitality should be utilized conservatively, and viably. In addition, amid extraction and generation of materials/ items, connection with environment is typically negative to the general public from numerous points of view. It incorporates water contamination, air contamination. Therefore, extraordinary care should be taken amid extraction, assembling, and utilization of a material/item. At end of its administration, the item might be reused rather than transferred. There are numerous advantages from reusing those include: to ration crude material assets and to moderate vitality required to separate and refine. Reusing likewise keeps away from natural contamination. Henceforth, last items should be planned and fabricated to such an extent that they are ecologically neighbourly, and simple to reuse with least utilization of crude materials and vitality. In the event of transfer into environment, items should be biodegradable. This demonstrates material lifecycle includes collaborations and trades among materials, vitality and the earth including the general public.

There is a requirement for better answers for the ecological issues. Therefore, there is extra cost to the item in the event that it is required to be created under conditions with the end goal that inconvenient ecological impacts are kept to least, i.e., green items cost more than expected items. Modern way to deal is to evaluate the ecological execution of items, that is, lifecycle examination/appraisal (LCA). LCA is clarified in detail in anticipated areas.

Our modern technologies and the manufacturing of the associated products impact our societies in a variety of ways—some are positive, others are adverse. Furthermore, these impacts are economic and environmental in type, and international in scope inasmuch as: (1) the resources required for a new technology often come from many different countries; (2) the economic prosperity resulting from technological development is global in extent; and (3) environmental impacts may extend beyond the boundaries of a single country.

13.3 RECYCLING ISSUES

Recycling and disposal assume an essential part in material determination and plan of cutting edge items. Reusing can be characterized as unflawed passage of material into its cycle after finishing of life as an item for interminable number of times. Ecologically, amicable means item after transfer into nature, must interface with it and get weakened so material comes back to a similar state from where it was extricated and put into lifecycle. For simplicity of transfer and reusing, item plan ought to facilitate destroying of framework. Distinctive building materials are reused and additionally biodegradable up to various degrees.

Metals and alloys have a tendency to get consumed up to some degree, i.e., biodegradable. Be that as it may, some of them are lethal. Then again, most metals and compounds are recyclable. Pottery/glasses are, be that as it may, are scarcely reused. It is on account of that crude materials are cheap, and reusing procedure is tedious and costly. Plastics and elastic are well known for their latency to nature and dampness. This same latency makes them hard to bio-corrupt. Along these lines, plastics are for the most part reused, and simply arranged through landfills. Thermoplastic polymers are effortlessly reused up on warming to higher temperatures. Then again reusing of thermoset plastics is a great deal more troublesome. Subsequently, these are generally arranged. Therefore, there is a pattern to utilize elective materials which are recyclable, for example, thermoplastic elastomers set up of conventions elastic.

Recycling of used products rather than disposing of them as waste is a desirable approach for several reasons. First of all, using recycled material obviates the need to extract raw materials from the earth, and thus conserves natural resources and eliminates any associated ecological impact from the extraction phase. Second, energy requirements for the refinement and processing of recycled materials are normally less for their natural counterparts; for example, approximately 28 times as much energy is required to refine natural aluminium ores as to recycle aluminium beverage can scrap. Finally, there is no need to dispose of recycled materials.

In many countries, environmental problems and issues are being addressed by the establishment of standards that are mandated by governmental regulatory agencies. Furthermore, from an industrial perspective, it becomes incumbent for engineers to propose viable solutions to the existing and potential environmental concerns. Correcting any environmental problems associated with manufacturing will influence product price. That is, manufacturing cost is normally greater for a "green" (or "environment friendly") product than for its equivalent that is produced under conditions wherein environmental issues are minimized. Thus, a company must confront the dilemma of this potential economic-environmental trade-off and then decide the relative importance of economics and of environmental impact.

One approach, implemented by industry to improve the environmental performance of products, is termed *lifecycle analysis/assessment*. With this approach to product design, consideration is given to the cradle-to-grave environmental assessment of the product, from material extraction to product manufacture to product use, and, finally, to recycling and disposal; sometimes this approach is also labelled "green design." One important phase of this approach is to quantify the various inputs (i.e., materials and energy) and outputs (i.e., wastes) for each phase of the lifecycle. In addition, an assessment is conducted relative to the impact on both global and local environments in terms of the effects on the ecology, human health, and resource reserves.

13.3.1 Metals

Metal alloys (i.e., Fe, Cu) to one degree or another experience corrosion and are also biodegradable. However, some metals (i.e., Hg, Pb) are toxic and, when landfilled, may present health hazards. Furthermore, alloys of most metals are recyclable; on the other hand, it is not feasible to recycle all alloys of every metal. In addition, the quality of alloys that are recycled tends to diminish with each cycle.

Product designs should allow for the dismantling of components composed of different alloys. Another problems of recycling involves separation of various alloys (i.e., aluminium from ferrous alloys) after dismantling and shredding; in this regard, some rather ingenious separation techniques have been devised (i.e., magnetic and gravity). Joining of dissimilar alloys presents contamination problems; for example, if two similar alloys are to be joined, welding is preferred over bolting or riveting. Coatings (paints, anodized layers, claddings, etc.) may also act as contaminants, and render the material non-recyclable.

Aluminium alloys are highly corrosion resistant and, therefore, non-biodegradable. Fortunately, however, they may be recycled; in fact, aluminium is the most important recyclable nonferrous metal. Since aluminium is not easily corroded, it may be totally reclaimed. A low ratio of energy is required to refine recycled aluminium relative to its primary production. In addition, there are a large number of commercially available alloys that have been designed to accommodate impurity contamination. The primary sources of recycled aluminium are used beverage cans and scrapped automobiles.

13.3.2 Glass

The one ceramic material that is consumed by the general public in the greatest quantities is glass, in the form of containers. Glass is relatively an inert material, and, as such, it does not decompose; thus, it is not biodegradable. A significant proportion of municipal landfills consist of waste glass; so also does incinerator residue.

In addition, there is not a significant economic driving force for recycling glass. Its basic raw materials (sand, soda ash, and limestone) are inexpensive and readily available. Furthermore, salvaged glass (also called "cullet") must be sorted by colour (clear, amber, and green), by type (plate versus container), and by composition (lime, lead, and borosilicate [or Pyrex]); these sorting procedures are time-consuming and expensive. Therefore, scrap glass has a low market value, which diminishes its recyclability. Advantages of utilizing recycled glass include more rapid and increased production rates and a reduction in pollutant emissions.

13.3.3 Plastic and Rubber

One of the reasons that synthetic polymers (including rubber) are so popular as engineering materials lies with their chemical and biological inertness. On the down side, this characteristic is really a liability when it comes to waste disposal. Polymers are not biodegradable, and, as such, they constitute a significant landfill component. Major sources of waste are from packaging, junk automobiles, automobile tyres, and domestic durables. Biodegradable polymers have been synthesized, but they are relatively expensive to produce. On the other hand, since some polymers are combustible and do not yield appreciable toxic or polluting emissions, they may be disposed of by incineration.

Thermoplastic polymers, specifically poly(ethylene terephthalate), polyethylene, and polypropylene, are the most amenable to reclamation and recycling, since they may be reformed upon heating. Sorting by type and colour is necessary. In some countries, type sorting of packaging materials is facilitated using a number identification codes; for example, a "1" denotes high-density polyethylene (HDPE). The recycled plastic is less costly than the original material, and quality and appearance are generally degraded with each recycle. Typical applications for recycled plastics include shoe soles, tool handles, and industrial products such as pallets.

The recycling of thermoset resins is much more difficult since these materials are not easily remould or reshaped due to their crosslinked or network structures. Some thermosets are ground up and added to the virgin moulding material prior to processing; as such, they are recycled as filler materials.

Rubber materials present some disposal and recycling challenges. When vulcanized, they are the thermoset materials, which makes chemical recycling difficult. In addition, they may also contain a variety of fillers.The major source of rubber scrap in the United States is discarded automobile tyres, which are highly non-biodegradable. Scrap tyres have been utilized as a fuel for some industrial applications (i.e., cement plants), but yield dirty emissions. Recycled rubber tyres that have been split and reshaped are used in a variety of applications such as automotive bumper guards, mud flaps, door mats, and conveyer rollers; and, of course, used tires may also be recapped. In addition, rubber tyres may be ground into small chunks that are then recombined into the desired shape using some type of adhesive; the resulting material may be used in a number of non-demanding applications such as place mats and rubber toys.

13.3.4 Composite Materials

Composites are inherently difficult to recycle because they are multiphase in nature. The two or more phases/materials that constitute the composite are normally intermixed on a very fine scale and trying to separate them complicates the recycling process. However, some techniques have been developed, with modest success, for recycling polymer-matrix composites. Recycling technologies will differ only slightly for thermoset-matrix and thermoplastic-matrix composite materials. The first step in recycling both thermoset- and thermoplastic-matrix composites is shredding/grinding, wherein the components are reduced in size to relatively small particles. In some instances, these ground particles are used as filler materials that are blended with a polymer (and perhaps other fillers) before fabrication (usually using some type of moulding technique) into post-consumer products. Other recycling processes allow for separating of the fibres and/or matrix materials. With some techniques the matrix is volatilized; with others it is recovered as a monomer. Of course, the recovered fibres have short lengths, as a result of the shredding/grinding process. In addition, fibres will experience a reduction of mechanical strength, the degree of which will depend on the specific recovery process as well as fibre type.

13.4 LIFECYCLE ANALYSIS AND ITS USE IN DESIGN

Materials extracted from the earth should have been dealt with for one prime reason - to moderate material and vitality. Be that as it may, materials, vitality and environment impacts are interrelated, i.e., everyone has effect on another either specifically/in a roundabout way. Modern way to evaluate the natural execution of items is named as lifecycle examination/appraisal (LCA). As such, LCA is the evaluation of ecological effect of an item over its lifecycle. The intricate communication between an item and nature is managed in the Lifecycle Assessment (LCA) technique. It is likewise known as ecobalance. LCA deliberately depicts and surveys all streams to and from nature, from a support to grave point of view.

Natural and environmental lifecycle of an item comprises all phases from crude material extraction to its waste administration. In ecological lifecycle evaluation, regular asset utilize and contamination emanation are depicted in quantitative terms. One essential explanation behind undertaking a LCA study is that there are developing worries about an assortment of natural issues as communicated by general sentiment, political bodies, and industry. Ecological concern might be identified with the long haul asset base of human social orders or might be more wellbeing related or it might be a sympathy towards the indigenous habitat accordingly. The quality of LCA is that it thinks about an entire item framework. Since an entire lifecycle is contemplated, LCA is not site particular. In this manner, ecological effect can't be demonstrated at an exceptionally point by point level. Prudent and social angles are excluded in LCA other than when utilized as a reason for examination. Hazard is another perspective not managed in LCA.

LCA is a technique for assessing the environmental aspects and potential impacts associated with a product by:

- compiling an inventory of relevant inputs and outputs of a product system;
- evaluating the potential environmental impacts associated with those inputs and outputs; and
- interpreting the results of the inventory analysis and impact assessment phases in relation to the objectives of the study.

LCA is not just item orientated; it is likewise quantitative and therefore apparently objective. Therefore, it was no more drawn out important to answer on basic dependable guidelines. With LCA came the idea that it was conceivable to quantitatively contrast options all together recognizing the naturally favoured alternative. Besides, it could manage ecological issues. Quantitative depiction of the natural effects of emanations and asset utilizing the three classes – assets utilize, human wellbeing, and environmental results – is a complicated issue. Natural issues are intricate and still not completely caught on. As an outcome, distinctive endeavours at creating models to portray natural effects have prompted diverse results.

LCA can be utilized as a part of a wide range of routes in item advancement. Lifecycle appraisal is a system for surveying the ecological perspectives connected with an item over its lifecycle. The most critical applications are: (1) analysis of the commitment of the lifecycle stages to the general ecological load, as a rule with the plan to organize enhancements on items or procedures, (2) comparison between items for inner or inward correspondences.

Product design and improvement have been viewed as an important range of uses of LCA. The point of convergence of LCA, the item, concurs with that of the item plan and improvement handle. This clarifies to some degree why LCA has been presented as an instrument for natural reception of item plans. The other part is that product development is seen as a decisive activity for achieving sustainability in industrial society – most of the environmental attributes are determined and built into the product during the design stage when materials are selected and constructions designed.

Concerning item plan, there is a need to see how an item impacts on the earth. To grow genuinely practical items, it must be conceivable to evaluate which plan arrangement is ecologically ideal. LCA apparatuses can help in this troublesome territory of eco-outline. Ten straightforward rules for eco-plan are:

1. Do not design products, but lifecycles.
2. Natural materials are not always better.
3. Underestimation of energy consumption.
4. Product life time improvement.
5. Do not design products, but services.
6. Minimum use of material.
7. Use recycled materials.
8. Make sure the product is recyclable.
9. Stupid interrogations, i.e., be suspicious about the established result.
10. Become a member O_2: an international network for sustainable design.

As the entire items lifecycle ought to be respected on the double, agents from outline, generation, showcasing, obtaining, and extend administration ought to cooperate on the eco-plan of another item as they have together the most obvious opportunity to foresee the comprehensive impacts of changes of the item and their natural effect. Ecological angles which should be investigated for each phase of the lifecycle can be viewed as:

1. Resources consumption (energy, materials, water or ground)
2. Air, water or the ground emissions.
3. Products related to waste.
4. Miscellaneous (noise and vibration)

Having made up a list on which stage of the lifecycle has which specific environmental issues, these viewpoints are assessed by ecological effects on the premise of a number of parameters, for example, reach out to ecological effect potential for development or capability of progress. The assessments can likewise consider promoting and advertising issues. As per this positioning the suggested changes are done and are inspected after a specific time. It must be done routinely to be aware of yearly audit of ecological enactment by national and worldwide agencies.

13.5 SUMMARY

For engineering materials to be studied we also need to emphasise on the various environmental issues. In many cases these engineering materials pollute the environment and in the processes of manufacturing them the environment gets degraded. We also need to know the fact that recycle, reuse and reproduce policy needs to be incorporated at every possible stage so that we can save the environment from getting degraded.

The economics of engineering is very important in product design and manufacturing. To minimize product cost, materials engineers must take into account the component design, what materials are used, and manufacturing processes. Other significant economic factors include fringe benefits, labour, insurance, profit, etc.

Environmental and societal impacts of production are significant engineering issues. In this regard, the material cradle-to-grave lifecycle is an important consideration. This cycle consists of extraction, synthesis/processing, product design/manufacture, application, and disposal stages. Materials, energy, and environmental interactions/exchanges are important factors in the efficient operation of the materials cycle. The earth is a closed system in that its materials resources are finite; to some degree, the same may be said for energy resources. Environmental issues involve ecological damage, pollution, and waste disposal. Recycling of used products and the utilization of green design obviate some of these environmental problems.

Recyclability and disposability issues were addressed in the context of materials science and engineering. Ideally, a material should at best be recyclable, and at least biodegradable or disposable. The recyclability and disposability of metal alloys, glasses, polymers, and composites were also discussed.

EXERCISE

1. Briefly explain the environmental considerations for engineering materials.
2. Summarize lifecycle analysis in details.
3. List the issues associated with recycling.
4. What is lifecycle model? Discuss in detail.
5. How can you justify that the 3R principles of recycle, reuse and reproduce save the environment from getting degraded?
6. Specify the recycle issues in metals and glass in detail.
7. How can various engineering materials be recycled without harming environmental conditions? Explain with examples.
8. Explain in detail the straightforward rules for eco planning.
9. What are the steps involved in LCA? Discuss.
10. Visualize the role of component design in the field of engineering materials.

Model Question Papers

Sample Question Paper I

1. (a) What is atomic packing factor of a crystal structure? Show that the atomic packing factor for the FCC crystal structure is 0.74.
   (b) Describe Fick's laws of diffusion.
   (c) Differentiate between Frenkel and Schottky defects.
   (d) Find the equilibrium concentration of vacancies in aluminium at 300 K. Enthalpy of formation of vacancies in aluminium, $\Delta H_f = 68$ kJ mol^{-1}; R = 8.314 Jmol^{-1}K^{-1}
2. (a) Draw the unit cells for BCC, FCC & HCP crystals. Calculate the effective number of atoms for each unit cell
   (b) How indexing of planes and directions are done? Explain.
3. (a) What are crystal defects? Explain in detail with neat sketches.
   (b) With neat sketches give the various types of crystal systems.
4. (a) What are dislocations? Explain with neat sketches.
   (b) What are the various sources for dislocations?
5. (a) Distinguish between slip and twinning.
   (b) Write notes on dislocation climb and cross slip.
6. (a) What is CRSS? Explain.
   (b) Write a note on plastic deformation of a single crystal.
7. (a) With a neat sketch explain the strain hardening of single crystal.
   (b) What are Sessile and Glissile dislocations? Discuss.

Sample Question Paper II

1. (a) Differentiate between slip and twinning with relevant diagrams.
   (b) What is APF? Calculate the APF for BCC system.
   (c) Write notes on dislocation climb and cross slip.
2. (a) State Gibbs phase rule. Explain all stages of iron-iron carbide with relevant diagrams.

(b) What do you mean by solid solutions? Elaborate TTT diagram of steel.

3. (a) Mention various types of polymers. Describe the processing and synthesis of polymers.

   (b) Give the properties and applications of different ceramics.

   (c) Name various heat treatment processes? Write advantages of undergoing various heat treatment processes.

4. (a) What are dislocations? With neat sketches explain the defects in engineering materials.

   (b) What are the designations of cast iron? Give the names, properties and uses of different cast irons.

   (c) Explain the true stress strain curve. What are various types of stress strain curves?

5. (a) Elucidate the tensile testing methodology for identifying the mechanical properties of engineering materials. List some of the properties of engineering materials those are identified using tensile testing.

   (b) Discuss the classification of engineering materials.

   (c) What is crystallization? Briefly cite the transition and melting of glass.

6. (a) Define the term creep. Summarize the generalized creep behaviour of engineering materials and effect of temperature on them.

   (b) Write down various crystal structures and properties of composites.

   (c) Explain how underground pipelines are protected using magnesium anode.

7. Attempt any four

   (a) Significance of phase diagrams?

   (b) Write a short note on structural composites and nano-composites.

   (c) Name various hardness testing methods. Explain any one in detail.

   (d) Cite various criteria for selection of NDT.

   (e) Make a distinction between case hardening and surface hardening heat treatment process.

Sample Question Paper III

1. (a) What is the difference between atomic structure and crystal structure? Calculate the radius of an iridium atom, given that it has an FCC crystal structure, a density of 22.4 g/cm, and an atomic weight of 192.2 g/mol?

   (b) What are point defects? Explain two types of point defects.

2. (a) Mention the primary conditions that favour the extensive substitutional solid solubility of an alloy system.

(b) Distinguish between the direction of the dislocation line, the Burger's vector and the direction of motion for both edge and screw dislocations.

3. (a) What are the differences between the states of phase equilibrium and metastability?

(b) Describe briefly the mechanisms of creep.

(c) Distinguish between ductile and brittle fracture.

4. (a) A sample of glass has a crack of half length 2 μm. The Young's modulus of the glass is 70 GNm^{-2} and the specific surface energy is 1 Jm^{-2}. Estimate its fracture strength.

(b) For a polymer-matrix fibre-reinforced composite, (i) compare the desired mechanical characteristics of matrix and fibre phases; and (ii) mention two reasons why there must be a strong bond between the fibre and matrix at their interface.

5. (a) Explain the differences between diamagnetism, paramagnetism and ferromagnetism.

(b) Why does the conductivity of a semiconductor change with impurity content? Compare this with the behaviour of metallic conductors.

6. Write any two of the following:

(a) Dislocation movement by slip

(b) Dislocation forest

(c) Materials classification

7. Compare interstitial and vacancy atomic mechanism for diffusion. Cite two reasons why interstitial diffusion is normally more rapid than vacancy diffusion?

Short Questions

1. The distance between (111) planes in FCC crystal structure is 2 A. Find the lattice parameter and atomic diameter.
2. A 45 kN force was applied on a Cu-Ni alloy tensile specimen having 2.5 mm diameter and 50 mm gauge length. Determine whether the specimen will undergo necking. Given: σ_x = 420 MPa and σ_y = 250 MPa.
3. What is *S-N* curve?
4. State Fick's first law of diffusion.
5. What is a peritectoid and monotectic reaction?
6. What is Jominy end-quench test?
7. Defme (a) T_g and (b) degree of polymerization.
8. What is stress corrosion cracking?

9. A steel has tensile strength of 1.6 GPa. A large tensile piece of such a steel has crack of length 7 mm in the interior and fractures at 0.6 GPa. Calculate its fracture toughness.
10. Define (a) curie temperature and (b) remanence of a magnetic material.
11. Define isomorphous system with examples.
12. Define Burger's vector.
13. Define the terms (a) susceptibility, and (b) permeability of a magnetic material.
14. Two samples A and B of a brittle material have crack length in the ratio 3: 1. What will be the ratio of tensile strengths of A and B?
15. What is stress corrosion cracking (SCC)
16. What are superalloys? Give examples.
17. Define glass transition temperature.
18. Define inelasticity and viscoelasticity.
19. State Griffith theory.
20. What is fatigue limit of a material?
21. What is corrosion fatigue?
22. What is meant by mobility?
23. What is the basic difference in the solidification (or melting) process between crystalline and noncrystalline structures?
24. What do you mean by annealing?
25. What is the mechanism behind the ultrasonic inspection of mechanical properties of materials?

Glossary

A

Abrasive. A hard and wear-resistant material (commonly a ceramic) that is used to wear, grind, or cut other material.

Absorption. The optical phenomenon whereby the energy of a photon of light is assimilated within a substance, normally by electronic polarization or by an electron excitation event.

Acceptor level. For a semiconductor or insulator, an energy level lying within yet near the bottom of the energy band gap, which may accept electrons from the valence band, leaving behind holes. The level is normally introduced by an impure atom.

Activation energy (*Q*). The energy required to initiate a reaction, such as diffusion.

Activation polarization. The condition wherein the rate of an electrochemical reaction is controlled by the one slowest step in a sequence of steps that occur in series.

Addition (or chain reaction) polymerization. The process by which monomer units are attached one at a time, in chain-like fashion, to form a linear polymer macromolecule.

Adhesive. A substance that bonds together the surfaces of two other materials (termed adherends).

Allotropy. The possibility of the existence of two or more different crystal structures for a substance (generally an elemental solid).

Alloy. A metallic substance that is composed of two or more elements.

Alloy steel. A ferrous (or iron based) alloy that contains appreciable concentrations of alloying elements (other than C and residual amounts of Mn, Si, S, and P). These alloying elements are usually added to improve mechanical and corrosion resistance properties.

Alternating copolymer. A copolymer in which two different repeat units alternate positions along the molecular chain.

Amorphous. Having a noncrystalline structure.

B

Bainite. An austenitic transformation product found in some steels and cast irons. It forms at temperatures between those at which pearlite and martensite transformations occur. The microstructure consists of α-ferrite and a fine dispersion of cementite.

Band gap energy (*Eg*). For semiconductors and insulators, the energies that lie between the valence and conduction bands; for intrinsic materials, electrons are forbidden to have energies within this range.

Bifunctional. Designating monomers that may react to form two covalent bonds with other monomers to create a two-dimensional chain-like molecular structure.

Block copolymer. A linear copolymer in which identical repeat units are clustered in blocks along the molecular chain.

Body-centred cubic (BCC). A common crystal structure found in some elemental metals. Within the cubic unit cell, atoms are located at corner and cell centre positions.

Bohr atomic model. An early atomic model, in which electrons are assumed to revolve around the nucleus in discrete orbitals.

Bronze. A copper-rich copper-tin alloy; aluminium, silicon, and nickel bronzes are also possible.

Burgers vector (b). A vector that denotes the magnitude and direction of lattice distortion associated with a dislocation.

C

Calcination. A high-temperature reaction whereby one solid material dissociates to form a gas and another solid. It is one step in the production of cement.

Capacitance (C). The charge-storing ability of a capacitor, defined as the magnitude of charge stored on either plate divided by the applied voltage.

Carbon-carbon composite. A composite composed of continuous fibres of carbon that are embedded in a carbon matrix.The matrix was originally a polymer resin that was subsequently pyrolyzed to form carbon.

Carburizing. The process by which the surface carbon concentration of a ferrous alloy is increased by diffusion from the surrounding environment.

Cathodic protection. A means of corrosion prevention whereby electrons are supplied to the structure to be protected from an external source such as another more reactive metal or a dc power supply.

Cation. A positively charged metallic ion.

Cement. A substance (often a ceramic) that by chemical reaction binds particulate aggregates into a cohesive structure. With hydraulic cements the chemical reaction is one of hydration, involving water.

Cementite. Iron carbide (Fe_3C).

Ceramic. A compound of metallic and nonmetallic elements, for which the interatomic bonding is predominantly ionic.

Ceramic-matrix composite (CMC). A composite for which both matrix and dispersed phases are ceramic materials. The dispersed phase is normally added to improve fracture toughness.

Cermet. A composite material consisting of a combination of ceramic and metallic materials. The most common cermets are the cemented carbides, composed of an extremely hard ceramic (e.g., WC, TiC), bonded together by a ductile metal such as cobalt or nickel.

Coulombic force. A force between charged particles such as ions; the force is attractive when the particles are of opposite charge.

Covalent bond. A primary interatomic bond that is formed by sharing of electrons between neighbouring atoms.

Creep. The time-dependent permanent deformation that occurs under stress; for most materials it is important only at elevated temperatures.

Crevice corrosion. A form of corrosion that occurs within narrow crevices and under deposits of dirt or corrosion products (i.e., in regions of localized depletion of oxygen in the solution).

Critical resolved shear stress ($^{\tau}$CRSS). The shear stress, resolved within a slip plane and direction, that is required to initiate slip.

Crosslinked polymer. A polymer in which adjacent linear molecular chains are joined at various positions by covalent bonds.

Crystalline. The state of a solid material characterized by a periodic and repeating three-dimensional array of atoms, ions, or molecules.

D

Defect structure. Relating to the kinds and concentrations of vacancies and interstitials in a ceramic compound.

Degradation. A term used to denote the deteriorative processes that occur with polymeric materials. These processes include swelling, dissolution, and chain scission.

Degree of polymerization (DP). The average number of repeat units per polymer chain molecule.

Dislocation line. The line that extends along the end of the extra half-plane of atoms for an edge dislocation, and along the centre of the spiral of a screw dislocation.

Dispersed phase. For composites and some two-phase alloys, the discontinuous phase that is surrounded by the matrix phase.

Dispersion strengthening. A means of strengthening materials wherein very small particles (usually less than 0.1 χm) of a hard yet inert phase are uniformly dispersed within a load bearing matrix phase.

Domain. A volume region of a ferromagnetic or ferrimagnetic material in which all atomic or ionic magnetic moments are aligned in the same direction.

Donor level. For a semiconductor or insulator, an energy level lying within yet near the top of the energy band gap, and from which electrons may be excited into the conduction.

Ductile-to-brittle transition. The transition from ductile to brittle behaviour with a decrease in temperature exhibited by BCC alloys. The temperature range over which the transition occurs is determined by Charpy and Izod impact tests.

Ductility. A measure of a material's ability to undergo appreciable plastic deformation before fracture. It may be expressed as percent elongation (%EL) or percent reduction in area (%RA) from a tensile test.

E

Edge dislocation. A linear crystalline defect associated with the lattice distortion produced in the vicinity of the end of an extra half plane of atoms within a crystal. The Burgers vector is perpendicular to the dislocation line.

Elastic deformation. Deformation that is non-permanent, that is, totally recovered upon release of an applied stress.

Elastic recovery. Non-permanent deformation that is recovered or regained upon the release of a mechanical stress.

Elastomer. A polymeric material that may experience large and reversible elastic deformations.

Electroluminescence. The emission of visible light by a *p-n* junction across which a forward-biased voltage is applied.

Electrolyte. A solution through which an electric current may be carried by the motion of ions.

Electromotive force (emf) series. A ranking of metallic elements according to their standard electrochemical cell potentials.

Electron configuration. For an atom, the manner in which possible electron states are filled with electrons.

Electronegative. For an atom, having a tendency to accept valence electrons. Also, a term used to describe non-metallic elements.

Electron energy band. A series of electron energy states that are very closely spaced with respect to energy.

Electroneutrality. The state of having exactly the same number of positive and negative electrical charges (ionic and electronic), that is, of being electrically neutral.

Electron state (level). One of a set of discrete, quantized energies that are allowed for electrons. In the atomic case each state is specified by four quantum numbers.

Electron volt (eV). A convenient unit of energy for atomic and subatomic systems. It is equivalent to the energy acquired by an electron when it falls through an electric potential of 1 volt.

Electropositive. For an atom, having a tendency to release valence electrons. Also, a term used to describe metallic elements.

F

Face-centred cubic (FCC). A crystal structure found in some of the common elemental metals. Within the cubic unit cell, atoms are located at all corners and face-centred positions.

Fatigue. Failure, at relatively low stress levels, of structures that are subjected to fluctuating and cyclic stresses.

Fatigue life (Nf). The total number of stress cycles that cause a fatigue failure at some specified stress amplitude.

Fatigue limit. For fatigue, the maximum stress amplitude level below which a material can endure an essentially infinite number of stress cycles and not fail.

Fatigue strength. The maximum stress level that a material can sustain, without failing, for some specified number of cycles.

Ferrite (iron). Body-centred cubic iron; also iron and steel alloys that have the BCC crystal structure.

Ferroelectric. A dielectric material that may exhibit polarization in the absence of an electric field.

Ferromagnetism. Permanent and large magnetizations found in some metals (e.g., Fe, Ni, and Co), which result from the parallel alignment of neighbouring magnetic moments.

Ferrous alloy. A metal alloy for which iron is the prime constituent.

Fibre. Any polymer, metal, or ceramic that has been drawn into a long and thin filament.

Fibre-reinforced composite. A composite in which the dispersed phase is in the form of a fibre (i.e., a filament that has a large length-to-diameter ratio).

Fibre reinforcement. Strengthening or reinforcement of a relatively weak material by embedding a strong fibre phase within the weak matrix material.

G

Galvanic corrosion. The preferential corrosion of the more chemically active of two metals that are electrically coupled and exposed to an electrolyte.

Galvanic series. A ranking of metals and alloys as to their relative electrochemical reactivity in seawater.

Glass–ceramic. A fine-grained crystalline ceramic material that was formed as a glass and subsequently crystallized.

Glass transition temperature (Tg). The temperature at which, upon cooling, a noncrystalline ceramic or polymer transforms from a supercooled liquid to a rigid glass.

Graft copolymer. A copolymer wherein homopolymer side branches of one monomer type are grafted to homopolymer main chains of a different monomer type.

Grain. An individual crystal in a polycrystalline metal or ceramic.

Grain boundary. The interface separating two adjoining grains having different crystallographic orientations.

Grain growth. The increase in average grain size of a polycrystalline material. For most materials, an elevated-temperature heat treatment is necessary.

H

Hall effect. The phenomenon whereby a force is brought to bear on a moving electron or hole by a magnetic field that is applied perpendicular to the direction of motion. The force direction is perpendicular to both the magnetic field and the particle motion directions.

Hardenability. A measure of the depth to which a specific ferrous alloy may be hardened by the formation of martensite upon quenching from a temperature above the upper critical temperature.

Hard magnetic material. A ferromagnetic or ferromagnetic material that has large coercive field and remanence values, normally used in permanent magnet applications.

Hardness. The measure of a material's resistance to deformation by surface indentation or by abrasion.

Hexagonal close-packed (HCP). A crystal structure found for some metals. The HCP unit cell is of hexagonal geometry and is generated by the stacking of close-packed planes of atoms.

High polymer. A solid polymeric material having a molecular weight greater than about 10,000 g/mol.

I

Integrated circuit. Millions of electronic circuit elements (transistors, diodes, resistors, capacitors, etc.) incorporated on a very small silicon chip.

Interdiffusion. Diffusion of atoms of one metal into another metal.

Intergranular corrosion. Preferential corrosion along grain boundary regions of polycrystalline materials.

Intergranular fracture. Fracture of polycrystalline materials by crack propagation along grain boundaries.

Intermediate solid solution. A solid solution or phase having a composition range that does not extend to either of the pure components of the system.

Intermetallic compound. A compound of two metals that has a distinct chemical formula. On a phase diagram it appears as an intermediate phase that exists over a very narrow range of compositions.

Interstitial diffusion. A diffusion mechanism whereby atomic motion is from interstitial site to interstitial site.

Isothermal transformation (T–T–T) diagram. A plot of temperature versus the logarithm of time for a steel alloy of definite composition. Used to determine when transformations begin and end for an isothermal (constant-temperature) heat treatment of a previously austenitized alloy.

Isotopes. Atoms of the same element that have different atomic masses.

Isotropic. Having identical values of a property in all crystallographic directions.

Izod test. One of the two tests that may be used to measure the impact energy of a standard notched specimen. An impact blow is imparted to the specimen by a weighted pendulum.

K

Kinetics. The study of reaction rates and the factors that affect them.

L

Laminar composite. A series of two-dimensional sheets, each having a preferred high-strength direction, fastened one on top of the other at different orientations; strength in the plane of the laminate is highly isotropic.

Large-particle composite. A type of particle-reinforced composite wherein particle-matrix interactions cannot be treated on an atomic level; the particles reinforce the matrix phase.

Laser. Acronym for *l*ight *a*mplification by *s*timulated *e*mission of *r*adiation—a source of light that is coherent.

Lattice. The regular geometrical arrangement of points in crystal space.

Lattice parameters. The combination of unit cell edge lengths and interaxial angles that defines the unit cell geometry.

Lattice strains. Slight displacements of atoms relative to their normal lattice positions, normally imposed by crystalline defects such as dislocations and interstitial and impurity atoms.

Leadframe. Part of an integrated circuit package. A frame to which may be connected an array of electrical wire leads that are attached to an integrated circuit chip.

Longitudinal direction. The lengthwise dimension. For a rod or fibre, in the direction of the long axis.

Lower critical temperature. For a steel alloy, the temperature below which, under equilibrium conditions, all austenite has transformed to ferrite and cementite phases.

Luminescence. The emission of visible light as a result of electron decay from an excited state.

M

Macromolecule. A huge molecule made up of thousands of atoms.

Magnetic field strength (H). The intensity of an externally applied magnetic field.

Magnetic flux density (B). The magnetic field produced in a substance by an external magnetic field.

Magnetic susceptibility (χm). The proportionality constant between the magnetization *M* and the magnetic field strength *H*.

Magnetization (M). The total magnetic moment per unit volume of material. Also, a measure of the contribution to the magnetic flux by some material within an *H* field.

Malleable cast iron. White cast iron that has been heat treated to convert the cementite into graphite clusters; a relatively ductile cast iron.

Martensite. A metastable iron phase supersaturated in carbon that is the product of a diffusionless (athermal) transformation from austenite.

Matrix phase. The phase in a composite or two-phase alloy microstructure that is continuous or completely surrounds the other (or dispersed) phase.

Matthiessen's rule. The total electrical resistivity of a metal is equal to the sum of temperature-, impurity-, and cold work-dependent contributions.

Melting point (glass). The temperature at which the viscosity of a glass material is 10 Pa-s (100 P).

Metal. The electropositive elements and alloys based on these elements. The electron band structure of metals is characterized by a partially filled electron band.

Metallic bond. A primary interatomic bond involving the non-directional sharing of nonlocalized valence electrons ("sea of electrons") that are mutually shared by all the atoms in the metallic solid.

Metal-matrix composite (MMC). A composite material that has a metal or metal alloy as the matrix phase. The dispersed phase may be particulates, fibres, or whiskers that normally are stiffer, stronger, and/or harder than the matrix.

N

n-Type semiconductor. A semiconductor for which the predominant charge carriers responsible for electrical conduction are electrons. Normally, donor impurity atoms give rise to the excess electrons.

Natural aging. For precipitation hardening, aging at room temperature.

Network polymer. A polymer produced from multifunctional monomers having three or more active covalent bonds, resulting in the formation of three-dimensional molecules.

Noncrystalline. The solid state wherein there is no long-range atomic order. Sometimes the terms *amorphous, glassy,* and *vitreous* are used synonymously.

Nonferrous alloy. A metal alloy for which iron is *not* the prime constituent.

Nonsteady-state diffusion. The diffusion condition for which there is some net accumulation or depletion of diffusing species. The diffusion flux is dependent on time.

Normalizing. For ferrous alloys, austenitizing above the upper critical temperature, then cooling in air. The objective of this heat treatment is to enhance toughness by refining the grain size.

O

Octahedral position. The void space among close-packed, hard sphere atoms or ions for which there are six nearest neighbours. An octahedron (double pyramid) is circumscribed by lines constructed from centres of adjacent spheres.

Ohm's law. The applied voltage is equal to the product of the current and resistance; equivalently, the current density is equal to the product of the conductivity and electric field intensity.

Opaque. Being impervious to the transmission of light as a result of absorption, reflection, and/or scattering of incident light.

P

Paramagnetism. A relatively weak form of magnetism that results from the independent alignment of atomic dipoles (magnetic) with an applied magnetic field.

Particle-reinforced composite. A composite for which the dispersed phase is equiaxed.

Passivity. The loss of chemical reactivity, under particular environmental conditions, by some active metals and alloys often due to the formation of a protective film.

Pauli exclusion principle. The postulate that for an individual atom, at most two electrons, which necessarily have opposite spins, can occupy the same state.

Pearlite. A two-phase microstructure found in some steels and cast irons. It results from the transformation of austenite of eutectoid composition and consists of alternating layers (or lamellae) of α-ferrite and cementite.

Periodic table. The arrangement of the chemical elements with increasing atomic number according to the periodic variation in electron structure. Nonmetallic elements are positioned at the far right-hand side of the table.

Peritectic reaction. A reaction wherein, upon cooling, a solid and a liquid phase transform isothermally and reversibly to a solid phase having a different composition.

Poisson's ratio. For elastic deformation, the negative ratio of lateral and axial strains that result from an applied axial stress.

Polar molecule. A molecule in which there exists a permanent electric dipole moment by virtue of the asymmetrical distribution of positively and negatively charged regions.

Polarization (P). The total electric dipole moment per unit volume of dielectric material. Also, a measure of the contribution to the total dielectric displacement by a dielectric material.

Polarization (corrosion). The displacement of an electrode potential from its equilibrium value as a result of current flow.

Polarization (electronic). For an atom, the displacement of the centre of the negatively charged electron cloud relative to the positive nucleus, which is induced by an electric field.

Polarization (ionic). Polarization as a result of the displacement of anions and cations in opposite directions.

Polarization (orientation). Polarization resulting from the alignment (by rotation) of permanent electric dipole moments with an applied electric field.

Polycrystalline. Refers to crystalline materials that are composed of more than one crystal or grain.

Polymer. A compound of high molecular weight (normally organic) the structure of which is composed of chains of small repeat units.

Q

Quantum mechanics. A branch of physics that deals with atomic and subatomic systems. It allows only discrete values of energy that are separated from one another. By contrast, for classical mechanics, continuous energy values are permissible.

Quantum numbers. A set of four numbers, the values of which are used to label possible electron states. Three of the quantum numbers are integers, which also specify the size, shape, and spatial orientation of an electron's probability density; the fourth number designates spin orientation.

R

Random copolymer. A polymer in which two different repeat units are randomly distributed along the molecular chain.

Recovery. The relief of some of the internal strain energy of a previously cold-worked metal, usually by heat treatment.

Recrystallization. The formation of a new set of strain-free grains within a previously cold-worked material; normally an annealing heat treatment is necessary.

Recrystallization temperature. For a particular alloy, the minimum temperature at which complete recrystallization will occur in approximately one hour.

Rectifying junction. A semiconductor *p-n* junction that is conductive for a current flow in one direction and highly resistive for the opposite direction.

Reduction. The addition of one or more electrons to an atom, ion, or molecule.

Resilience. The capacity of a material to absorb energy when it is elastically deformed.

Resistivity. The reciprocal of electrical conductivity, and a measure of a material's resistance to the passage of electric current.

Resolved shear stress. An applied tensile or compressive stress resolved into a shear component along a specific plane and direction within that plane.

Reverse bias. The insulating bias for a *p-n* junction rectifier. Electrons flow into the *p* side of the junction.

S

Sacrificial anode. An active metal or alloy that preferentially corrodes and protects another metal or alloy to which it is electrically coupled.

Screw dislocation. A linear crystalline defect associated with the lattice distortion created when normally parallel planes are joined together to form a helical ramp. The Burgers vector is parallel to the dislocation line.

Secondary bonds. Interatomic and intermolecular bonds that are relatively weak and for which bonding energies are relatively small. Normally, atomic or molecular dipoles are involved. Examples of secondary bonding types are van der Waals forces and hydrogen bonding.

Selective leaching. A form of corrosion wherein one element or constituent of an alloy is preferentially dissolved.

Self-diffusion. Atomic migration in pure metals.

Self-interstitial. A host atom or ion that is positioned on an interstitial lattice site.

Solid solution. A homogeneous crystalline phase that contains two or more chemical species. Both substitutional and interstitial solid solutions are possible.

Solid-solution strengthening. Hardening and strengthening of metals that result from alloying in which a solid solution is formed. The presence of impurity atoms restricts dislocation mobility.

Solidus line. On a phase diagram, the locus of points at which solidification is complete upon equilibrium cooling, or at which melting begins upon equilibrium heating.

T

Tape automated bonding (TAB). One type of integrated circuit lead frame design in which the substrate is a thin and flexible polymer film. Electrical contact between the IC chip and lead frame is via very thin and narrow copper fingers.

Temper designation. A letter-digit code used to designate the mechanical and/or thermal treatment to which a metal alloy has been subjected.

Tempered martensite. The microstructural product resulting from a tempering heat treatment of a martensitic steel. The microstructure consists of extremely small and uniformly dispersed cementite particles embedded in a continuous ferrite matrix. Toughness and ductility are enhanced significantly by tempering.

Thermal tempering. Increasing the strength of a glass piece by the introduction of residual compressive stresses within the outer surface using an appropriate heat treatment.

Thermally activated transformation. A reaction that depends on atomic thermal fluctuations. The atoms having energies greater than an activation energy spontaneously react or transform.

Thermoplastic (polymer). A semicrystalline polymeric material that softens when heated and hardens upon cooling. While in the softened state, articles may be formed by moulding or extrusion.

U

Unit cell. The basic structural unit of a crystal structure. It is generally defined in terms of atom (or ion) positions within a parallelepiped volume.

Unsaturated. A term describing carbon atoms that participate in double or triple covalent bonds and, therefore, do not bond to a maximum of four other atoms.

Upper critical temperature. For a steel alloy, the minimum temperature above which, under equilibrium conditions, only austenite is present.

V

Viscoelasticity. A type of deformation exhibiting the mechanical characteristics of viscous flow and elastic deformation.

Viscosity. The ratio of the magnitude of an applied shear stress to the velocity gradient that it produces; that is, a measure of a noncrystalline material's resistance to permanent deformation.

Vitrification. During firing of a ceramic body, the formation of a liquid phase that upon cooling becomes a glass-bonding matrix.

Vulcanization. Nonreversible chemical reaction involving sulphur or other suitable agent wherein crosslinks are formed between molecular chains in rubber materials. The rubber's modulus of elasticity and strength are enhanced.

W

Weld decay. Intergranular corrosion that occurs in some welded stainless steels at regions adjacent to the weld.

Welding. A technique for joining metals in which actual melting of the pieces to be joined occurs in the vicinity of the bond. A filler metal may be used to facilitate the process.

Whisker. A very thin, single crystal of high perfection that has an extremely large length-to-diameter ratio. Whiskers are used as the reinforcing phase in some composites.

White cast iron. A low-silicon and very brittle cast iron, in which the carbon is in the combined form as cementite; a fractured surface appears white.

Whiteware. A clay-based ceramic product that becomes white after high-temperature firing. Whiteware include porcelain, china, and plumbing sanitary ware.

Y

Yielding. The onset of plastic deformation.

Yield strength. The stress required to produce a slight yet specified amount of plastic strain; a strain offset of 0.002 is commonly used.

Index

T

U

V

W

Y

Z